Genetics of Natural Populations

The Continuing Importance of Theodosius Dobzhansky

Professor Theodosius Dobzhansky in his Columbia University laboratory (January 22, 1962), smiling and relaxed, surrounded by his ever present fly-bottles.

Genetics of Natural Populations

*The Continuing Importance
of Theodosius Dobzhansky*

Edited by Louis Levine

*New York
Columbia University Press*

Columbia University Press
New York Chichester, West Sussex
Copyright (c) 1995 Columbia University Press
All rights reserved

Library of Congress Cataloging-in-Publication Data

Genetics of natural populations: the continuing importance of
    Theodosius Dobzhansky / edited by Louis Levine.
        p.   cm.
        Includes bibliographical references and index.
        ISBN 0-231-08116-2
        1. Population genetics. 2. Evolution (Biology) 3. Dobzhansky,
    Theodosius Grigorievich, 1900–1975. I. Levine, Louis, 1921–   .
    QH455.G466   1995
    575.1—dc20   94-29691
                                                              CIP

●

Casebound editions of Columbia University Press books are printed on permanent
and durable acid-free paper.

Printed in the United States of America

c 10 9 8 7 6 5 4 3 2 1
p 10 9 8 7 6 5 4 3 2 1

Contents

PART 4. Genetic Load and Life Cycles

PART 5. Speciation and Mating Behavior

PART 6. Molecular Studies

Preface

The idea for this volume had its origin in discussions with Jeff Powell regarding my desire to organize a workshop in honor of Theodosius Dobzhansky, as part of the program of the Seventeenth International Congress of Genetics, which took place in August 1993 in Birmingham, England. Since Professor Dobzhansky's death in 1975, many symposia and workshops have been held in his memory. A goodly number were retrospective and stressed the importance of his ideas and work in molding the thinking and research of evolutionary genetics during the middle half of the present century. Now, almost twenty years since his passing, the time has come to assess his influence on current thought and research. Such an undertaking would call for chapters devoted to discussions of his intellectual legacies and their relevance to modern ideas on evolution and, in addition, chapters devoted to current research that had its origin with him or his writings.

The title for this volume was provided, in part, by Dobzhansky himself from the title of his famous series of forty-three papers ("Genetics of Natural Populations"), which have become the symbol of his devotion to the study of the evolutionary process. It was necessary to add only a descriptive phrase, indicating the focus of the book.

A most gratifying response was forthcoming from those contacted for this project. Even when an individual was not able to join the group, the expression of love for Dodik (as he was affectionately called) and indebtedness to him were truly moving. The twenty-four of us, and coauthors as indicated, who have contributed to this volume consist of colleagues who worked with Dobzhansky at various times, geneticists from other universities who did research in his laboratory, his own Ph.D. students whom he considered his "intellectual offspring," and the Ph.D.s of his students, namely, Dobzhansky's "intellectual grandchildren." Within these chapters one will often find personal statements, because to interact with Dobzhansky the scientist always meant interacting with Dobzhansky the human being, and all who did so were affected by the experience.

A number of biographical accounts of the Professor as well as listings of his publications have been printed. They are noted at the end of this preface for those who may wish to learn more about this remarkable man.

The chapters that make up this volume fall into six groups. In the first group, Reflections and Reactions, one will find reviews covering the accomplishments of Dobzhansky at the various universities in the United States where he taught, did research, and produced Ph.D.s. Also included are discussions of his importance in the development of research programs on evolutionary genetics in other areas of the world. This is followed by chapters on Dodik's Intellectual Legacies. There is an evaluation of his advancement of science both through the addition of new information to existing bodies of knowledge and through his "inspirational breakthroughs." Additionally, this section offers reviews of the effects of his contributions to current ideas on genetic variation in populations, coadaptation, balancing selection, chromosomal polymorphism, and neo-Darwinism. The last chapter in this section presents arguments for considering Dobzhansky a theoretician rather than a laboratory and field investigator.

In the third group, Chromosomal Polymorphisms, a series of chapters reflect the area of current research that has its origin in Professor's pioneering field studies of *Drosophila pseudoobscura*. Included are investigations of *D. robusta*, *D. pseudoobscura*, *D. mediopunctata*, and *D. subobscura*. A consideration of Genetic Load and Life Cycles follows. These interrelated areas of study were the focus for much of Dobzhansky's research, and the results obtained were critical in the formulation of his ideas on the genetic structure of populations.

The fifth section, Speciation and Mating Behavior, contains chapters detailing current research that has its foundation in one of Dobzhansky's great contributions to evolutionary thinking, namely, his biological definition of species. Also discussed is mating behavior, which often proves to be the reproductive isolating mechanism that keeps species genetically independent. The final group of chapters deals with Molecular Studies. In his research, Dobzhansky worked at the limits of resolution of the techniques available to him. Unfortunately, the techniques did not include nucleotide base sequence analysis or the ability to detect transposable elements. As the authors (both Dobzhansky "grandchildren") of these final chapters point out, current molecular studies were inspired by Professor's earlier work and are designed to examine the evolutionary process, seeking to understand its intricacies much as he had done.

As part of the editorial process, it was agreed that the contributors to this volume would also serve as reviewers of related chapters, thereby reducing the amount of overlap of material that might otherwise occur. In the spirit of collegiality, reviewers almost uniformly sent their comments directly to the authors. Eleven of us were able to attend the Genetics Congress, where there were direct discussions between "authors" and "reviewers," to the benefit of all.

Columbia University Press has had a long history of publishing books by Dobzhansky: *Genetics and the Origin of Species* (1937), *The Biological Basis of Human Freedom* (1956), *Genetics of the Evolutionary Process* (1970), *Human Culture: A Moment in Evolution*, with E. Boesiger, edited and completed by B. Wallace (1983); and books about Dobzhansky, e.g., *Dobzhansky's Genetics of Natural Populations*, edited by R. C. Lewontin et al. (1981). With respect to the present volume, they have graciously agreed to continue the tradition. Special mention must be made of Mr. Ed Lugenbeel and Ms. Laura Wood for their encouragement and prodding, both of which have been greatly appreciated.

I take much pleasure in thanking all the contributors to this volume for their wholehearted cooperation. They have all agreed that royalties from this volume are to go to the Theodosius Dobzhansky Memorial Fund, administered by The Society for the Study of Evolution, which awards grants annually to a "young scientist for support of field or laboratory research in evolution."

I must add a special word of thanks to Jeff Powell for the help I received from him at every stage of this project. However, whatever shortcomings it may have are solely my responsibility. Finally, I would like to say that for me a very meaningful reward of this undertaking has been a greatly increased appreciation of the impact of Theodosius Dobzhansky on the lives of so many human beings.

Louis Levine

READINGS ON DOBZHANSKY

Ayala, F. J. and T. Prout. 1977. Theodosius Dobzhansky: 1900–1975. *Social Biology* 23:101–7.

Ehrman, L. 1977. Theodosius Grigorievich Dobzhansky: 1900–1975. Scientist and Humanist. *Behav. Genet.* 7:3–10.

Glass, B., ed. 1980. *The Roving Naturalist: Travel Letters of Theodosius Dobzhansky.* Philadelphia: American Philosophical Society.

Land, B. 1973. *Evolution of a Scientist: The Two Worlds of Theodosius Dobzhansky.* New York: Thomas Y. Crowell.

Levene, H., L. Ehrman, and R. Richmond. 1970. Theodosius Dobzhansky up to now. In M. K. Hecht and W. C. Steere, eds., *Essays in Evolution and Genetics in Honor of Theodosius Dobzhansky,* pp. 1–41. New York: Appleton-Century-Crofts.

Lewontin, R. C., J. A. Moore, W. B. Provine, and B. Wallace, eds. 1981. *Dobzhansky's Genetics of Natural Populations. I–XLIII.* New York: Columbia University Press.

Genetics of Natural Populations

The Continuing Importance of Theodosius Dobzhansky

Professor Theodosius Dobzhansky with a number of his students at the XIII International Congress of Genetics, August 1973. Front row left to right are Sergey Polivanov, Theodosius Dobzhansky, Eliot B. Spiess, and Francisco J. Ayala. Middle row: Jeffrey R. Powell, Wyatt W. Anderson, Abd El-Khalek M. Mourad, and Louis Levine. Back row: K. Sankaranarayanan, Timothy Prout, Rollin C. Richmond, and George R. Carmody.

PART ONE

Reflections and Reactions

1

Mentor, Colleague, Friend

Chia-Chen Tan

My story begins with a special species of ladybird beetle known as *Harmonia axyridis*, which shows great variation in its elytral color patterns. In the fall of 1930 I was admitted to Yecheng University in Peking (Beijing) and became a graduate student of Professor J. C. Li who was the first Chinese Ph.D. student of T. H. Morgan. At the suggestion of the entomologist professor Chen-fu Wu, the chairman of the biology department, I began to work on the variations and inheritance of the color patterns of this species of beetle.

After a year and a half, I wrote up my master's thesis, which was later split into three separate papers. One of them, on the inheritance of the color patterns of the beetle, the core of my thesis, was sent by J. C. Li to T. H. Morgan at The California Institute of Technology (Cal. Tech.) with the intention that it be published in the United States. Morgan immediately passed on the manuscript to Dobzhansky who had previously studied and published on the geographical distribution of different color patterns of the same species, based on collections in the Lake Baikel and Irkutzk area of the Asiatic part of the Soviet Union.

After reviewing my manuscript, Dobzhansky was so pleased with my work that he immediately wrote me a very warm letter together with some comments and constructive remarks on the manuscript. Following further correspondence and revision, the paper was approved by Morgan and Dobzhansky and was published in *The American Naturalist* (Tan and Li 1934). Meanwhile my relationship with Dobzhansky developed to such an extent that I was accepted to work for the Ph.D. degree with him in Morgan's laboratory.

I arrived at the San Pedro Harbor of Los Angeles after a twenty-day sea voyage. I recollect the vivid sight of Dobzhansky who came personally to meet me at the dock and made all the necessary arrangements for my living accommodation. It is no exaggeration to say that the ladybird beetle is what brought us together. His ardent enthusiasm and straightforwardness left a deep impression on me. Under the guidance of this good teacher and friend, my scientific life reached an important turning point.

I had been accepted as T. H. Morgan's Ph.D. student, but because of Morgan's busy administrative duties, I was actually assigned to study and work under the joint sponsorship of Sturtevant and Dobzhansky, and Dobzhansky was directly responsible for the supervision of my thesis research work. I was, in fact, Dobzhansky's first Ph.D. student. Dobzhansky and I got along very well. He was always very thoughtful and kind to me. This may have been because, aside from the fact that we had a common interest in the ladybug, we also had a more or less similar cultural background. He had a keen interest in the Chinese culture, both ancient and modern. At the same time, both of us were foreign residents in the United States.

During my three years at Cal. Tech. I was associated with *Drosophila* and *Drosophila* geneticists practically every day. And just at that time the giant salivary chromosome technique was being used increasingly in genetic studies. Dobzhansky first taught me how to make smear preparations of *Drosophila* salivary gland chromosomes and I immediately used this technique in combination with classical genetic methods in my Ph.D. thesis, entitled "The genetical and cytological maps of the autosomes in *Drosophila pseudoobscura*" (Tan 1936a, b). While Sturtevant and Dobzhansky began to search for the inversion differences on the third chromosome of different geographical strains of *Drosophila pseudoobscura*, I made use of the giant salivary chromosome technique to study the internal structural differences of chromosomes between what were then referred to as two races of *D. pseudoobscura* but later described as two separate species, *D. pseudoobscura* and *D. persimilis* (Tan 1935).

Continuing with this line of research, I collaborated with Dobzhansky in the study of genetic differentiation between *D. pseudoobscura* and *D. miranda*, two phylogenetically more distant species that are crossable to each other but produce sterile offspring (Dobzhansky and Tan 1936). In the meantime I worked with Sturtevant in studies on the genetic differences between two more widely separated noncrossable species, *D. pseudoobscura* and *D. melanogaster*, by comparative studies of the genetic maps of the two species to identify the homology of the genes (Sturtevant and Tan 1937).

In 1937, when I was about to return to China, Dobzhansky's book *Genetics and the Origin of Species* had just been published. This publication may be considered the most important book on evolutionary theory in the twentieth century. It successfully integrated Mendelism and Darwinism to explain the origin of species in terms of genetic principles. The book has laid the foundation for the modern synthetic theory of evolution. I have treasured this book, not only for personal reasons, because my work on the ladybugs and *Drosophila* is quoted in numerous places, but more importantly, because I brought a copy back to China, which was later used as the textbook for the course entitled "Experimental Evolution," which I offered to my students during World War II, when the university at which I taught was evacuated to the interior far west of China. During the wartime I kept up a correspondence with Dobzhansky who constantly encouraged me to overcome the difficult conditions of research and daily life.

Dobzhansky and Dunn helped make arrangements through the Rockefeller Foundation for me to spend one year of sabbatical leave at Columbia University during 1945–46. I arrived in New York City in May, after more than two months of travel, having flown over the Himalaya Mountains and waited for American air transports in India and Egypt. I was very excited to rejoin Dobzhansky and his family after such a long and dangerous journey and to celebrate together the end of World War II.

I spent the summer of 1945 in the biological laboratory at Cold Spring Harbor, where I started to work on the genetic mechanism of sexual isolation between *D. pseudoobscura* and *D. persimilis*. My work on sexual isolation was in reality a part of the joint program of Dobzhansky and Mayr on this aspect of *Drosophila* evolution. The design of my experiments followed Dobzhansky's scheme in his experiments on the study of hybrid sterility factors in *D. pseudoobscura-persimilis* hybrids. The results of my experiments showed that numerous genes exist mainly distributed on the X-chromosome and chromosome 2, which are responsible for distinguishing the mating behavior between the two species (Tan 1946).

In my days at Columbia I found Dobzhansky's scientific achievements, both experimental and theoretical, coming to a climax and his influence spread far and wide. His *Genetics and the Origin of Species* was revised and was widely accepted with great excitement by the biological community of the day.

When I met Dobzhansky again, in Stockholm during the eighth International Congress of Genetics in 1948, I was facing a critical choice, whether to return to China or not. It was difficult because I learned that at a recently concluded Lenin Agricultural meeting held in the Soviet Union,

Lysenko, under Stalin's banner, had formally denounced classical genetics as politically bourgeois and reactionary and had put in its place his theory of genetics as the official socialist and proletarian genetics. I was afraid that the Chinese government would follow the Soviet Union in dealing with teaching and research in genetics. The last time I saw Dobzhansky was in the winter of the same year in New York City when I was on my way back from Europe to China. I set out on my journey with mixed feelings. I was encouraged by Dobzhansky to be brave enough to seek the truth. I have always believed firmly in the motto of Cal. Tech.: "Truth makes you free," and I lived accordingly during the years following.

After 1948, I lost contact with those abroad, except through newspapers and journals, for thirty long years. In 1978 I was invited to attend the fiftieth anniversary of the founding of the biological division at Cal. Tech. by T. H. Morgan. I availed myself of this opportunity to make a special trip to Davis to mourn my most respected and beloved teacher and friend Theodosius Dobzhansky, who had passed away there three years earlier. I owe so much to Dobzhansky that no words can express my feelings. As I have said elsewhere (Tan 1989):

> This great man left mankind a very generous heritage. It will be inscribed together with his name on the monument of the history of biological sciences.

REFERENCES

Dobzhansky, Th. and C. C. Tan. 1936. Studies on hybrid sterility, III: A comparison of the gene arrangement in two species, *Drosophila pseudoobscura* and *Drosophila miranda. Zeit. ind. Abst. Ver.* 72:88–114.

Sturtevant, A. H. and C. C. Tan. 1937. The comparative genetics of *Drosophila pseudoobscura* and *D. melanogaster. Genetics* 34:415–32.

Tan, C. C. 1935. Salivary gland chromosomes in the two races of *Drosophila pseudoobscura. Genetics* 20:392–402.

———. 1936a. Genetic maps of the autosomes in *Drosophila pseudoobscura. Genetics* 21:796–807.

———. 1936b. The cytological maps of the autosomes in *Drosophila pseudoobscura. Z. Zellforsch.* 26:119–43.

———. 1946. Genetics of sexual isolation between *Drosophila pseudoobscura* and *Drosophila persimilis. Genetics* 31:558–73.

———. 1989. Remembering Theodosius Dobzhansky. *Genome* 31:230–32.

Tan, C. C. and J. C. Li. 1934. Inheritance of the elytral color patterns of the ladybird beetle (*Harmonia axyridis* Pallas). *Am. Natur.* 68:252–65.

2

Recollections of a Coauthor and Close Friend

G. Ledyard Stebbins

I first met Dr. Dobzhansky in 1936, when I visited T. H. Morgan's laboratory at Cal. Tech., a year after I had arrived in Berkeley to join the genetics department there, as a junior geneticist and research worker on a grant held by its chair, E. B. Babcock. After Morgan had told me about his new staff member, a true genius who had, nine years earlier, arrived from Russia, he took me to the laboratory, where I saw a young man of moderate build, with a full head of close-cropped hair, with his wife Natasha, examining *Drosophila* chromosomes. Their objective was to find out whether the physical distance between mutant loci did or did not correspond with "map distances," as calculated from crossover frequencies, based upon the assumption that crossover frequency was the same throughout the length of the chromosome. Of course, they found no correlation, owing to the presence of various positions of "hot spots" of high crossover frequency within a limited chromosomal segment. Shortly afterward, salivary chromosomes became available for a much more detailed and accurate attack on the same problem. The research of Calvin Bridges on salivaries made Dobzhansky's results obsolete and almost forgotten.

For the next few years, I was busy with research on *Crepis* and other plants in collaboration with Babcock and neither saw Dobzhansky nor paid much attention to his research. When I joined the faculty as assistant professor of genetics and was asked to organize a course on organismic evolution for seniors and graduate students, I looked over his newly published *Genetics and the Origin of Species*, but it contained too much on *Drosophila* and too little on plants to serve as a text, although I highly recommended it as an outside reference. Meanwhile a graduate student who was about to get his Ph.D., I. Michael Lerner, had known Dobzhansky for a number of years, shared with him Russian as their native language plus a mutual interest in population genetics, and attracted visits to Berkeley from Dobzhansky while he was still at Cal. Tech. From their discussions, some of which were in English, I learned about aspects of evolution previously unfamiliar to me, particularly the ideas of Sewall Wright, who at that time was Dobzhansky's hero. From Michael, I obtained this remarkable story.

When Dobzhansky was awarded a fellowship to come to the United States and work in Morgan's laboratory, and was given permission by the authorities at Leningrad to do so, he had to obtain a passport and visa to enter. When he reached the U.S. Embassy at Riga, Latvia (at that time the United States and the Soviet Union did not have diplomatic relations with each other), he was asked "What kind of visa do you want?" His typically naive reply was

> *"What do you have?"*
> *"It depends on what you do."*
> *"I want to learn from Professor Morgan the latest knowledge about U.S. genetics, particularly how to work with the fly Drosophila."*
> *"So you want to study? Then we'll issue a student visa."*

Equipped with this permit, he went to New York and Columbia University in 1927 but shortly afterward accompanied Morgan's group to Pasadena and Cal. Tech. After the specified three-year period was over, the Dobzhanskys learned that in the Soviet Union, genetics had come under a cloud, owing to the machinations of a young communist named T. D. Lysenko, who had the backing of the education commissars and of Stalin himself. They wisely decided that if they returned, no career for them would be open. After hearing of their plight, Morgan used funds available to him to support them for a year in his laboratory, then took steps to obtain for them a visa of immigration, using the subterfuge of sending them to spend a few weeks with the chair of the zoology department at the University of British Columbia (UBC), a close friend.

Reentry to the United States, however, turned out to be an almost insurmountable problem. When Dobzhansky showed the U.S. consul at Vancouver his student visa, he was asked the usual questions, including:

> *"Have you taken gainful employment while in the United States?"*
> *"Well, for the past year I have been working for Professor Morgan."*
> *"What! Didn't you read the lines on your visa that states specifically that you cannot be employed in our country if you enter on a student visa?"*

"I guess I didn't notice, since I didn't read English easily." (All this through Lerner as interpreter, who at that time was at UBC and had been introduced to Dobzhansky by the chair of the zoology department).

"Well, since you have already broken our laws, we can't give you any kind of visa for reentry."

This shock and news threw them into despair. Since the Dobzhanskys had overstayed their leave, a return to the Soviet Union would certainly mean banishment to Siberia and perhaps worse. They would have to leave Canada in a few days and had nowhere to go. In desperation, Dobzhansky telephoned to Morgan in Pasadena. Morgan told him not to worry: he surely would find a way out.

Morgan immediately telephoned to the office of Dr. Millikan, president of Cal. Tech., asking urgently for an immediate reply. He was told that Millkan was in Washington.

"Get on the telephone, find him, and tell him to phone me at once."

After a bit of sleuthing, Millkan's administrative assistant discovered that he was on President Hoover's yacht, cruising on the Potomac River. Realizing that Morgan was an indispensable chair on her faculty, she persisted until she got through, and Millkan telephoned Morgan. Said Morgan: "One of the most brilliant and promising young scientists in my group is in deep trouble with the U.S. Immigration Service. Do what you can to help us out."

According to this story, which in detail may be somewhat apocryphal, Millkan went to the U.S. President, who telephoned the Secretary of State, and from there the message went to the head of the Immigration Service, then to the consulate in Vancouver, and the Dobzhanskys secured immigration visas.

When Dobzhansky started to investigate the distribution of inversions in the third chromosome of *Drosophila pseudoobscura*, he emphasized aspects of population genetics that were relevant to some of Sewall Wright's hypotheses, and he also constructed a phylogeny of the different inversions. Since neither of these topics appeared to me to be relevant to the principles of evolution that I recognized in plants, I at first took little notice of them. Soon, however, he interpreted differences in terms of geographic and ecological distribution. Of special interest was the inversion content of populations from the desert margin in southern California: Andreas Canyon near Palm Springs, Piñon Flats at 900 meters near the foothills of San Jacinto Mountains, and Idylwild, at 1800 meters in these mountains themselves. I clearly saw with him that here was an unusual opportunity to study Darwinian natural selection in a species in which hypotheses could be tested under controlled conditions.

Ten years earlier, when as a Harvard graduate student I had heard my professor of general physiology declare that "Evolution is good fodder for newspaper Sunday supplements, but it isn't science. You can't experiment with two million years." As an eager young Darwinian evolutionist, I was irked and pricked by this statement and looking for chances of giving a lie to it. Dobzhansky's research on *Drosophila* inversions appeared to give me my answer. From then on, I was his ardent follower throughout his life.

The summer of 1944 marked the beginning of a period of close, intimate, and highly profitable relationship with Dobzhansky. That summer he decided to examine populations of *Drosophila pseudoobscura* and *D. persimilis* along a transect of locations established by H. M. Hall and J. C. Clausen for their experimental investigations of plants, from Jackson in the foothills, to Mather at a middle altitude, and up to timberline five miles north of Tioga Pass, just northeast of the boundary of Yosemite National Park. Since this transect was easily accessible from my home in Berkeley, I asked and was granted hospitality at the cabin belonging to the Carnegie Institution of Washington that he occupied, and I looked forward eagerly to sitting at the feet of the great evolutionist and absorbing knowledge from him.

I soon learned, however, that one could never sit at the feet of Dobzhansky when he was at a field station. When he was not sleeping or eating, he was either setting out baited traps for *Drosophila*, catching them in bottles, preparing and examining their squashed chromosomes, or taking his "leisure" by mounting a horse and riding rapidly in a chosen direction. The only possible way of communicating with him was to mount another horse and ride equally rapidly in the same direction. Fortunately, I had spent teenage years at a boarding school where horseback riding was compulsory, and so I had no difficulty in keeping up with him on the gentle nag chosen for me.

The first of these rides was to a fine pristine meadow surrounded by tall pine trees and carpeted with bunch grasses belonging to species that I was studying intensively in my efforts to breed a species or race that would provide better forage for cattle, at least during the beginning and end of the long dry summers characteristic of interior California. At once I spotted scattered tufts of woodland wildrye grass (*Elymus glaucus*), dozens of low bristly mats of squirrel tail (*Sitanion hystrix*), and an exceptionally large number of plants that were intermediate in stature and appearance. These intermediate plants were like the sterile hybrids that I had seen elsewhere and had synthesized by crossing in my garden plot. I rode my horse to each type of grass, leaned down and picked a few stems, and while my horse was busy munching others, examined the technical characteristics of each one, as well as checked fertility vs. sterility via the presence or absence of developing seeds. Then I rode up to Dobzhansky, while both of us were astride our mounts, and explained what I had found. At once, his eyes glowed: "Stebbins," he said, "You are the first person who has ever seen, collected, and identified a species hybrid from the back of a horse." For a veteran horseman who several years before had ridden 4,000 kilometers in central Asia doing research on the evolution of the domestic horse, this was of prime importance and greatly raised his estimate of my capability.

Toward the end of the summer, I visited Mather again, and we saw Cottonwood Meadow when the grass seeds were ripe. He had been complaining about the situation, saying that it was hard for him to believe that natural selection could permit the formation of so many sterile hybrids. "Of course, *Drosophila* orders things much better." I countered with the fact that flies have evolved by selection for elaborate courtship patterns that keep males and females belonging to different species apart from each other, whereas grass plants stay in one place all the time, and while their flowers are open, their feathery stigmas cannot help receiving wind-borne pollen from all directions.

To show him clearly that my story was correct, I asked him to look at seed-bearing heads of both parental species, so that he could be sure they were full of plump seeds. I asked him then to look at hybrids, which were easy for him to recognize, and said:

"I'll give you a bottle of beer for every seed that you find on one of those plants."
"You're a piker," said he, "When I was in Central America with Michael White, he offered me a bottle of champagne for each individual of a rare mantid that he wanted for chromosome counting."
"All right, I'm no piker. It's a bottle of champagne for each seed."

Then he and his daughter Sophie, who was with us, really went after them in earnest. While I watched, to make sure that they did not sneak in a few parental seeds by mistake, they spent nearly an hour ransacking more than 100 hybrid plants, thrashing and beating them out, and ended up with 28 seeds out of a possible 10,000 to 15,000. Well, we did not immediately get drunk on champagne, but in 1946, while I was staying at their home and delivering the Jesup Lectures, we had a pretty good party in honor of the event.

Perhaps at least in part because of our discussions at Mather, I received from L. C. Dunn, chair of Dobzhansky's department, an invitation to present a series of Jesup Lectures on plant evolution at Columbia University during the autumn of 1946. Dobzhansky insisted on my accepting the invitation proffered by him and his wife Natasha to stay in their apartment while in New York. This was an exceptional honor, from which I did my best to profit and contribute in turn as much as I could. The endless discussions, day after day, while we were walking to and from the campus, plus daily lunches with graduate students and often also with John Moore, then zoology professor at Columbia's Barnard, did much to help me formulate my ideas about evolution. Occasional visits to his friend, biochemist Alfred Mirsky at the Rockefeller University, and weekends at the Cold Spring Harbor Laboratory, whose director was Milislav Demerec, at that time transferring his interest from *Drosophila* to microbial genetics, expanded my horizons even more.

My intimacy with the Dobzhansky family taught me things about human genetics and culture. At that time the English plant cytogeneticist C. D. Darlington was insisting in published papers and books that the ability to pronounce the words of a particular language, specifically the English diphthong "*th*," has a genetic basis. In fact, he postulated a genetic linkage between the A blood group phenotype and the ability to pronounce the English "*th*." When he heard contrary reverberations from Dobzhansky and others, he and English friends spread around the following apocryphal conversation between Dobzhansky and Ernst Mayr:

> *Doby (an affectionate nickname): "Ernst, you know zat Darlington's idea is silly! Why, anyone can pronounce ze 'th.'"*
> *Mayr: "Yes, dat's right."*

This was, of course, correct with respect to Doby and Ernst, both of whom learned their English as adults. But when I was in the Dobzhansky's apartment, I heard their daughter Sophie, then a girl of thirteen, talking with her parents. While both parents pronounced "*th*" and other English sounds in the manner caricatured by Darlington, and had done so ever since Sophie was a small child, she spoke English with a typical New York accent, hardly different from mine, a native New Yorker.

After 1951, when Dobzhansky's interests became transferred from *Drosophila pseudoobscura* and its temperate relatives to the tropical American species *D. willistoni* and its relatives, I saw less of him. Nevertheless, at various meetings, particularly the International Congress of Genetics at Montreal in 1958, we discussed, often passionately, the issues of population genetics that emerged from his population cage experiments. The major address presented by H. J. Muller at Montreal, which vigorously supported the classical theory: predominance of genomes that consist of "wild type" alleles at most loci, and attacked with equal vigor the "balance theory" of Dobzhansky, which promoted the widespread adaptive value of polymorphism at many loci, including heterotic fitness, was particularly distasteful to Dobzhansky since it appeared to contradict the results of his experiments. During this period, he worked out the concept of balancing selection as he presented it in chapter 5 of his 1970 book *Genetics of the Evolutionary Process*. His concept of heterotic balance has been severely criticized recently, but his suggestion that a balance based upon diversifying selection "is possibly even more important in nature than the heterotic type" has much to be said for it. This is especially true in light of recently revealed biodiversity in microhabitats, as well as the discovery that mutations and small alterations of chromosome structure are far more ubiquitous than they were believed to be by Muller and other classicists. The most important lesson to evolutionists of molecular genetics as it has been developed recently is that a genotype consisting of rarely mutated "wild type" alleles is virtually impossible to maintain. It is certain to be altered by neutral mutations, by stabilizing natural selection, or by modifiers of slightly deleterious mutations that provide a stable equilibrium via interaction with various balancing factors.

My friendship with Dobzhansky became very close after his retirement from Rockefeller University in 1970, which had been preceded by the death of his wife Natasha in 1969. He had approached Robert Allard, then chair of our genetics department, about the possibility of moving to Davis, along with his research associate Francisco Ayala. Allard's persuasiveness with the chancellor and his staff turned that possibility into reality, and in the fall of 1970, Dobzhansky became adjunct professor, and Ayala, associate professor at the University of California, Davis. Both were supplied with newly equipped laboratories for continuing their research.

Personally, this gave me an exceptional opportunity to repay a longstanding debt of hospitality. During the spring term, Doby would have to live alone while preparing for housing during the following autumn, unless somebody invited him to stay with them. I explained to my wife Barbara that here was an excellent opportunity to repay Dobzhansky for his hospitality toward me in 1946. She agreed, though with some reservations. When he came, however, his easy smile and relaxed informality, plus their common interest in art, particularly Renaissance Italian painting, stimulated a close, warm friendship that lasted until his death.

3

Dobzhansky at Columbia

Howard Levene

Theodosius Dobzhansky first came to New York in 1927 as a Rockefeller fellow. He joined Thomas Hunt Morgan as an associate at Columbia where he remained until the fall of 1928 when Morgan moved to the California Institute of Technology and took all the people in the laboratory with him, including Calvin Bridges, A. H. Sturtevant, and Dobzhansky. Accordingly, the first Columbia year was only part of a year and there is little to say about it except that he was getting used to the swing of things in this country. At that time the zoology department was very crowded on the top floor of Schermerhorn Hall. Not until a year later was the administration of the university persuaded to build a Schermerhorn extension and thereby provide a good deal more space.

Dobzhansky was very disappointed in the facilities of Columbia, which he complained were small, dirty, and inadequate, and I am sure Cal. Tech. was a great improvement. When he arrived at Cal. Tech. he moved to a small house where he remained happily for a number of years, working primarily with Sturtevant. A number of research projects were undertaken there, and he also had two graduate students: Poulson and, from China, C. C. Tan. He also developed a habit of frequent collecting trips in southern California and around Yosemite.

During most of the Cal. Tech. days Dobzhansky worked closely with Sturtevant, but at some point near the end of his stay, a rift developed between them. Partially for that reason, when L. C. Dunn invited Dobzhansky to come to Columbia in 1940, he was happy to accept. He came in the fall of 1940, and in the spring of 1941 he gave the first course of a long-time series on the material from *Genetics and the Origin of Species*. During the first year he covered about the first half of the book. The second year, in the spring of 1942, he gave the second half of the course on the second half of the book.

I was at that time a beginning student at Columbia and attended the course. Personal history here might be interesting. I had decided to come to Columbia to study both mathematical statistics and genetics and evolution, and I spoke with L. C. Dunn, the chair of the zoology department. Since there was no statistics department, we agreed that I would be accepted as a student in the zoology department but would be free to take as many, or as few, zoology courses as I wanted and to mainly take courses in statistics and mathematics. I spoke to Dunn in May 1941, and at that time he told me Dobzhansky was away in California, but he did write and tell Dobzhansky about this promising student who was interested in both Dobzhansky's work and mathematical theory.

In the fall Dunn signed me up for various courses. Dobzhansky was still in California. He arrived a few days after the term began, and I went to see him and said my name was Howard Levene. "Levene, I am delighted to see you. I have a lot of problems for you," and that was my introduction. It was also my introduction to a habit that was continued over the years, of Dobzhansky's leaving for California and fieldwork just as soon as classes were over and then arriving, usually a few days after the term started, innocently saying "Oh! has the term already begun?" I always suspected he knew very well he was late. In any event he would arrive later, and we had to arrange for the first few classes to be covered either by Dunn or by some advanced graduate student of Dobzhansky's.

In his Columbia days most of the graduate students referred to him as the professor, while his good friends, who included Dunn, John Moore, Francis Ryan, and myself, referred to him as "Dodik," which came, I believe, from his first name, Theodosius. His wife, Natasha, also referred to him, of course, as "Dodik" with a soft middle 'd.' Near the end of his years at Columbia, and then at Rockefeller, the students began to refer to him as "Doby," an abbreviation of his last name rather than his first, but his oldest friends continued to use "Dodik."

In the spring of 1942 I attended the second part of his class on evolution and population genetics. There was a large and important student body made up largely of postdoctoral fellows, research associates, and other senior people as well as graduate students. The former included, among others, Salome Glucksohn Schoenheimer, Verne Bryson, Taylor Hinton, James Neal, Marie Holt, who was the assistant in the laboratory, and myself. By that time Dodik and I knew each other a little better, and he said "You will, of course, do the lectures discussing the mathematics of change under natural selection, since this is something I am not too familiar with and I always have some trouble with it." So I devoted two or three lectures to that, and then, since the course was run as a seminar and each person gave a talk on some relevant paper, I was given papers by Carl Epling on *Oenothera organensis,* which is a complicated story of an isolated group.

In the fall of 1942, I attended the repeat of the first part. The attendance now was much reduced since the older people had already been there and, since World War II had started, there were fewer students; the students included Harriet Taylor (later Harriet Taylor Ephrussi), Evelyn Meisel Witkin, and one or two others. Bruce Wallace, a fellow student, left before the end of the term to join the Air Force.

In 1943 Dobzhansky made the first of many foreign trips. He had recently found seasonal variation in populations of *Drosophila* in California, apparently due to natural selection under different climatic conditions. The obvious thing was to look at what happens in the tropics in rain forests where conditions were constant throughout the year, and Brazil seemed a perfect place for this. Accordingly, he tried to get to Brazil, but this was during World War II when travel was not that easy. However, he finally got grant money to enable him to go, and he arranged for André Dreyfus at the University of São Paulo to be his host. Dobzhansky and Dreyfus hit it off very well and became firm friends until a few years later when, unfortunately, Dreyfus died.

During the stay in Brazil, in addition to his productive collecting in the forest, Dobzhansky lectured, and a number of prominent young biologists attended his lectures. Many of them were induced to go into the general field of *Drosophila* population genetics both in the laboratory and in the field, and as the years went on, a number of them visited Columbia as postdoctoral fellows for a year and sometimes longer. They provided Dobzhansky with a broad base of former students in Brazil, on whom he could count for assistance at various places. Prominent among them were Pavan, DaCunha, Cavalcanti, Cordeiro, Malogolowkin, and Kerr.

Bruce Wallace returned to Columbia after several years in the Air Force and defended his dissertation in the fall of 1947. Since I had defended in the spring of 1947 I was eligible to be on Bruce's defense committee and because I was already advising him on statistical aspects of his dissertation I was asked to join in. Bruce was very nervous during the examination and came in with white lips from chewing antacid tablets. I believe he had slept very little the night before. Nevertheless he did a good job defending. Dobzhansky considered Bruce his best student. A few years later Dunn was a visitor at Harvard for a term, and he came across a very brash and bright student named Richard Lewontin and persuaded him to come to Columbia. There Richard worked with Dobzhansky. The facilities on the eighth floor of Schermerhorn extension left something to be desired for the *Drosophila* people. There were walk-in incubators, a general student laboratory where the genetics course was given, and a number of small offices usually shared by two people and equipped with a big table, chairs, some shelves, and nothing else. The students according to their needs received a dissecting microscope and occasionally a high-powered microscope for studying chromosome preparations plus an etherizing bottle. They all shared the incubator. There was also a kitchen for the preparation of food. Boris Spassky had to help all the students when they had problems with the flies, and he was in charge of all the stocks, so they had to go to him for whatever stocks they were going to be working with. Dick Lewontin shared his small office with Timothy Prout. Dick had a fairly small project studying the orange mutant and was able to complete the laboratory work fairly easily, but Tim had a much bigger project involving a number of strains of wild chromosomes and requiring intercrossing all of them to explore the viability of the hybrid, as well as looking at the viability of all the homozygotes. They engaged in eternal conversations on genetics and other subjects, which were very fascinating and led to the increased education of both participants but seriously interfered with making the crosses and doing the counting, which Tim had to do. There was some question whether Tim would ever complete his degree if Dick did not finish his and move on, so that Tim could work full time.

During the time Dick was finishing his studies Dobzhansky had become very interested in the problem of genetic homeostasis, namely, the ability of heterozygous chromosomes to produce flies that develop well under many different environmental conditions. Homozygotes for particular wild chromosomes might be good under some conditions but usually not so good under others. This variation was part of the general picture of coadaptation that Dobzhansky and Wallace developed and that they held to be a very important aspect of population genetics.

Dobzhansky's usual definition of homeostasis was that it involved a very small variance in viability. His definition did not, however, say much about the means, and the definition did not appeal to Lewontin, who argued that under this definition the most homeostatic chromosome would be one having zero variance and that a recessive lethal fit this description perfectly. So under Dobzhansky's definition a recessive lethal was perfectly homeostatic. This use of reductio ad absurdum did not, however, appeal at all to Dobzhansky, who tended not to like this kind of mathematical reasoning, and so he would tell his visitors "this crazy fellow thinks that lethals are homeostatic." At his defense Dick was not nervous, but he did get into an argument with Dobzhansky over the definition of genetic homeostasis. When, finally, Dick's time was up, Dunn, who was acting as chair, said "Alright, Mr. Lewontin, would you please step out while we discuss the exam?" Dick said, "What? Is the two hours up already?" Lewontin is the only student I ever saw who felt that the defense examination was too short instead of too long.

The visitors to the laboratory usually came for six months to a year and were expected to work with *Drosophila* while they were there. The idea was that working on a project would produce a paper. The younger and less experienced visitors were usually happy to ask the great man to suggest a topic to them and then they would do it. Some of the more senior ones would arrive with a problem in mind, and on arrival Doby would discuss it with them and often tell them that these projects were unsuitable. Either stocks were not available or, in order to get useful results, too large an experiment would have to be done for one person to manage, or else the time required to complete the experiment would be longer than the person's expected stay at Columbia.

In these cases the professor, having been thinking about the matter for some time, would then suggest a suitable topic that would fit in with the visitor's interests and that could actually be done in the time available. After an agreement was made for such a project he would take the person in to Spassky, tell Spassky what was going to be done, and then Spassky would provide any necessary apparatus, primarily dissecting microscopes, etherizing bottles, and such things, and would then either immediately, or quickly, provide the flies with which the experiment would be started.

During their visit Dodik would frequently stop by each of the laboratories to see how the experiment was going and would often look at some of the flies under the microscope and at some of the data, in order to see whether any problem existed or whether the project looked as if it would come through in time. When the experiment was over, the visitor would write it up and then present the professor with a draft that was usually not in terribly good shape. It was done under the pressure of time, and frequently not in good English, since some of the visitors did not have a great familiarity with the language. In such cases Doby would, over the course of a day, rewrite the paper in good English and then present it to the author. In spite of the fact that Doby usually suggested the topics, was totally involved in their development, and, in numerous cases, actually wrote the paper, he never permitted himself to be listed as a coauthor.

As mentioned earlier, Boris Spassky played an essential role in the laboratory. Boris was born in Russia but attended Charles University in Czechoslovakia where he graduated with a degree in forestry. He came to the United States in 1930 and attended the University of California at Berkeley where he received a master of science degree. During these years, Boris met and worked for Dobzhansky. After a short period with the U.S. Forest Service, Boris returned to Dobzhansky's laboratory at the California Institute of Technology. Shortly thereafter Dobzhansky moved to Columbia and brought Boris and his wife Natasha with him. By coincidence Boris and Dodik had wives whose name was Natalie and whose father's name was Peter. As a result they were both Natasha Petrovna, and when they got together to talk they were constantly calling each other by that name. Both Natashas, from time to time, worked in the laboratory, and close to the time of the move to New York, each of them gave birth to a daughter.

Boris was a large man. He had a slow method of speech and was a little bit naive in some ways. For instance, on one occasion when Louis Levine was working on the last stages of his dissertation, I dropped by to see the professor but he was away. When he came back Boris told him "the real Levene was here to see you," distinguishing me from the student, and for several years after that everybody around the laboratory was calling Lou, Pseudo-Levine. Boris's height was particularly helpful when, on occasion, exterminators would come to the eighth floor of Schermerhorn, where they would spray some insecticide for insect parasites in the mouse laboratory at the other end of the hall. They would then typically try to spray the entire floor so that there would be no chance of the bugs' coming back, but Boris would hear of their being there and would appear bearing down on them and saying loudly "keep away from this end of the hall. Very important work is going on and it would be destroyed." Since on those occasions he looked very forbidding, the exterminators would beat a hasty retreat.

Dobzhansky had a tiny office at Columbia that opened into a quite large laboratory room where he counted flies and looked at chromosomes. A quarter of that laboratory was occupied by Boris. In addition there was a tiny room next door where there would be an assistant to do other chores, primarily the chromosome preparations. Starting in 1949, that person was Ogla Pavlovsky. Olga and her husband Vadim were trained—he in chemistry and she in biology—in the Ukraine, but when the Nazis invaded the Ukraine, the Pavlovskys tried to go to the south, reaching the Crimea, which was also overrun by the Nazis, and they were finally captured. Ogla was able, however, to get a position in a German laboratory as an assistant, and so the couple managed to survive. Eventually they were able to get away and arrived in New York where they met Dodik who hired her to fill the position of laboratory assistant and chromosome preparator.

In 1962 Dunn had to retire, because he had reached the age of sixty-eight. At first Dodik pushed for Bruce Wallace as his replacement, since he had always wanted Bruce as his colleague and ultimate replacement. However, there was no enthusiasm for Bruce in the department, and so Dodik then suggested Richard Lewontin. While there was some positive response to Dick, particularly from John Moore, it became clear that not enough support existed and that the department would probably appoint a biochemical geneticist as Dunn's replacement. This would leave Dodik isolated in the department, and so he decided to leave. A number of opportunities opened, particularly at the University of Chicago and the Rockefeller Institute (not yet Rockefeller University). I pushed for Rockefeller, since he would still be in New York and it would be easy to collaborate. At Rockefeller he would have lots of space and could keep it when he retired. Also, I think staying in New York, where he had so many friends, appealed to him, and in any event he moved to Rockefeller.

It just so happened that in 1961–62 activities at his Columbia laboratory had reached an all-time high, with three postdoctorals: Lee Ehrman, Chana Malogolowkin (on a long-term visit), and Angela Solima from Italy, and six Ph.D. candidates: Francisco Ayala and Marvin Druger, who were nearly finished; Sergei Polivanov and David Weisbrot, who were well along; and George Carmody and Seymour Kessler, who still had quite a way to go. After the move I was the faculty member in charge of all these people, with Chana in charge of operation of the laboratory and all the students visiting Dodik from time to time to report on their work. After two years of this arrangement, we had to leave Schermerhorn to free the space for Dodik's successor, Geoffrey Zubay, and Chana returned to Brazil. Angela had already married Tom Simmons and moved to Washington. At about this time, one of a row of University-owned private homes on 117th Street was remodeled to provide laboratory facilities. The two remaining students, Carmody and Kessler, moved there and finished, with Lou Levine coming by from City College to run the laboratory. Shortly after, the private houses were torn down to make way for the School of International Affairs.

Around 1960 Doby visited Spain, and while in Madrid he met a graduate student named Francisco Ayala, who had an undergraduate degree in physics and a doctor of theology degree with dissertation in Latin and who had become a Dominican priest and was now studying genetics. Doby was much impressed with him and urged him to come to Columbia for a Ph.D. Ayala came to Columbia to work on his dissertation and became exceedingly popular with the other graduate students. His dissertation, which I read carefully for correction, as I did for the other students, was the only dissertation I ever saw that was written in perfect English, with no grammatical errors, with no difficulties in seeing what he was getting at, and with very little need for change. Also, unlike the other students, Ayala was a good conversationalist with broad interests and was often invited to the Dobzhansky home for evening get-togethers with other friends.

While a student Ayala lived in a Dominican monastery on the upper West Side where his duties included saying mass at the crack of dawn, and he would then come and spend the day in the laboratory. On graduating he did not want to return to Spain since he was exceedingly anti-Franco and also against the hierarchy of the Spanish Roman Catholic Church, which was mostly pro-Franco. He accordingly obtained a teaching position at Providence College, which was a Dominican college in Providence, Rhode Island, and spent several years there while he was getting more disenchanted with the Catholic Church, with the Dominicans, and with Spain. Eventually he resigned from the Dominicans and the priesthood. He was in touch with Dobzhansky about his situation, and Doby suggested that he come to Rockefeller, where Dobzhansky had in the meantime moved, and become his associate there, which he did, and they continued in that sort of relationship with Francisco as assistant professor at Rockefeller University until 1971, when the tables were turned.

Dobzhansky had gone to Rockefeller, for one thing, because people could retain their laboratories after they retired. But times had changed and space was getting tight. Even before the time had come for Dobzhansky to retire in 1970, the provision was that, upon retirement, he could have an office and a very small laboratory but not the half floor that he had been having for himself and associates. At this point Dodik's very old friend Michael Lerner at Berkeley had a brilliant idea. Davis had an opening in their genetics department, and if they hired Francisco as an associate professor with tenure, they could get Dobzhansky thrown in practically for free and would get all the benefits of the great man's being at Davis along with the younger, dynamic Ayala, who was making increasingly important contributions to the field of population genetics. Both Francisco and Dodik approved of the plan, and this idea was pushed ahead, and so the tables were turned. For the rest of his life Dodik was, at least officially, simply an associate of Francisco, who was the boss. Of course, Dodik had the freedom to do what he wanted.

With Dobzhansky's leaving Rockefeller, Boris Spassky decided to retire, ending a most pleasant and fruitful association of close to thirty-five years. Olga Pavlovsky was able, however, to remain with the professor, and in Davis, the Pavlovskys had a house adjoining Dodik's so that he could spend lots of time with them and have most of his dinners there. He very much missed Natasha, who had died in 1969. Unfortunately, about three years after they moved to Davis, Olga developed breast cancer and had to return to New York, ending a twenty-five-year period of devoted service.

Davis was a very attractive place since it had a large genetics department with great activity in many fields and also some very good friends such as Ledyard Stebbins and Herman Spieth. There were many other people at Davis with whom Dodik became very friendly.

In 1968, well before the move to Davis in 1971, Dodik and Wyatt Anderson were on a trip to Colombia when Natasha phoned to tell me that he had gone to a doctor for a routine checkup and the doctor had found that he had lymphoma. The diagnosis was that he had some two years to live and that chemotherapy might help but would not be a cure. Dobzhansky did not admit his illness to me for some time and eventually said he had simply assumed that I knew. He would regularly have to go for a white blood cell count, which tended to get higher and higher, and in some cases they reached a level that would put most people in a hospital, but because he was tremendously strong he would ignore it and go about his business. It was not until a good time later that the disease began to catch up with him, probably in 1974, certainly in November, when he and Francisco came to New York for an international scientific meeting at the Waldorf-Astoria run by the Reverend Moon. He told me his principal reason for coming was to see his daughter Sophie for the last time. He was definitely in poor health and spent a good deal of time in his room. As soon as he and Francisco had participated in the session they were scheduled for, he left the meeting and went back to Davis. After that his health continued to decline, but he continued to go to his office regularly, until one morning when he did not feel well enough to get up for breakfast. Michael Andregg, his graduate student who was living next door, phoned Francisco to come with his car and drive them to the hospital. Francisco came right over and drove them to the hospital, but sadly, Dodik died on the way.

4

Resistance and Acceptance: Tracing Dobzhansky's Influence

Costas B. Krimbas

In his 1937 book _Genetics and the Origin of Species_ Theodosius Dobzhansky indicated the way to transform the study of evolution into an experimental scientific program. This book was instrumental in stating clearly the synthetic theory of evolution, and by this formulation Dobzhansky played a key role in the history of modern evolutionism. In his version of the synthetic theory he combined the knowledge and techniques of Morgan's school of genetics with the conceptual constructs of Chetverikov's Russian school and the mathematical models of population genetics and evolution of Sewall Wright. From 1937 on, Dobzhansky devoted himself in building up, enlarging, and promoting this new experimental field, population and evolutionary genetics. In the United States he was subsequently recognized as its leader. But his importance was also acknowledged outside the country. Thus J. B. S. Haldane once said that the only reason to visit the United States was to see Dobzhansky! broodcast

Dobzhansky's work and ideas were disseminated outside the United States, not only through his books and scientific papers, but also by his

The narrative is based on personal reminiscences and information I have requested from several students, collaborators, and other scientists (a few others have not replied): Mrs Berthe Boesiger, Drs. H. Burla, R. Falk, S. Lakovaara, L. Levine, R. C. Lewontin, D. Marinkovi'c, E. Nevo, G. Pasteur, C. Petit, J. R. Powell, A. Prevosti, K. Sankaranarayanan, A. Saura, D. Sperlich, L. Van Valen, and B. Wallace. I thank them all.

lectures during his numerous travels and by his foreign students and collaborators. A special mention should be made of South America, a privileged area for his influence: Dobzhansky in 1948–49 spent an entire sabbatical year in Brazil; made numerous friends, students, and collaborators; and promoted an ambitious experimental program. At the end of his life he extended his activity to Mexico. His students and collaborators Louis Levine, Jeffrey R. Powell, and Wyatt W. Anderson have continued along with their American and Mexican colleagues this program of *Drosophila* research in Mexico.

However, the situation has not been the same for Europe. Of course, the transfer of an idea is not a passive process, similar to the diffusion of a soluble chemical into a solvent according to a random walk of molecules. It is more like the success (or failure) to establish a new colony or the germination of a plant grain following its dispersal. In the latter case the subsequent growth and establishment of an individual plant are subject to and depend on environmental conditions and vagaries, on systematic factors, and eventually on random events. The persons responsible for the transfer of an idea generally display specific attributes and have in some sort been selected for this; the channels and ways of transfer are not random. Following the initial transport, the final acceptance, and incorporation of ideas, the ways of looking at things, problematics, concepts, and knowledge or their rejection, marginalization, and oblivion depend on several conditions defining their receptivity. These include the interest shown for them and the resistance they encounter. The factors controlling receptivity are related to the intellectual history of the country, to the temporary prevailing intellectual trends, and to the existing social structures and groups.

Europe in this text will be used in an extended sense, including Israel and Egypt. It is an area comprising several countries with markedly different and important intellectual traditions, countries having been subjected during World War II to different traumatic experiences that marked them. Dobzhansky's influence could reach Europe in a significant way only after the war, since his ideas on evolution were widely presented shortly before its declaration. Thus, tracing the fate of Dobzhansky's influence in different European countries may provide a good illustration of the various processes and situations encountered when a novel and powerful idea is disseminated.

In England Dobzhansky left few direct influences. John Beardmore, coming from Thoday's laboratory, spent the entire year of 1958 at Columbia University working with him as a postdoctoral fellow. Beardmore later became professor at Groeningen and thereafter at South Wales. He is probably the person who carried most of Dobzhansky's tradition in England, but he should mainly be considered as a product of the English local tradition. Moreover, Philip Sheppard, from Liverpool, a member of the Fisher-Ford school, felt it necessary to spend a year with Dobzhansky, but apparently he was not much influenced by him. In 1959–60, during the year of Dobzhansky's sabbatical leave, John Maynard Smith, a student of Haldane, was selected to replace him at Columbia University. This was probably due to the fact that Maynard Smith had published several papers showing important heterotic effects in hybrids between pure strains of *Drosophila subobscura*. From this stay at Columbia, Maynard Smith hardly received any influence from Dobzhansky. I have also mentioned the favorable attitude that J. B. S. Haldane harbored for him; Haldane extended his friendly approach also to Bruce Wallace, the most Dobzhanskian of Dobzhansky's students. In 1965 Dobzhansky was elected foreign fellow of the Royal Society.

In spite of this honor, English geneticists did not appreciate Dobzhansky *a sa juste valeur.* For them he was another successful evolutionist, nearly on the right track, but they considered themselves as constituting the group of the frontal advance that derived directly from Darwin; they were the direct descendents from the God. The dominant Fisher-Ford school probably also had some differences to settle with Dobzhansky, owing to his association with Sewall Wright, the evolutionist who first (with the exception perhaps of Chetverikov) discovered the importance of genetic drift. It is significant that Dobzhansky was elected to the membership of the Royal Society only after he had been a hard selectionist for a number of years. Some members of the Fisher-Ford school still retain some hostility toward him. Not so for E. B. Ford, who for years was his friend. Ford in his biographical memoir of Dobzhansky recognizes that he *"was the most distinguished geneticist in the U.S.A. and one of the most distinguished in the world (italics mine)."* This posthumous praise reflects neither any special influence Dobzhansky might have left on England nor the acceptance that he was instrumental in forming and propagating the neo-Darwinian theory of evolution.

Along with England we should examine the Netherlands. These two countries are closely connected in such a way that sometimes it seems legitimate to consider the Netherlands as an intellectual extension of England. The presence of Beardmore in Groeningen helped in creating a Dobzhanskian tradition. His successor W. Van Delden has also worked with Dobzhansky. In addition the problematic and work of Professor

W. Scharloo does not seem completely alien to that of Dobzhansky. In Leiden the genetics chair was occupied by Krishnaswami Sankaranarayanan, of Indian origin but naturalized Dutch, for years a student of Dobzhansky. However, Sankaranarayanan's interest is focused on radiation genetics and thus is far from an evolutionary problematic.

A country with a very strong and important local tradition does not accommodate easily to newcomers. This was the case for England. France is a slightly different story. Here too a strong tradition, more ancient than the English one, prevailed in evolutionary studies starting with Buffon, followed by Lamarck, and ending with Geoffroy Saint Hilaire. The presence of Lamarck's figure inhibited the expansion of the Darwinian tradition to this country. France has been the most anti-Darwinian of the European countries, and before World War II Lamarckism prevailed among French evolutionists. Thus Dobzhansky had to contend with the French anti-Darwinism and eventually Bergsonism. This is meant to describe his ambiguous relations with several French evolutionists, e.g., P.-P. Grasse and A. Vandel. Both displayed a friendly behavior toward him but at the same time did not become convinced by his arguments and did not accept his ideas. On the other hand among the Darwinists only Philippe L'Heritier could be considered truly friendly to him.

Georges Teissier apparently disliked Dobzhansky; he seemed to think that Dobzhansky in some way stole from him and from L'Heritier the credit for inventing population cages and subsequently experimenting with them. Dobzhansky did in fact use these *demometres*, as L'Heritier, their inventor, called them, but did expressly mention that they were devices already used by L'Heritier. If the papers of Dobzhansky have been more widely read than those of the French scientists, this was mainly due to the fact that in these cages Dobzhansky had followed the evolutionary fate of naturally occurring inversions, not laboratory mutants, as the French scientists did.

Boris Ephrussi, probably for different reasons, also seemed antagonistic to Dobzhansky; I remember in 1958 when I met him at Columbia University in New York (he had been my teacher in Paris) and we started a friendly conversation. When asked, I replied that I was there working with Dobzhansky, and he instantaneously became icy, turned his back on me, and left! This might explain why Dobzhansky never became a member of the French Academy and why he was bestowed the honorary degree at the Sorbonne too late in life. It was only after the death of Teissier, in 1972, that he obtained the degree and was invited to lecture at the Collège de France (both in 1975)!

In spite of this, Dobzhansky had more followers in France than in England. Maxime Lamotte, a self-made French evolutionist and professor at Ecole Normale in Paris, was one of his first contacts. Dobzhansky invited him to participate in the 1959 Cold Spring Harbor Symposium he was organizing (with C. S. Coon, L. C. Dunn, G. L. Stebbins, B. Wallace, and M. Demerec), in order to confront Philip Sheppard, also invited. Lamotte had published his thesis, a major work on the population genetics of color morphs of land snails, and had suggested that drift played an important role in determining their geographic distribution. On the contrary Sheppard was an advocate of natural selection.

Claudine Petit was probably the first French collaborator of Dobzhansky. Dobzhansky at every occasion spoke of the great discovery Petit had made, the rare-male effect (a form of frequency-dependent sexual selection giving to the rare males an advantage proportional to their rarity). This form of selection explained the retention of polymorphism in a population without requiring a differential survival of genotypes, deaths of individuals. The support of Dobzhansky was crucial for Petit, a student of Teissier, and helped her academic recognition in France. Later Petit worked in New York in Dobzhansky's laboratory together with Lee Erhman, an American student of Dobzhansky, who also studied the same phenomenon. Petit became professor at Paris 7 University and passed the Dobzhanskian tradition to her students, especially Dominique Anxolabéhère and Georges Perriquet. Petit was also instrumental in the election of Costas Krimbas, another student of Dobzhansky, to Paris 6 University and the decision to give an honorary degree to R. C. Lewontin, one of Dobzhansky's most important students.

However, the closest collaborator of Dobzhansky in France, his local representative and friend, was Ernst Boesiger, born Swiss but naturalized French, who worked in Montpellier. Boesiger was deeply interested in philosophical and humanistic questions and in this he fitted well with Dobzhansky's later years' interests. On the other hand Boesiger was, in my opinion, neither a great experimenter nor very critical in interpreting results; he never questioned or challenged the views of his mentor. The fact that in the French milieu Boesiger was somewhat isolated did not permit a rapid expansion of Dobzhansky's influence. But Boesiger has translated Dobzhansky's texts in French, and they produced together, in 1968, in French, a book of evolutionary essays. They also left unfinished a book on human culture and its evolutionary interpretation. This last text was completed by B. Wallace and published in 1983, several years after their deaths. In Belgium, an intellectual satellite of France, Dobzhansky also had a student, M. J. Heuts, now in Louvain.

Thus in France Dobzhansky did not much affect the older generation, which remained Lamarckian, and he had a moderate reception by the Darwinians of his age group. He had, however, a greater influence on the younger generation of evolutionists.

Germany and Austria present a different story. The Nazi social Darwinism, Nietzscheism, and racism, which led to the genocide of several ethnic groups, including the Jews, made population genetics and evolutionism fields of study that German biologists avoided discussing after the war was over and their country was occupied by the allied forces. It is significant that only the Felix Mainx group in Vienna was active in population genetics research and studied naturally occurring inversions in *Drosophila subobscura*. Of this group Diether Sperlich, later professor at Tübingen, spent a few months in 1964 in Dobzhansky's laboratory. Actually in respect to the name, Sperlich's chair in Tübingen is the only one of population genetics in Germany and Austria! Sperlich had many students (among them W. Pinsker, now professor in Vienna) and promoted Dobzhansky's ideas and concepts, together with some originating from his old professor, Mainx: the idea that the selective mechanism responsible for the retention of inversions resided in "position effects." Another German student of Dobzhansky was W. Drescher, now in Bonn, working on honeybees.

The favorite species of the Mainx group was *Drosophila subobscura*, which soon became the European counterpart of the North American *Drosophila pseudoobscura*, the favorite material of Dobzhansky and his American students. Apart from the Vienna and later the Tübingen group, several other groups worked with this material: Burla's group in Zurich, Prevosti's group in Barcelona, Krimbas's group in Athens, Marinkoviç's in Yugoslavia, and Lakovaara and Saura in Scandinavia. The work with *subobscura* was initiated in England before the war by C. Gordon, J. B. S. Haldane, H. Spurway, and their students. Most of this research was, however, focused on the study of the formal genetics of this species. By contrast the Athens, Barcelona, Beograd, Scandinavian, Vienna, and Zurich groups performed work oriented to the same direction as that of Dobzhansky and his collaborators and used similar techniques.

Hans Burla met Dobzhansky first in São Paulo, Brazil, during 1948–49 and, as he says, he became imprinted with his enthusiasm and tried to imitate his ease in approaching demanding projects. Later Burla became professor in Zurich. In this respect note that Boesiger had failed to obtain a position in Switzerland, perhaps because Adolf Portmann, a comparative anatomist and professor in Basel, who also represented evolutionism in Switzerland, apparently did not like him. Actually Portmann did not seem to like Dobzhansky either, because Portmann was too much of an idealistic philosopher to accommodate with a mechanistic evolutionism, as he saw Dobzhansky's! But in Zurich Professor Hadorn accepted Burla, who got an assistantship. Burla and his students (especially Walter Goetz and Hans Jungen) followed Dobzhansky's program but did not readily accept all the "master's decisions." Apparently Burla found Dobzhansky authoritative and dogmatic. In Switzerland the scientific structure is very individualistic; biology varies from place to place and no schools are formed nationwide. The intellectual interchange between universities in different parts of the country is restricted, despite common annual meetings. Everyone sticks to his or her personal brand of biology. It is a reflection of the political structure of Switzerland as seen in its federalism, in which power is regional and decentralized. Thus Burla's influence was restricted to Zurich and to his own group.

In Scandinavia Dobzhansky had some good contacts. His friends there were Arne Muntzig and his successor Ake Gustafsson, professors in Lund. He also knew Otto Mohr, from Oslo, a human geneticist, and Esko Suomalainen from Helsinki. In Denmark he was on good terms with Oivind Winge and Morgens Westergaard, both in Copenhagen. In Stockholm, at the 1948 Congress of Genetics, Westergaard urged Dobzhansky to take his "genius" student Ove Frydenberg to the United States. Subsequently Frydenberg worked with Dobzhansky in Brazil but ended as one of his most heartfelt enemies: he was accused of destroying, by releasing the wrong strain of flies, a monumental natural experiment in the island of Angra dos Reis. Frydenberg shared the blame with Maghalaes, a Brazilian geneticist. Probably Frydenberg was also overcritical of Dobzhansky's experimental designs. Later Frydenberg became professor in Aarhus and had some students.

In the 1970s two Finnish geneticists, Seppo Lakovaara, professor in Oulu, and his assistant Anssi Saura worked in Dobzhansky's laboratory at Rockefeller University. A few years later Saura returned and worked again with Dobzhansky. Saura is now professor in Umeå and pursues research on Dobzhansky's program using *D. subobscura*. Both Saura and Lakovaara have contributed considerably in establishing Dobzhansky's legacy in Sweden and Finland. Dobzhansky became a foreign member of the Swedish, the Danish, and the Finnish Academies of Sciences.

The third group in succession that started working with *D. subobscura*, especially in inversion polymorphism, after those at Vienna/Tübingen and at Zurich, was the Athenian. Costas Krimbas spent more than two years working as a postdoctoral in Dobzhansky's laboratory (1958–60) and subsequently was elected professor at the Agricultural University of Athens. It was at a time in Greece when the prewar generation of professors of German and French intellectual derivation were giving the path to those who studied in Anglo-Saxon countries. Thus the moment was favorable for Krimbas, because at that time everything American enjoyed a high esteem. Krimbas and his students (the Greeks E. Zouros, S. Tsakas, and M. Loukas; the Portuguese A. Brehm; and the Iranian M. Khadem) continued to work on the program Dobzhansky had proposed. Another Greek who spent a postdoctoral in Dobzhansky's laboratory, then at Rockefeller University, was Costas Kastritsis. Kastritsis got his Ph.D. in Texas with W. S. Stone and then, influenced by Krimbas, went to Dobzhansky's laboratory. After leaving New York Kastritsis took positions in Texas and subsequently was elected professor at the University of Thessaloniki. Kastritsis with his students translated into Greek Dobzhansky's *Genetics of the Evolutionary Process*.

Eleutherios Zouros was originally a student of Krimbas and later of R. C. Lewontin, who was at that time professor in Chicago. Zouros subsequently held positions of professor in Halifax and in Crete Universities. Thus the Dobzhansky tradition became dominant in Greece, and this was due to a weak or nonexistent local tradition in this area and to the favorable timing when Dobzhansky's ideas were brought to the country. In the 1960s the old guard of German- and French-educated intellectuals started being replaced by those of English and American education. Actually Krimbas was in between the two generations, having studied in Switzerland and in France before coming to Dobzhansky's laboratory.

The relations among the European scientists belonging to the same Dobzhanskian tradition formed a net-like pattern. A. Pentzou-Daponte and her professor at Thessaloniki University, A. Kanellis, collaborated first with D. Sperlich and then with E. Boesiger. This was another case of transition from the German intellectual tradition (Kanellis worked with N. Timofeeff-Ressovsky in Berlin before and during the war) to the American one through Sperlich, an Austrian, and Boesinger, a Swiss-German, as intermediates. Bruce Wallace spent half of a sabbatical year at Krimbas's laboratory in 1965 and another half at Groeningen. In 1983 he spent two months with Mourad's group in Alexandria, Egypt. In 1986–87 he also spent twelve months in Sperlich's laboratory in Tübingen. Sperlich spent some sabbatical leave time in J. Powell's laboratory at Yale, and Krimbas in Lewontin's laboratory at Harvard.

Countries in the periphery, or rather the semiperiphery (according to I. Wallerstein's concepts), like Greece, were more likely to adopt new ideas coming from the metropolitan center, the United States, and marginalize the influences they had received from predecessor metropolitan centers. This is the case for several countries at the margins of the European metropolitan centers. Thus Yugoslav scientists were influenced directly by Dobzhansky or through F. Ayala, a former student of Dobzhansky and his associate for a long time, or through J. Powell, his nearly last student. Dragoslav Marinković, from Beograd, played the role of the *chef de file* in this case. He led a group of scientists working with *D. subobscura* including M. Andjelković, V. Kekić, M. Milosević, M. Milanović, and N. Tucić. Some members of this group collaborated in faunistic surveys of *Drosophila* with G. Baechli, an associate of H. Burla; others with Sperlich; and others with W. Anderson, another student of Dobzhansky in the United States. In Yugoslavia the change in intellectual tradition was from the German to the American one (and the Russian). Sperlich and Baechli played the role of intermediates, both belonging by birth to a Germanic tradition but being active members of the American one.

Two Egyptian geneticists worked with Dobzhansky at Columbia University toward the end of the 1950s: Professor A. (=Abd el Azim) O. Tantawy came on a sabbatical leave; Abd el Khalek Mourad, later professor also in Alexandria, obtained his Ph.D. with Dobzhansky. Before their arrival Dobzhansky had visited Egypt and had extensively collected *Drosophila* flies with these collaborators and with the help of Professor G. S. Mallah. Another Egyptian, M. M. Dawood, studied with T. Prout in California, a former student of Dobzhansky, and collaborated with Monroe Strickberger, also a former student of Dobzhansky. Dawood years later occupied the position of Minister of Agriculture in Egypt.

The move from the dominant English tradition in Egypt to the American one was swift in these areas of knowledge; even Khishin, a former student of Charlotte Auerbach, seemed more affected by the Dobzhanskian heuristics than by the English tradition, according to R. Falk. In Egypt Dobzhansky was widely accepted also because of his pro-Arab and his anti-Israeli attitude. Some charged him with being an anti-Semite, which is a gross overexaggeration. The truth is that he never visited Israel; either he refused to go or he avoided going. Perhaps at that time visiting both Israel and Arab countries was incompatible even for an American. Anyway he disapproved of Israel's politics toward the Palestinians.

The late Elizabeth Goldschmidt, professor of genetics in Jerusalem, worked in Dobzhansky's laboratory in 1950–51. She seems not to have enjoyed her stay there, having had a difficult time with him. Goldschmidt, in collaboration with E. Stumm-Zollinger from the Vienna group, studied the inversion polymorphism of *D. subobscura* populations in Israel. She did not leave a permanent influence on Israeli evolutionism, her only student, Tuviah Kushnir, having been killed in the Israeli-Arab war. After her suicide she was replaced by R. Falk, a student of J. Crow, and thus belonging to the rival group of H. J. Muller, which challenged the views of Dobzhansky as far as the origin of the genetic load is concerned and the forces maintaining the genetic polymorphism in natural populations.

Thus Dobzhansky would not have left any trace in Israel if it had not been for the presence of Eviator Nevo. Nevo, a heretic among the Israeli geneticists, is a hard panselectionist, considered by Falk more Dobzhanskiist (sic) than Dobzhansky! For a time Nevo worked at R. C. Lewontin's laboratory in Harvard and now holds the position of professor at the University of Haifa. Nevo has many collaborators and students and produces massive results and numerous papers.

The Egyptian case fits well with the model that countries at the margins are more prone to accept an influence. Israel, on the contrary, should not be considered a marginal country because it has an important scientific diaspora located at the metropolitan center, which is in close contact with the indigenous scientists. Furthermore, as I explained above, political reasons made the adoption of Dobzhansky's heuristics very difficult in Israel. Chana Malogolowkin-Cohen, a collaborator of Dobzhansky in Brazil, moved later to Israel. She did collaborate with Sperlich, but her influence in Israel was minimal.

Italy is a country with a long and venerable biological tradition but with a rather recent one in genetics and modern evolutionism. Three distinguished Italians have independently and concurrently brought the new ideas: Adriano Buzzati-Traverso, from Pavia (brother of Dino, the novelist); Claudio Barigozzi, from Bologna; and Giusepe Montalenti from Naples and later Rome. All of them appreciated Dobzhansky and were his friends. Dobzhansky frequently visited Italy and took an active part in several conferences organized by the Italians. He became a rather early foreign member of the Academia Nazionale dei Lincei. The good reception of his ideas is also probably due to the fact that Buzzati-Traverso before and during the war was a close collaborator with N. Timofeeff-Ressovsky; together they conceived a monumental experimental plan to study the effect of genetic drift by observing *Drosophila* populations in many Italian localities. This plan was never realized, but Buzzati-Traverso initiated studies of *obscura* group species and especially of *D. subobscura*. His close friend and taxonomist, Francesco Pio Pomini, who described several of the European species, was killed in 1941 at the Albanian front.

The Italian scientific milieu was not virgin and had already received positive influences from the very same school that also inspired Dobzhansky, namely, the Russian group of Chetverikov and this via Timofeeff-Ressovsky. In Brazil Dobzhansky worked with two Italians, Bruno Battaglia and Florence Padoa. Battaglia worked again with him at the Columbia University laboratory, in 1958. Dobzhansky appreciated Battaglia's proof that some color morphs of a copepod were retained by what he called a balanced selection. Battaglia became professor at Padova and director of the oceanographic station at Venice. He exerted an important influence in scientific affairs in his country. However, no active *Drosophila* research following the heuristic of Dobzhansky is to be encountered now in Italy.

The last but not least country to mention is Spain. Antonio Prevosti, originally an anthropologist, soon became interested in *D. subobscura* population genetics. Prevosti's early contacts were with the English school. In 1953 he met Dobzhansky in Pavia, at the "Symposium on Genetics of Population Structure" organized by A. Buzzati-Traverso just before the Ninth International Genetics Congress in Bellagio. This contact and the reading of *Genetics and the Origin of Species* left a vivid impression on Prevosti. He never worked in Dobzhansky's laboratory, but he was very much attracted by his naturalistic and humanistic views. Prevosti's work fits well with Dobzhansky's program. Ten years later Prevosti started the study of the inversion polymorphism of *D. subobscura*. From Prevosti derives the major part of the Spanish workers on population genetics and evolutionism. A. Fontdevila, A. Lattore, L. Serra, R. de Frutos, F. Mestres, M. Aguade, M. Papaceit, and many others are his students, who continue and extend Prevosti's investigations. Prevosti also had close contacts with some South American scientists, D. Brncic and S. Koref-Santibanez, who were also closely connected to Dobzhansky.

Dobzhansky also had two Spanish students, Eduardo Torroja, now in Madrid, and Francisco Ayala. Ayala cannot any longer be considered as belonging to the Spanish contingent. He has become an American citizen and has pursued his remarkable carrier in the United States, in close contact with Dobzhansky until the latter's death. However, the Spanish origin of Ayala and the ties he retains with the Spanish scientists increased the influence of Dobzhansky's ideas in Spain. Ayala also had Fontdevila working in his laboratory.

The acceptance of Dobzhansky in Spain fits rather well with the hypothesis that countries with a weak tradition in modern evolutionism

TABLE 4.1.

"European" Students, Collaborators, Friends, and Correspondents of Dobzhansky
(Those marked by + are dead.)

1. Students and Collaborators
England
    John Beardmore (Groeningen, S. Wales)
    +Philip Sheppard (Liverpool)
    John Maynard Smith (London, Brighton)
France
    +Ernest Boesiger (Montpellier)
    Claudine Petit (Paris)
    (Georges Pasteur, Montpellier via C. D. Kastritsis)
Belgium
    M. J. Heuts (Louvain)
Holland
    Van Delden (Groeningen)
    Krishnaswami Sankaranarayanan (Leiden)
Spain
    Francisco Ayala (New York, Davis, Irvine)
    Edouardo Torroja (Madrid)
    (Antonio Fontdevila, Barcelona via F. Ayala)
Italy
    Bruno Battaglia (Padova)
Switzerland
    Hans Burla (Zurich)
Denmark
    +Ove Frydenberg (Aarhus)
Germany/Austria
    Diether Sperlich (Wien, Tübingen)
    S. Koref-Santibanez (Santiago, Chile; Berlin)
    Willie Drescher (Bonn)
Finland/Sweden
    Seppo Lakovaara (Oulu, Finland)
    Anssi Saura (Oulu, Finland; Umeå, Sweden)
Yugoslavia
    Dragoslav Marinković (Beograd)
Greece
    Costas B. Krimbas (Athens)
    Costas D. Kastritsis (Thessaloniki)
Israel
    +Elizabeth Goldschmidt (Jerusalem)
    Chana Malogolowkin-Cohen (Tel Aviv)
    (Eviator Nevo, Haifa via R. C. Lewontin)
Egypt
    +Abd el Khalek Mourad (Alexandria)
    A. O. Tantawy (Alexandria)

2. Friends and Correspondents
England
    +E. B. Ford (Oxford)
    +J. B. S. Haldane (London)
France
    M. Lamotte (Paris)
    +P. Grasse (Paris)

Philippe L'Heritier (Paris, Clermont Ferrant)
Germany
    +Bernard Rensch
Sweden
    +Ake Gustafson (Lund)
    +Arne Muntzing (Lund)
Finland
    Esko Suomalainen (Helsinki)
Norway
    +Otto Mohr (Oslo)
Denmark
    +Oivind Winge (Copenhagen)
    +Mogens Westergaard (Copenhagen)
Austria
    +Felix Mainx (Wien)
Italy
    +Adriano Buzzati-Traverso (Parma, Napoli, Roma)
    C. Bargozzi (Bologna)
    Montalenti (Roma)
Spain
    Antonio Prevosti (Barcelona)
Poland
    Gajewski (Warsaw)

would be more prone to accept and incorporate the new ideas of this field. In this respect note that the dictatorship in Spain had a negative effect on scientific studies. Many important Spanish scientists left for Latin America and it was difficult to get a position in the Spanish universities if the Phalange or the Catholic Church did not support the candidate. From this few scientists escaped, among them Prevosti. Thus the dictatorship had a negative effect on the scientific tradition in this country (and in Portugal).

As I pointed out above, most of the European evolutionists working with *D. subobscura* were either former students and collaborators of Dobzhansky or were strongly influenced by him. They constituted a group whose members exchanged information and ideas. Burla in 1970 organized in Zurich the first *"Subobscura* Conference" (European Conference of Population Genetics of *Drosophila subobscura*). The only scientist present in this conference not working with *D. subobscura* was Dobzhansky. Thus the *subobscura* work was symbolically put under the aegis of the old master! A second meeting was organized in Barcelona (Sitges) by Prevosti in 1982 and a third in Tübingen by Sperlich in 1984.

The European group, including the scientists of Spanish obedience working in the Canary Islands (A. Gullon, V. M. Cabrera, J. M. Larruga, and A. M. Gonzalez), have produced an important amount of work and contributions to the field of modern evolutionism. Their achievements among others include first the proof that in nature the formation of clines of gene arrangement frequencies is due to selection. They were able to ascertain this through a careful study of the South America colonizing populations of *D. subobscura*. Their other accomplishments include the invention of a method for estimating the effective population size from temporal allozyme frequency changes, the careful study of linkage disequilibria in natural populations between allozyme markers and inversions, the discovery of a predicted missing link in inversion phylogenies, the establishment of species phylogenies of the *obscura* group, and the use of molecular markers in population genetics studies of *subobscura*. Thus Dobzhansky's legacy matured and produced fruits in continental Europe.

It is significant that this fruition mainly took place in Spain, Germany, Switzerland, Greece, Yugoslavia, and Sweden. They are countries that were out of the main track of biological research during the period before World War II or (for Germany and Sweden) out of the main track of modern evolutionary work. By contrast, England, France, and Italy did not participate in that, following their own local traditions. Thus, apart from random events, the pattern of our original working hypothesis seems verified, that in Europe a virgin terrain was most favorable for receiving and incorporating the new heuristics of Dobzhansky.

PART TWO

Intellectual Legacies

5

The Legacies of Theodosius Dobzhansky

Bruce Wallace

I have been assigned the task of identifying and expounding those aspects of Dobzhansky's work that constitute his legacy to evolutionary genetics or those that are likely to constitute a lasting contribution to the still slowly emerging field of population biology. In carrying out this task, I shall not attempt merely to list Dobzhansky's many contributions to evolutionary and human biology; Ayala (1985) and Levene, Ehrman, and Richmond (1970) have prepared excellent accounts of Dobzhansky's many accomplishments. Indeed, these existing articles make my task today more—not less—difficult. What can I say that is new? Or, stated differently, how can I shed new light on well-known items of Dobzhanskiana?

Hand-drawn diagrams of an onion and a Christmas cactus (figure 5.1) that have yellowed with age while hanging on the wall above my desk will provide me with a context within which to discuss Dobzhansky's work. The two plants I sketched were meant to remind me of the manners in which science can progress. My onion (through an admitted inversion of botanical fact) represents progress in science achieved by adding one thin, cautious layer at a time to an already established body of knowledge. This, in my opinion, is the progress favored (or, at least, fostered) by granting agencies that demand research proposals containing (1) a thorough review of the past, (2) a detailed account of the investigator's own objectives, and (3) a virtual guarantee that these objectives can and will be met.

The onion model described here corresponds to Thomas Kuhn's (1964) view that most scientific research is cumulative. Many seminars begin today with slides or transparencies that state (as if neither the speaker nor the audience could otherwise understand) the objectives of the research,

FIGURE 5.1

Sketches of an onion and of a Christmas cactus that are mentioned early in the text.

the hypotheses to be tested, and the means by which these hypotheses have been put to the test. Unfortunately, the layer of knowledge being added by the speaker is generally so thin that deviant results (unless they suggest an actual reversal of expectations) are not normally significant. Hence, in Kuhn's terminology, no one cumulative-type study is likely to overturn a prevailing paradigm. Switching metaphors, it is a rare straw, indeed, that breaks the camel's back.

I have dwelt on my onion model because much of Dobzhansky's research must be acknowledged as having been cumulative in precisely that sense. "Science," he would tell his students, "is 90 percent perspiration and 10 percent inspiration." One kept busy! One kept one's hands dirty! Dirty hands, in Dobzhansky's estimation, rendered one's inspirational (theoretical) thoughts realistic. In my own case, it was a study on microdispersal (Wallace 1970) that prompted Dobzhansky to exhort me over a period of several years: "Bruce, you *must* work on migration!" He invoked ethics far beyond the usual work ethic in buttressing his position. Much of Dobzhansky's scientific life was spent, each of us who knew him must admit, in confirming Kuhn's views regarding cumulative science.

The Christmas cactus in my ancient sketch represents my view of genuine, substantial scientific progress: Although each blossom of a Christmas cactus emerges only from the terminal edge of one of the plant's many joints, the position of that joint on the plant seems to be haphazard, if not random.

The Christmas cactus serves to represent the 10 percent of Dobzhansky's effort that was inspirational. Here one finds the basis of his enduring interests. Here one also finds his legacies—insights that altered the course of both his own otherwise routine activities and those of others. And here, unlike the case of added-on layers of onion, one finds intellectual advances that are sufficiently large to reveal associated errors. Sadly, one must acknowledge that, in science, Christmas cactus blossoms slowly become transformed into onions. The stimulating novelties of today become the hardened cores upon which layers of cumulative knowledge will be added tomorrow.

The first of Dobzhansky's inspirations that led him to depart from tradition was his view of geographic variation in coccinellid beetles. Local populations of "ladybird" beetles are polymorphic; that is, individual members of these populations differ both in color patterns and in the sizes of their many markings. Not only is there variation within populations, however, but (as a rule) there is also conspicuous variation among populations that inhabit different geographic areas. This geographic variation consists, for the most part, of differences in the frequencies of morphs that occur in many, widely dispersed local populations. In brief, geographic variation does not involve diagnostic morphs in this or that local population; on the contrary, frequencies of morphs change, often clinally, over broad areas.

Tradition would have led Dobzhansky to delineate areas within which local populations resembled one another more closely than any one of them resembled populations in other areas. The clusters of populations so identified would then be named—perhaps as subspecies through the use of Latin trinomials. And there the matter would have been dropped. One's work was finished!

Dobzhansky did otherwise. He saw geographic variation as an integral part of a species's composition. Just as a local population encompasses all the genotypes that may be generated in agreement with Hardy-Weinberg expectations, so the species encompasses interpopulation variation. That is the reality! To be complete, then, one's work must include studies of (and explanations for) the origin of the observed variation, the maintenance of this variation in local populations, and the diversity that arises among geographically separated populations. Although the point is seldom stressed, one must also explain the cohesiveness of the species's genetic endowment that exists despite tremendous genetic variation.

A second insight that follows from the first is that which Dobzhansky had with respect to the origin of species. Granted that each of two species consists of a geographically widespread assemblage of dissimilar populations, what then becomes the relationship between the two species themselves? Here, again, one encounters a Christmas cactus blossom, a legacy: A species is "that stage of the evolutionary process at which one actually or potentially interbreeding array of forms becomes segregated into two or more separate arrays which are physiologically incapable of interbreeding" (Dobzhansky 1935).

Speciation is a *stage* of what may or may not be a more or less continuous process but, nevertheless, a decisive stage. Within any one species, migrant individuals moving among the many local populations carry with them their genetic wares, which can be incorporated into the genetic fabric of each resident population. Speciation has occurred, however, when populations of species B *always* reject the wares offered them by migrants of species A—and vice versa. An event—speciation—has occurred that, aside from possible competitive interactions among individual members, puts two clusters of local populations on independent evolutionary paths. Before speciation occurs, the abortive evolutionary paths of local populations repeatedly diverge only to coalesce once more; after speciation, subsequent coalescing becomes impossible.

These two legacies lead us to a third: isolating mechanisms. Much of Dobzhansky's research effort of the mid-1930s (continued by C. C. Tan in the mid-1940s) involved classical genetic analyses of the sterility exhibited by *Drosophila pseudoobscura* x *D. persimilis* (known then as races *A* and *B* of *D. pseudoobscura*) hybrid males. Exciting as the results of these particular tests were, I believe the *legacy* that Dobzhansky has left us in this area of research is a three-parted one: (1) that species are reproductively isolated, (2) that this isolation requires *at least* a two-locus genetic basis, and (3) that the "final touches" (premating) to this isolation are wrought by natural selection as reinforcements that follow the reestablishment of contact between populations that have surpassed the species stage in their evolutionary process.

Ayala (1985) points out that the term *isolating mechanisms* was introduced by Dobzhansky (1937) and that it remains standard terminology among evolutionary biologists. The insight that led Dobzhansky to postulate (at least two) complementary genes as the heritable basis for reproductive isolation appears to be withstanding the tests of time (for example, see Hutter, Roote, and Ashburner 1990). The reinforcement of isolating mechanisms following the reestablishment of once spatially isolated species has not been fully accepted (see Mayr 1988:361), but the consistently greater sexual isolation observed between pairs of sympatric strains than between pairs of allopatric strains of the same species suggests that Dobzhansky's intuition was correct (see Ehrman 1965; Wasserman and Koepfer 1977). This intuition has been supported as well by the observation that *Drosophila* flies from populations of hybrid origin (*D. pseudoobscura* and *D. persimilis*) improve over a period of two years their ability to recognize members of their own individual populations (Wallace, Timm, and Strambi 1983).

Here, then, are three legacies, bearing upon the heart of evolutionary biology, that have been left us by Dobzhansky:

1. Populations are genetically heterogeneous collections of dissimilar individuals, and species are heterogeneous collections of genetically dissimilar populations. (Hence, the oft-cited expression: The most elementary step in evolutionary change is a change in gene frequency.)

2. Speciation is a stage in the evolutionary process. Populations that have passed that stage can no longer exchange genes; their evolutionary paths are now essentially independent (except that each may be a component of the other's environment).

3. The exchange of genes between species is prevented by isolating mechanisms whose initial genetic basis is provided by two or more complementary genes but whose elaboration is the result of natural selection acting in quantifiable, comprehensible ways.

A memorial service for Dobzhansky was held on the Columbia University campus during 1976. The featured guest speaker was E. B. Ford, the eminent ecological geneticist from Oxford University. After praising Dobzhansky as an individual of tremendous talents, Ford reminded his audience of some melancholy truths: In a relatively brief time, all those who knew Dobzhansky personally would also be dead, but his publications would remain. Next, however, all those who were influenced by his writings would also be gone and Dobzhansky would then have passed out of history. The sequence of events sketched by Ford was correct for its time, brought about largely by the admonition "not to read anything more than fifteen years old." As G. Evelyn Hutcheson remarked shortly before his death: "People know that I am famous, but they don't know what I am famous for."

A legacy, in my view, is a contribution that defies Ford's view of unremitting attrition through time. A legacy is like an epidemic disease that spreads and persists even though both the original victims and the original causative organisms vanish. In this sense, I believe Dobzhansky's *Genetics and the Origin of Species* among all his books is a legacy. That particular title may not persist (indeed, Dobzhansky himself abandoned it in its last revision), but the message it carried lives on. *Genetics and the Origin of Species* was the means by which Dobzhansky presented what have been identified earlier as three legacies and, as the conveyance of that message, has itself become a legacy. I believe that none of Dobzhansky's other books, interesting or profound as they may have been in their time, is a legacy in the same sense as his classic text. It brought sense and logic to an otherwise completely muddled branch of biology. Some, including Stephen Jay Gould (see Dobzhansky 1982:xxxiv), claim that the first edition of *Genetics and the Origin of Species* began the modern synthesis.

In the course of his experimental work, Dobzhansky encountered many situations that demanded interpretations other than those provided by existing conventional wisdom, by the prevalent paradigm in Thomas Kuhn's terminology. *Coadaptation, homeostasis*, and *euheterosis* are but three of the terms that he either invented or stressed in an effort to understand his laboratory and field observations. I consider it unlikely that these concepts will spread with sufficient rapidity to forestall their eventual disappearance, à la E. B. Ford's prediction.

One final Dobzhansky legacy remains, however: his students and their students. Dobzhansky produced more than forty doctoral students and trained a comparable number of postdoctoral fellows. For Dobzhansky and his wife Natasha, students were not merely students, they were friends— even family members. Many of us learned how to entertain in the Dobzhansky apartment—how to drink, how to eat, and how to carry on a lively, civilized conversation. We met each other—across many ages—not only in the laboratory but also in the Dobzhansky home. We met professorial colleagues (L. C. Dunn, Marcus Rhoades, and Alfred Mirsky come immediately to my mind) as well as older postdoctoral fellows (C. C. Tan, B. Sonnenblick, and Dietrich Bodenstein, for example); later, we met the younger generation (R. C. Lewontin, Timothy Prout, Francisco Ayala, Rollin Richmond, and Jeff Powell). Long before electronic computers made networking a catch phrase, Dobzhansky's students established a worldwide network of friends, colleagues, and lasting friendships. Even today, for example, Danko Brncic in Santiago, Chile, responds warmly to the news that my daughter will arrive at Viña del Mar in order to study the blood physiology of penguins. Of all the legacies left by Dobzhansky, his encouragement of friendship and collegiality among his many students and colleagues may be the most important. Those of us contributing to this book attest that Dobzhansky's training has persisted until now, at least. Future meetings of our students—as local assemblages, perhaps, but with cordially received, interassemblage migrants—will determine whether a true legacy is in the making—one that is of Dobzhansky even though it may not bear his name. The generous sharing of intellectual ideas, of suggestions regarding investigative procedures, of research materials, and of hospitality in the laboratory and at home are the hallmarks of a lasting Dobzhansky legacy.

If I may, I would like to turn now to a topic that was of considerable concern to Dobzhansky, a topic commonly referred to as the "neutralist-selectionist" controversy. Because scientists are human beings, scientific intercourse is not always carried out on a lofty intellectual plane, devoid of emotion, and subject always to rational thought. Persons hate being wrong. They hate even more being told that they are wrong, especially so if indeed they are. Skins are thin; personalities are complex; minds and recollections are selective.

Throughout his scientific career, Dobzhansky studied genes and traits that are subject to *natural selection*. He surely acknowledged whenever appropriate that chance events occur, but even here he pleaded ignorance when attempting to identify a neutral phenotypic trait. The morphologies, physiologies, and behaviors of organisms were, in his opinion, the outcome of evolutionary change, change wrought mostly—if not entirely—by natural selection.

In 1968, Kimura published an article in which he proposed that the recently discovered molecular variation involved genetic differences that were not subject to natural selection. The molecular variation that had already been demonstrated in numerous species of organisms had, in Kimura's opinion, no effect on the relative fitnesses of their carriers. One should be careful to note that it is *existing* genetic variation that Kimura claimed to be selectively neutral! Deleterious mutations, in Kimura's view, do not contribute to observable genetic variation; such mutations are eliminated from populations by a purifying or cleansing selection.

R. C. Lewontin (1974:200) summarized Kimura's arguments as follows:

The neoclassical argument is made up of two complementary parts. First, it attempts to show that the balance theory is untenable because it involves internal contradictions, and then attempts to show that the neoclassical theory is compatible with the data. Both parts of the argument are essential. To show only that the neoclassical view is compatible with the facts, to show that it is a sufficient theory, is not enough, for then the two theories stand side by side with no way to choose between them. It is a cornerstone of the neoclassical argument that the balance view is irreconcilable with all the important known facts.

My own role in the developing debate, and one that Dobzhansky seemingly accepted, was to argue that the basis on which Kimura eliminated natural selection was false: genetic loads as commonly calculated need not threaten a population's existence; indeed, the calculations may be irrelevant to that end. And, as for the "cost of evolution," many persons questioned its supposed consequences. I (1988) have even called the cost of evolution a computational artifact. There is no need to review the copious literature on the neutralist-selectionist controversy here; interested persons can consult Kimura's (1991) review article and my own *Fifty Years of Genetic Load* (Wallace 1991a).

What may be less well known are my recent efforts to resolve the apparent differences between the neutralist and selectionist positions. Through time, both parties to this dispute have shifted their positions. Therefore, it is important for future discussions to identify and focus on what the remaining differences are. I identify my present position by acknowledging that individual components of fitness are often poorly correlated with one another. Some, in fact, such as number and size of eggs, are, of necessity, negatively correlated. Because components of fitness are numerous, have optimal values, and are often poorly correlated, total fitness cannot be highly correlated with any one component. The matter is made complex (and interesting) by every component's possessing an intermediate optimum value beyond which further change reduces total fitness. Paradoxically, this assertion applies even to percent egg-hatch where the maximum, 100%, may be achieved only by substantially prolonging the time required for embryonic development. Even at the molecular level, unseemly haste results in errors.

The average correlation between total fitness and its individual components becomes zero if even a small average negative correlation ($-1/n$, where n equals the number of fitness components) exists between the various components and subcomponents. Recognition of this fact can be discerned in Lush (1951). It was made explicit by Dickerson (1955) and "rediscovered" by Reeve, Smith, and Wallace (1990).

In an attempt to identify a population (1) within which the individual components of fitness were known for each member of the population and (2) for which pairwise tests of "total" fitness had been performed among the population's individual members, I turned to professional prizefighters. For every major (championship or near-championship) bout, some fifteen aspects (age, weight, height, reach, etc.) of each contender are listed in the sports section of major newspapers (e.g., *New York Times*). In the opinion of gamblers and others interested in boxers and boxing, these numerical data bear in important ways upon the outcomes of fights. The considerations I have discussed here concerning total fitness and its components suggest otherwise. The data (Wallace 1991b) are overwhelmingly on my side: winners and losers of important prizefights do not differ systematically with respect to the carefully noted components of fitness. That these *are* components of fitness is attested, however, by the stabilizing selection exhibited in their regard in heavyweight bouts.

This brief account can be concluded by the following statement: If total fitness is not correlated with its components and subcomponents, it cannot be correlated with the genes responsible for those components. This statement is true even with respect to genes exhibiting pleiotropic effects. The genetic differences responsible for variation with respect to fitness components (recall that no genetic basis exists for total fitness; this relationship is *assigned* to genes, not *caused* by them) must be molecular. It is an accident of history (and a result of technical simplicity) that allozyme, protein, and DNA variation were revealed before the causal connection between quantitative characters and molecular variation was established. Thus, one can concede that the bulk of all genetic variation must appear to be neutral, even if it underlies selectively important traits. As a result of this conclusion, one sees that future discussions of what is normally regarded as the neutralist-selectionist controversy must be carried out on a new plane and against a new backdrop. In my opinion (see Wallace 1993), much of the controversy vanishes when one notes that "neutral" genetic variation includes that which "appears to be neutral."

REFERENCES
Ayala, F. 1985. Theodosius Dobzhansky. *Natl. Acad. Sci. Biographical Memoirs* 55:163–213.

Dickerson, G. E. 1955. Genetic slippage in response to selection for multiple objectives. *Cold Spring Harbor Symp. Quant. Biol.* 20:213–24.

Dobzhansky, Th. 1935. A critique of the species concept in biology. *Philos. Sci.* 2:344–55.

———. 1937. Genetic nature of species differences. *Am. Natur.* 71:404–20.

———. 1982. *Genetics and the Origin of Species* (with an introduction by Stephen Jay Gould). New York: Columbia University Press.

Ehrman, Lee. 1965. Direct observation of sexual isolation between allopatric and between sympatric strains of the different *Drosophila paulistorum* races. *Evolution* 19:459–64.

Hutter, Pierre, John Roote, and Michael Ashburner. 1990. A genetic basis for the inviability of hybrids between sibling species of *Drosophila*. *Genetics* 124:909–20.

Kimura, Motoo. 1968. Evolutionary rate at the molecular level. *Nature* 217:624–26.

———. 1991. Recent development of the neutral theory viewed from the Wrightian tradition of theoretical population genetics. *Proc. Natl. Acad. Sci.* 88:5969–73.

Kuhn, Thomas S. 1964. *The Structure of Scientific Revolutions.* Chicago: University of Chicago Press.

Levene, Howard, Lee Ehrman, and Rollin Richmond. 1970. Theodosius Dobzhansky up to now. *Evol. Biol.* (Suppl.):1–41.

Lewontin, R. C. 1974. *The Genetic Basis of Evolutionary Change.* New York: Columbia University Press.

Lush, J. L. 1951. Effectiveness of selection: Summary. *J. Animal Sci.* 10:18–21.

Mayr, E. 1988. *Toward a New Philosophy of Biology.* Cambridge: Belknap Press of Harvard University Press.

Reeve, Russell, Eric Smith, and Bruce Wallace. 1990. Components of fitness become effectively neutral in equilibrium populations. *Proc. Natl. Acad. Sci.* 87:2018–20.

Wallace, Bruce. 1970. Observations on the microdispersion of *Drosophila melanogaster. Evol. Biol.* (Suppl.):381–99.

———. 1988. In defense of verbal arguments. *Persp. Biol. Med.* 31:201–11.

———. 1991a. *Fifty Years of Genetic Load: An Odyssey.* Ithaca, N.Y.: Cornell University Press.

———. 1991b. The manly art of self-defense: On the neutrality of fitness components. *Quart. Rev. Biol.* 66:455–65.

———. 1993. Toward a resolution of the neutralist-selectionist controversy. *Persp. Biol. Med.* 36:450–59.

Wallace, B., M. W. Timm, and M. P. P. Strambi. 1983. The establishment of novel mate-recognition systems in introgressive hybrid *Drosophila* populations. *Evol. Biol.* 16:467–88.

Wasserman, M. and H. R. Koepfer. 1977. Character displacement for sexual isolation between *Drosophila mojavensis* and *Drosophila arizonensis. Evolution* 31:812–23.

6

Four Decades of Inversion Polymorphism and Dobzhansky's Balancing Selection

Timothy Prout

No one would question that when Dobzhansky started his work on *D. pseudoobscura* inversions in the 1930s he did so because he wanted to study evolution in natural populations. The survey of inversion frequencies continued for four decades (Anderson et al. 1991) and very little evolution was observed. Does this mean the project was a failure? The answer is that it was not a failure, because "studying evolution" means two very different classes of activity. As used above, *evolution* means change, and very little change was observed. The other use of the word means studying the implications of Darwin's theory of natural selection, which is the cause of evolution. The Darwinian theory is that at any time in the past there was additive genetic variation for fitness (no tautology). The principle of uniformitarianism leads to the prediction that populations should exhibit such variation at any time, including today. It is this prediction that played a major role in the modern synthesis of which Dobzhansky was a leader.

Many academic departments today have evolution in their titles. The departments contain population and quantitative geneticists who are directly concerned with the Darwinian prediction; other persons such as evolutionary ecologists study adaptations or reproductive isolation and then attempt to reconstruct the past Darwinian mechanism; others are systematists who also infer past evolution without direct concern for the Darwinian mechanism, except those studying the molecular clock. It is a rare person, however, who is directly observing and documenting evolutionary change.

Dobzhansky's primary interest was pursuing the Darwinian prediction of the presence of fitness-related variation in natural populations. Early on he was not interested in the inversions, because he regarded them as neutral variation, but as soon as hints of nonneutrality arose, his serious investigations started. It is fair to say that he did detect natural selection in the form of seasonal variation, and his studies of experimental populations provided indirect evidence for selection in the wild. Using chomosomes extracted from his surveys he developed the idea of "coadapted gene complexes" (Dobzhansky et al 1977). The coadaptation in his case resulted in inversion heterosis, a stimulating idea at the time. A summary of this research of Dobzhansky's together with modern developments is given by Anderson (1989). The point here is not to attempt to recapitulate Dobzhansky's many contributions, but rather to identify the goal of the inversion studies, namely, the confirmation of the Darwinian prediction that natural populations contain, to use a Dobzhansky phrase, "the raw material" for evolution and also the documenting of the natural selection process.

The discoveries of Dobzhansky and others at the time led naturally to the concept of "balancing selection," another Dobzhansky phrase, which means any mode of selection that maintains variation, such as heterosis between coadapted inversions. This immediately led to the dilemma—"you cannot have your variation and eat it too" (evolve). Solutions to the dilemma were offered (see below); also another school proposed that no dilemma existed, because the variation due to balancing selection was not the raw material for evolution: the raw material consists of the rare variants at mutation-selection equilibrium. The point is that Dobzhansky's inversion studies and similar work today were and are on equilibrium populations and focus on the mechanism of evolution rather than track evolution itself.

Most of the documented cases of evolution itself are the results of drastic human perturbations of the environment such as industrial melanism, heavy-metal tolerance, resistance to antibiotics and pesticides, and some others noted by Williams (1992). The amount of evolutionary change in these cases is not great in terms of the amount of genetic change. Industrial melanism involved an allelic substitution at a single locus, and recent studies of insecticide resistance show that in many cases a single allelic substitution is the cause. These case are certainly interesting, but they are small genetically and have been extremely rapid on an evolutionary time scale.

One might ask how much "natural" evolution might be expected during the forty years of the Dobzhansky inversion survey. Much has been learned since the start of the survey in 1936. An example of such knowledge is illustrated in figure 6.1, which depicts the evolution of size in the radiolarian protozoan *Pseudocubus vema* (adapted from Kellogg 1975 by Futuyma 1986). This spans 2.5 Myr, which is approximately the same time span

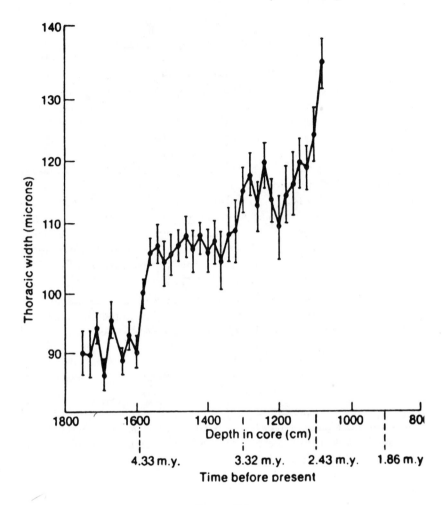

FIGURE 6.1.
Change in size of the radiolarian protozoan *Pseudocubus vema* for which the fossil record is highly detailed. Adapted from Futuyma 1986, after Kellogg 1975.

estimated by Aquadro et al. (1991) from the molecular analysis of the amylase locus for the *D. pseudoobscura* inversion tree (figure 6.2). In figure 6.1 the interval between the sample points is ca. eighty thousand years. A

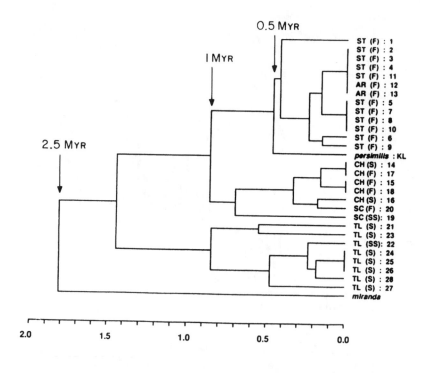

FIGURE 6.2.

Dendrogram of restriction map haplotypes for the *amy* gene region of *D. pseudoobscura, D. persimilis,* and *D. miranda* obtained by the unweighted pair group method with arithmetic mean. Each chromosome is identified by gene arrangement, *Amy* allozyme, and strain number. The scale of distances measures half the nucleotide sequence divergence between sequences. From Aquadro et al. 1991.

sample of forty years from this would hardly show anything and would certainly reveal no information concerning the trend of increasing size during the 2.5 Myr. One might say that size and inversions are different kinds of traits. The reply is that at the basic genetic level a change in inversion frequency is analogous to the change in allele frequency at a polygenic locus underlying size. A change in frequency of an allele or an inversion defines the minimum amount of evolution. Over the 2.5 Myr of *Pseudocubus* size evolution, many alleles at many loci probably went to fixation in sequential replacement. Similarly, inversion evolution means a new inversion becoming fixed within a species, as exemplified by the Hawaiian *Drosophila*, in which branches of an inversion tree predominantly become sequentially fixed in different species along each branch (Carson and Kaneshiro 1976).

The question, then, is this: is the four-decade survey of inversion polymorphism in *D. pseudoobscura* an evolutionary glimpse of the polymorphic transition from, say, Santa Cruz to one of the descendant inversion's eventual sweeping through the species? For example, the three-decade survey showed that there had been an increase in Tree Line on the West Coast, and the fourth-decade survey showed its continued increase (Anderson et al. 1991). Is Tree Line destined to prevail throughout the species and thus complete a Hawaiian *Drosophila* type of inversion evolution?

There is evidence against such an interpretation. Aquadro et al. (1991) (figure 6.2) applied the time calibration of Caccone, Amato, and Powell (1988) and found that the origin of the tree is about 2 Myr, and, more important, the polymorphism itself may be 1 Myr old. One might expect that during this time period a directional trend would have been inferred as shown in the *Pseudocubus* size evolution. More relevant are the Hawaiian *Drosophila*, in which, in some cases, it has been possible to infer that the time required for inversion evolution is less, and perhaps considerably less, than 1 Myr (Carson 1976). Furthermore the branch to *D. persimilis*, which, according to Aquadro et al. (1991), originated 1/2 Myr ago, generated a new tree of polymorphic inversions, rather than fixed a new inversion derived from *D. pseudoobscura* as a Hawaiian *Drosophila* would. This historical evidence, together with the field and laboratory studies of natural selection initiated by Dobzhansky and continued by others (Anderson 1989), strongly suggests that this third-chromosome system is governed by balancing selection, a Darwinian mechanism that prevents evolution, in this case for an evolutionarily relevant period of time.

Whether Dobzhansky expected to observe evolution during the surveys is not known, but what is clear is that his main interest was the study of the Darwinian mechanism, which is the cause of evolution. That this mechanism in its balancing mode can stop evolution so that the process can be studied in detail can be regarded as fortunate. (If one wanted to understand the workings of the internal combustion engine, which makes an automobile move, it would be more convenient to perform the study while the motor was idling, including, incidentally, studying the braking device that prevents movement).

This is not to say that balancing selection cannot be a phase that is a prelude to evolution. Some have proposed, Dobzhansky included, that evolution could occur by one set of balanced genotypes being replaced by another set. Another, simpler idea is that the variation controlled by balancing selection could provide the raw material for evolution when a change in the environment favored one of the variants, which would then proceed to fixation (evolve). The response to directional selection in this case would proceed faster than that favoring a rare variant formerly at mutation-selection equilibrium, as in the cases of industrial melanism and insecticide resistance.

The most active research today on natural selection is the study of quantitative characters (Roff 1992; Stearns 1992). Quantitative genetic variation for fitness-related traits is nearly as prevalent as allozymes (Mousseau and Roff 1987), and the use of the Lande-Arnold (1983) type of procedure has shown natural selection on fitness components in many cases. Again, it is assumed that balancing selection is operating, and therefore, the traits are not evolving at the evolutionary moment. The mechanism of balancing selection on quantitative traits is not so simple as for simple Mendelizing traits, and the anticipation is that the developing theory will be applied to natural populations.

Dobzhansky was one of the leaders of the modern synthesis, one of his major contributions being the initiation of experimental studies of the Darwinian process. This led inevitably to the discovery of the balancing selection phenomenon, which is extremely fortunate because it is apparently very widespread and conveniently provides ample time for detailed studies of a process that is actively preventing evolution but that in a different mode drives evolution.

REFERENCES

Anderson, W. W. 1989. Selection in natural and experimental populations of *Drosophila pseudoobscura.* In Proc. 16th Int. Cong. Genet. *Genome* 31:239–45.

Anderson, W. W., J. A. Arnold, D. G. Baldwin, A. T. Beckenbach, C. J. Brown, S. H. Bryant, J. A. Coyne, L. G. Harshman, W. B. Heed, D. E. Jeffery, L. B. Klaczko, B. C. Moore, J. M. Porter, J. R. Powell, T. Prout, S. W. Schaeffer, J. C. Stephens, C. E. Taylor, M. E. Turner, G. O. Williams, and J. A. Moore. 1991. Four decades of inversion polymorphism in *Drosophila pseudoobscura* Proc. Natl. Acad. Sci. 88:10367–71.

Aquadro, C. F., A. L. Weaver, S. W. Schaefer, and W. W. Anderson. 1991. Molecular evolution of inversions in *Drosophila Pseudoobscura:* The amylase gene region. *Proc. Natl. Acad. Sci.* 88:305–9.

Caccone, A., G. D. Amato, and J. R. Powell. 1988. Rates and patterns of scn DNA and mtDNA divergence within the *Drosophila melanogaster* subgroup. *Genetics* 118:671–83.

Carson, H. L. 1976. Inference of the time of origin of some *Drosophila* species. *Nature* 259:395–96.

Carson, H. L. and K. Kaneshiro. 1976. *Drosophila* of Hawaii: Systematics and ecological genetics. *Ann. Rev. Ecol. Syst.* 7:311–45.

Dobzhansky, Th., F. J. Ayala, G. L. Stebbins, and J. W. Valentine. 1977. *Evolution*, p. 115. San Francisco: Freeman.

Futuyma, D. J. 1986. *Evolutionary Biology*, p. 294. New York: Sinauer.

Kellogg, D. E. 1975. The role of phyletic change in the evolution of *Pseudocubus vema* (Radiolaria). *Paleobiology* 1:359–70.

Lande, R. and S. J. Arnold. 1983. The measurement of selection on correlated characters. *Evolution* 37:1210–26.

Mousseau, T. A. and D. A. Roff. 1987. Natural selection and heritability of fitness components. *Heredity* 59:181–97.

Roff, D. A. 1992. *The Evolution of Life Histories, Theory and Analysis.* New York: Routledge, Chapman, and Hall.

Stearns, D. A. 1992. *The Evolution of Life History Traits.* New York: Oxford University Press.

Williams, G. C. 1992. *Natural Selection*, pp. 128–29. New York: Oxford University Press.

7

Dobzhansky and the Deepening of Darwinism

Hampton L. Carson

We must emphasize how amazingly close modern genetics has come to the Darwinian understanding of evolution. To the brilliant thesis of Darwin arose the brilliant antithesis of studies of mutation and to some degree Mendelism. At one time it appeared that this point of view utterly excluded Darwinism. . . . And now we have a synthesis so harmonious that it is necessary to declare that ever since Darwin we have only been deepening Darwinism

—Alexander Serebrovsky 1925 (translated and quoted by Adams 1980)

The majority of species of organisms are distinguishable by morphological or physiological characters. Despite the skepticism of some systematists and straw-splitting evolutionists (see Gould and Lewontin 1979), most such characters may also be shown to play an adaptive role in the life of the species. In turn, these same characters dominate the systematic descriptions; they are the characters that Darwin associated with species and that formed the centerpiece of his theory of natural selection.

Using a "natural history" approach, Darwin focused sharply on the origin of key adaptive characters through natural and sexual selection, as it acts upon variability existing in species's populations. Modern genetics has shown that the genetic components of that variability are the raw materials that empower the selective process through direct effects on reproductive potential of individuals ("Darwinian fitness"). Attention to this grand central generalization of biology has, however, repeatedly gone in and out of focus during this century. Here I recount a few of these shifts, including the signs of recovery from the current blurred condition.

The astonishing contemporary accomplishments of molecular biology have provided a way to recapitulate the history of existing populations; at long last it is possible to bring hard, quantitative genetic data to bear on the problems of phylogeny and systematics. Curiously, however, phylogeny can be pursued without much attention being given to the modus operandi of evolutionary change that Darwin promoted. The lure of phylogenetic tree construction is now so great that evolutionists tend to be carried away from populational thinking about causal factors of evolution.

The rediscovery of Mendelism in the first decade of this century led to intense preoccupation in the laboratory with disparate research topics that often appeared to the naturalist as arcane: linkage, crossing-over, gene action, chromosome aberrations, mutation research, and polyploidy, to name only a few. But the new methods and concepts produced for the early geneticists such intriguing data that investigation of their relationship to selection and evolutionary process tended to be ignored or at least set aside. This caused an early blurring of evolutionary process much like the present one that accompanies the current fashion in molecular biology. This was especially true of Morgan's research group at Columbia University over the five extraordinarily productive years from 1910 to 1915. Genetics as a science was indeed necessarily done at that time with specially prepared laboratory materials; natural populations and evolutionary process were out of both sight and mind.

By 1920, despite the paroxysm of revolution, news of the new findings in *Drosophila* genetics penetrated several Russian laboratories, most notably that of Nikolai Kol'tsov, director of the Institute of Experimental Biology at Moscow University. At about this time Kol'tsov appointed Sergei Chetverikov to his staff. A forty-year-old entomologist, naturalist, and an intense student of Darwinian theory, Chetverikov came to the institute with no knowledge of genetics; he was exposed there for the first time to some of the accomplishments of Morgan's group with *Drosophila*. Indeed, H. J. Muller from Morgan's laboratory had visited Moscow not long before and had personally delivered thirty-two mutant stocks of *Drosophila melanogaster* to the institute's Anikovo genetics station. Between 1922 and 1927, Chetverikov assembled a small group of keen workers who invented and developed with him a truly new empirical science, the genetics of populations. Unlike the Morgan group, the Russian group's genetic work was guided directly in the direction of Darwinian evolution in nature, leaving formal laboratory genetic manipulations and elucidation of details to others.

Chetverikov's classic 1926 paper sets forth an early statement of the evolutionary synthesis with clarity and vigor, although its conclusion is not couched in such grandiose terms as the quotation from Serebrovsky, given above. For the first time, Darwinian evolution and genetics had been brought into focus together. Almost right away, however, communist theoreticians perceived genetic determination as hostile and reactionary. No attempt was made to reconcile what was basically a misunderstanding of the roles of genetics and environment in the evolutionary synthesis. Within a few years, the group at Kol'tsov's laboratory was dispersed under this political pressure. In 1929, Chetverikov was denounced, arrested, imprisoned, and banished; Timofeeff-Ressovsky went to Germany and continued writing and working. Although later released from prison, Chetverikov never carried out much further research in population genetics.

Within five years after the Chetverikov paper, however, the evolutionary focus of genetics burst on the world with the publication of the books of Fisher (1930) and Haldane (1932) and the incisive short theoretical paper of Wright (1932). But these three statements arose from genetic, statistical, and mathematical considerations of the genetic system, thus differing from the naturalist approach of Chetverikov. Darwinian emphasis on natural populations was not accented and the biological world had yet to be convinced that genetics and Darwinism were compatible. Thus it fell, a few years later, to another naturalist-turned-geneticist, Theodosius Dobzhansky, to integrate genetics and natural history.

As has been widely recognized, Dobzhansky's *Genetics and the Origin of Species* (1937) brought the evolutionary synthesis into sudden brilliant focus. Its success was due to the author's appreciation of both the genetical-mathematical and natural history aspects of evolutionary change. Although Dobzhansky was deeply influenced by the population genetics of the Chetverikov group and visited their laboratory often, he never participated directly in the work there. Like Chetverikov, however, he had an entomological background and his early papers on coccinellid beetles showed a strongly developed Darwinism and a continuing interest in the study of the natural history and systematics of populations.

A perusal of the chronology of Dobzhansky's publications reveals how uniquely qualified he was to articulate these two disparate approaches to the most fundamental problem of biology. At the University of Leningrad from 1924 to 1927, he published about twenty papers on coccinellid beetle populations; this was work begun at the University of Kiev. When he went to Morgan's laboratory at Columbia University in 1927, his studies of natural populations were temporarily replaced by strictly defined laboratory genetic problems. The writing of *Genetics and the Origin of Species* was the instrument that set the future course not only for his own research but for that of dozens of other young scientists in his laboratory and around the world. The work on *Drosophila pseudoobscura* was already off and running by 1936. Before trying to articulate some of Dobzhansky's major contributions to the synthesis, I would like to digress and discuss some of the attitudes present at that time.

The Crucial Factor in the Evolutionary Synthesis: Adjustment of the Genome to the Environment by Natural Selection

The first approaches to the synthesis of Darwinism and genetics were quite optimistic, simplistic, and naive. Regrettably, this mind-set still persists in some circles. For example, the early Russian workers in particular had shown that wild populations of *Drosophila melanogaster* carried a wealth of "visible" mutant genes existing as recessive variants in heterozygous condition. This spawned the early and somewhat improbable hypothesis that the differences between species, indeed the adaptive characters themselves, might have become established through simple fixation of a naturally selected group of such genes in a population. Only later did it become clear, both in *Drosophila* and in other forms, that the oligogenes recognized at that time were best looked upon as major genes of individually large effect. Basically they are dysgenic departures from the normal continuously varying wild type. As raw material for evolutionary change, they were no more probable candidates than the deformities and inborn errors of metabolism that were turning up in the human species.

Abandoning of this naive view of genome adjustment to the environment gave way quickly in the 1940s to the crucial experiments and theoretical writings of Kenneth Mather (e.g., Mather 1949). Mather and his group performed experiments showing that the application of artificial selection to populations could cause gradual phenotypic change over a number of generations, clearly implying that genes of cryptic effect were far more important than the oligogenes in genetic adjustment to the environment. Many systematic observers were not impressed, however, since the metrics employed by the experimenters dealt with such characters as chaeta number in *Drosophila*. Although well adapted to statistical analysis, such a character is difficult to relate to the easily conceivable demands of natural selection.

The findings of the *Drosophila* work on quantitative inheritance came as no surprise to the animal and plant breeders, who for many years had been altering genomes selectively, using these same methods. Nevertheless, they had not developed them in any experimental material that would yield to easy analysis over many generations. The practical breeders, moreover, had not chosen to stress the obvious evolutionary aspects of their work. In fact, artificial selection by man had played a strong role in Darwin's original analysis of selective forces and has continued to do so in current work. There remains no doubt that artificial selection closely mimics natural selection. Its role is especially striking when directed toward certain fitness-enhancing characters where it can be shown to impinge primarily on genes of individually small effect. There has always been a tendency in genetics to simplify the effects of genes by thinking of each gene as having a product producing a neat and easily recognizable effect on the phenotype, independent of all other genes. The problem is that although we can often recognize the effect of a single gene, this does not mean that the effect we use to recognize it is the only effect it has. Dealing with more than one gene at a time, both theoretically and practically, has always been so difficult that reversion to "single-gene thinking" ("one gene, one effect") has repeatedly been used as a refuge from complexity. A view of the phenotype as dependent on the interaction of gene products through a network of effects was advocated for more than fifty years by Sewall Wright (see 1982 for a more recent strong statement of this concept).

Dobzhansky was not impressed with the polygene concept. In discussions with him during the summer of 1950 at Mather, California, for example, he made it clear to me that he felt that in population genetics we should not fail to use the most powerful tool of modern genetics: individual gene or chromosome segment recognition. He insisted that this was not offered by the polygene method. His disinclination to use the term *polygene* may also stem from disagreement with Mather over the notion that the polygenes were physically located in the heterochromatic areas of the chromosomes. Mather himself later abandoned this view, but the term *polygene* has persisted as a useful one for emphasizing genes of individually small effect or genes that may modify the effects of large-effect genes. Since the phenotype is the product of the genotype as a whole, every character is affected by numerous genes and is thus basically polygenic. On the other hand, the attraction of genetic variants like inversions was very great precisely because each entity could be monitored individually, making it unnecessary to depart from the basic single-gene models of Fisher and Wright.

Electrophoresis

The tendency to embrace "single-gene thinking" in population studies of natural populations had a major (and perhaps unfortunate) reprieve during the electrophoresis era of the 1960s and 1970s. Focus on natural variation was set off by the landmark paper of T. R. F. Wright (1963). The method vaulted into the void left by disillusion with the large visible recessives as the raw materials of evolution. The vast amount of heterozygosity for allozymes present in natural populations led many to hope that here at last was a method that would permit the identification of polygenes, allowing a qualitative analysis of the genetics of quantitative traits.

Sad to report, however, this great new hope was not realized nor was it possible to relate any fitness parameters to the interaction of the products of these genes. These biochemical variants turned out indeed to be largely neutral to selection, not contributing to the genetic basis of crucial adaptive differences either within or between intraspecific populations or species. Unfortunately, a whole generation of ecologists appear to have inherited the view that the genome is made up of simple allozyme-like variants. Like inversions, these loci have served famously as genetic markers for conditions in the genome or events that take place there. Indeed, as markers, they have signaled the triumph of Mendelism as the underlying universal mechanism basic to the genetic system of sexually reproducing organisms. This is no small contribution. They have a fascinating molecular evolution of their own that constitutes a genomic behavior bearing little relationship to the adaptive evolution of Darwin and Dobzhansky.

The dawn of DNA studies in the 1980s abruptly brought investigation of genome variability for the first time to the most extreme analytical level, that of the very structure of the gene itself. Analysis became so elegant, detailed, and precise at the level of the code that, as previously in evolutionary biology, the exact contribution of the genes to fitness and the evolutionary future of the individuals in populations again eluded us.

Let us step back for a moment and examine the phenomenon of Dobzhansky, the chromosome inversion, and the theory of coadaptation that emerged and see how close the work came to dealing with the interaction of gene products that impinge on the phenotype.

Chromosomal Inversions

Inversions captured Dobzhansky's interest at an early time primarily for two reasons. First, they appeared to be wholly natural genetic variants that not only could be characterized geographically in a species but also showed Mendelian behavior, making them perfectly suited to the basic manipulations of theoretical population genetics. Second, most inversions embrace very large sections of the genome, making them, potentially at least, of major importance. What appealed to Dobzhansky most was that inversions could be used to monitor the rules of mutation, recombination, and selection in both natural and laboratory-constructed populations. This fertile prospect consumed not only Dobzhansky's career but that of many other workers, especially those working on *Drosophila* and other species with favorable giant chromosomes.

The research on inversions in a number of species of *Drosophila* has recently been summarized in several major reviews (e.g., Ashburner, Carson, and Thompson 1981–1986; Krimbas and Powell 1992). In perusing these accounts, one may ask the question: what has this work contributed to an understanding of the basic Darwinian problem of the genetic analysis of those particular adaptive characters that are unique to differentially adapted populations and species? As one who has devoted a lifetime to such study, it is somewhat embarrassing for me to admit that the contribution in this area appears to have been very small. An inversion is basically a change in gene order. Although sometimes very large in extent, an inversion appears to be initially cast randomly into the genome of a species by mutagenesis, particularly that mediated by mobile genetic elements. The ease of demonstration of inversions in *Drosophila*, their indelible character, and great size relative to the base pair sequences of the DNA make them genome markers par excellence. Using them, one can look beyond the single gene locus to the behavior of substantial segments of the genetic system as it is manipulated by the stochastic processes of meiosis and recombination at sexual reproduction.

The value of the inversion method is increased by virtue of the fact that most inversions in the euchromatic regions of the chromosome may be clearly diagnosed as monophyletic. The two "breaks" that accompany each such inversion may be visually mapped with exquisite precision in the giant polytene chromosomes. Natural inversions, unlike some artificially induced ones, are almost all viable in the homokaryotypic condition. Chromosomal variants that are inviable are presumably screened out by normalizing selection soon after the original inductive event.

Balanced Inversion Polymorphism

One of the most remarkable features of natural inversions is their strong tendency to be retained in natural populations; thus they appear to be tenacious, indelible units apparently held in the population largely by balancing selection. Clearly, many are elements of long-standing historical importance to the genome of the species that carries them. Evidence is very strong that, following Fisher, selective advantage of the heterokaryotype is the primary explanation for their persistence. Although somewhat controversial, few seriously question the balance hypothesis in this instance. Invoking heterokaryotypic superiority in fitness has held up well as a formal explanation of the stability of sectional chromosomal heterozygosity in nature.

It is one thing to recognize that large sections of the genome of a species may exist in a state of balance but quite another to explain what properties these genetic variants may have that make natural selection favor them. When initially recognized in salivary gland chromosomes, inversions were found to interfere with the occurrence of recombination by crossing-over. They are thus properly looked upon as effective blockers of recombination among the many genes that must exist within these large linear sections of chromosome. Why and how blockage of crossing-over could contribute to fitness remained a perplexing puzzle.

Coadaptation

What has clearly become one of the major continuing influences of Dobzhansky's work long after his death pertains specifically to this theoretical topic. In the third edition (1951:122) of *Genetics and the Origin of Species*, after reviewing some early population-cage experiments with *Drosophila pseudoobscura*, he proposed the following fertile hypothesis:

The chromosomes with different gene arrangements carry different complexes of genes (arising ultimately through mutation). The gene complexes in the chromosomes found in the population of any one locality have been, through long continued natural selection, mutually adjusted, or 'coadapted,' so that the inversion heterozygotes possess high adaptive values. The gene complexes in different localities are not coadapted by natural selection, since heterozygotes for such foreign gene complexes are seldom or never formed in nature. Heterosis is, therefore, an outcome of a historic process of adaptation to the environment.

Dobzhansky refined the concept of coadaptation in his incisive summary at Cold Spring Harbor in 1955, in which he made the famous contrast between the "classical" and "balance" views of population genetic structure. Wallace (1968, 1991) has expanded the theory of coadaptation and has reviewed the experimental work that supports it. Coadaptation may be defined as the selection process by which harmoniously collaborating products of genes of relatively small individual effect become accumulated in the gene pool of a local population or deme. As Mather (1953) has argued, the integration of the effects of such genes may be of two kinds; the first involves a balance among alleles leading to overdominance and through it to a balanced polymorphism ("relational balance"). The other is a balance among different genetic loci in a single chromosome and is referred to as "internal" (or epistatic) balance.

I have long felt that the views of Dobzhansky and Mather on genetic coadaptation have a great deal in common. Neither investigator chose, however, to point out these similarities in explicit terms, preferring to maintain separate terminologies that have tended to obscure their similarity. For example, Dobzhansky rarely used the word *polygene*, apparently for the reasons mentioned earlier.

Although Dobzhansky employed discontinuous genetic variants in his research, already in the first edition of *Genetics and the Origin of Species* (1937) he makes a strong case that continuous variability in geographical races consists of many genes of small effect, citing the work of Sumner (e.g., 1932) on genetic variation within species of *Peromyscus*. Dobzhansky concluded (1937:60) that "The genetic basis of continuous variation is probably similar to that of discontinuous variation."

Mather, strongly influenced by his background in quantitative inheritance and plant and animal breeding, brought about an overdue communication between the community of practical breeders and the *Drosophila* genome engineers. Until the publication of the first edition of his *Biometrical Genetics* in 1949, there had been a strong tendency for those concerned with continuous variation and artificial selection to develop an independent approach to genetics. This had remained separate from the more precise single-locus orientation and analytical genetics of the Morgan school. Indeed, this new concordance of views was already anticipated by a little-known paper (Mather and Dobzhansky 1939). This was their only coauthored work, a classical study of the biometrical differences between what would soon be recognized as the sibling species *Drosophila pseudoobscura* and *D. persimilis*. Looking back at the work of these two great pioneers, proceeding mostly in physical and intellectual isolation from each other, we may see how well the continuing work of one tended to mesh with and corroborate the work of the other. The result was a gratifying clarification of the nature of the genetic structure of populations.

Coadaptation of Relatively Inverted Segments of the Genome

That the inversion heterokaryotype reduces very greatly the amount of effective recombination between two relatively inverted sections was well established at an early time (Sturtevant and Beadle 1936; Carson 1946). As Dobzhansky argued, this process converts such a region of the chromosome essentially into a linear block of genes that may be inherited as a unit. In a 1959 review, he adopted the Darlington and Mather (1949) term *supergene* to refer to such sections of the genome that may embrace many gene loci. The initial stochastic inversion event is thought to have "captured" or stabilized a preexisting beneficial association between the included genes. The inversion thus prevents crossing-over from disassociating adjacent loci, whose products interact to contribute to the ease with which natural selection can perpetuate the combination.

The event of "capture" can, however, only be part of the story. After the inversion has increased in frequency, owing perhaps to heterotic properties of the captured chromosome sector, a substantial number of homokaryotypes for both the inverted sequence and the original sequence will be formed in the population. Individuals carrying either of these homokaryotypes will, unlike the heterokaryotype, show unencumbered recombination over the length of the relevant chromosomal section. Subsequent mutation and further meiotic recombination thus permit many slightly differing haplotypes of both original and derived sequence to be formed. In turn, these can form a variety of heterokaryotypes that are presented to the selective process. Accordingly, refinement of coadaptation can proceed over many generations in various sectors of the genome. What may have been at first a crude heterotic effect may now become selectively more attuned to the demands of the environment.

Coadaptation in the Absence of Inversions

In order for "capture" to occur, coadaptation theory favors the idea that at least some coadaptive process must occur in a section of the genome that originally lacks inversions. Indeed, it is possible that well-developed coadaptation between sequentially unaltered segments within a genome could be a common condition even in populations or entire species that lack inversions. The peculiarities of the distribution of inversions in nature now appear to be very frequently tied to the presence of chromosomal breaks mediated by certain active transposable elements. Of course, crossing-over is expected to produce recombinations that destroy, by chance, gene combinations that give favorable gene associations. Such a process within the genome may, however, be counteracted by the continual monitoring of fitness by the process of selection. Thus selection might continue to strongly favor eugenic combinations of genes that function well in enhancing the fitness of the individual. Such a result might lead to the development and maintenance of a heterotic combination or balance even in the absence of a specific crossover-reducer, such as an inversion, translocation, or other nuclear event that locally inhibits chiasma formation.

Although the theory outlined above is a logical extension of what little we know about the details of the modus operandi of the genetic system, hard data are mostly lacking in systems that lack inversions. *Drosophila melanogaster*, for example, in which the genome may be easily marked, demonstrates relatively few blocks of genes that show epistatic properties. But one should note that the very process of introducing and accumulating markers is likely to have a perturbing effect on the very fitness systems that one may wish to study. Genes that directly contribute to fitness in an unmarked genome from nature are difficult to monitor. Thus, unmarked cryptic epistatic blocks of polygenes may be important objects of natural selection without being recognized as such. Methods to identify such systems and to distinguish between relational and epistatic balances are badly needed to extend Dobzhansky's coadaptation hypothesis. As discussed below, the use of apparently benign natural markers, such as RFLPs (restriction fragment length polymorphisms) as used by Paterson et al. (1988), for example, is a promising approach.

Selection and Coadaptation

Evolution appears to result from a dual process: (1) the principal directive force is Darwinian natural selection that maximizes fitness and (2) selection elicits Dobzhanskian genetic coadaptation at the DNA level in the genome. The detailed workings of the individual gene locus, its regulation and transcription, provide only partial information about the role that gene is playing, in conjunction with others, as contributors to fitness in the Darwinian sense. Thus, the very fact that molecular methods must deal with one gene, or a portion of a gene, at a time is simultaneously both a power and a weakness. Natural and sexual selection demand we consider not just the products of a single gene but also the interactions between gene products.

Artificial selection, so close in its operation to natural selection, remains today, as so often in the past, the most powerful single approach to the analysis of the directive forces that bring about evolutionary change. Selection can operate on the genotypic state (i.e., cause evolution to occur) only through first impinging on phenotypes. To understand evolution we must develop methods to enable us to look in back of the phenotype to the interaction of a number of genes.

Very recently, quantitative geneticists have developed the method of marker-assisted selection (MAS). As indicated earlier, genetics has developed a wealth of genetic variants that, even if rarely important in adaptive evolutionary advance themselves, are quite easily mapped, manipulated, and inserted into the genome at prearranged points. Specific areas of the chromosome that appear from indirect evidence to be responding to selection in some undetermined way may be targeted for study. The effects of selection on such a genome section may enable the quantitative geneticist to identify specific genetic alterations that have been brought about by selection.

A relatively recent paradigm of this approach is the work on *Drosophila melanogaster* by Shrimpton and Robertson (1988a,b). Using many of the classical "visible" marker genes on chromosome three of this species, they were able to pick up and analyze in detail a single particular third chromosome (designated "C") from a line that had been strongly selected for polygenic effects that increase the number of bristles, principally the sternopleurals. Chromosome C was divided into five sections, by use of mutant markers to construct homozygotes for each section. Each homozygote was then tested for bristle effects. The results of this experiment are not very different from earlier findings that the number of chromosome sections that can affect sternopleural bristle number is relatively low. For example, Thoday and Thompson (1976) recognized only nine major sternopleural bristle regional effects on all three major chromosomes, far less than the hundreds of small-effect genes that were once postulated. Indeed, these findings suggest that quantitative genetics must recognize the existence of certain major gene-like effects whose expression is altered through the effects of polygenic modifiers found widely distributed in the genome. Wright (1977:463), in his discussion of the shifting balance theory, outlines this possibility for the amelioration of the deleterious effects of an engineered gene of large effect. The "transilience" concept introduced by Templeton (1980) invokes a similar process without relying on interdeme selection for the spreading of such a newly evolved coadaptation. Indeed, there is much in common among the theories of shifting balance, coadaptation, and transilience.

The comment above is made merely to suggest that polygene effects are indeed amenable to genetic manipulation using marker genes. Now that RFLPs and even more sophisticated molecular markers are available, it should be possible to pursue the process of coadaptation at a more sophisticated level. A growing literature is available that makes use of molecular markers of various sorts to assist practical selection procedures for study of quantitative trait loci (QTLs). Fairly recent examples are the work of Helentjaris et al. (1986), Weller, Soller, and Brody (1988), and Paterson et al. (1988) on *Lycopersicon;* and Edwards, Stuber, and Wendel (1987) on maize. Lande and Thompson (1990) have given a provocative recent account dealing with QTLs and the uses of MAS to analyze them.

When such results are obtained, one must approach their interpretation with regard to adaptive changes in the organism with caution, recalling the Gould-Lewontin warning against an overindulgence in panselectionism with regard to specific characters. Wright (1982) has, however, eloquently argued that selection operating on the improvement of relative fitness must impinge primarily on a network of developmental pathways so that the recognition of a genetic change as adaptive must be tempered by the realization that we are not dealing with single-gene effects. It is now time for the evolutionist to enter this field with a renewed sophistication and attempt to relate quantitative characters and evolutionary change.

The intellectual curiosity and drive of Theodosius Dobzhansky led the way for us into the morass of evolutionary genetics. He stayed around long enough for us to profit immensely from his pioneering approaches. More than making any single key discovery, he led us along the tangled path in what now appears to emerge as the most profitable direction. Thus hopeful signs appear that a new breakthrough is coming in the understanding of the genetic basis of what Darwin would have considered adaptive change. The stage is set for us to continue the deepening of Darwinism evoked not so long ago by Serebrovsky and so elegantly embellished by Dobzhansky.

REFERENCES

Adams, M . B. 1980. Sergei Chetverikov, the Kol'tsov Institute, and the evolutionary synthesis. In E. Mayr and W. B. Provine, eds., *The Evolutionary Synthesis*, Cambridge: Harvard University Press.

Ashburner, M., H. L. Carson, and J. N. Thompson Jr., eds. 1981–1986. *The Genetics and Biology of Drosophila*. Vols. 3a–3e. London: Academic Press.

Carson, H. L. 1946. The selective elimination of inversion dicentric chromatids during meiosis in the eggs of Sciara impatiens. *Genetics* 31:95–113.

Chetverikov, S. S. 1926. On several aspects of the evolutionary process from the viewpoint of modern genetics. *Zhurnal eksperimental'noi biologii*, ser. A, 2 (1):3–54 (in Russian). See translation by M. Barker, I. M. Lerner, ed. 1961. *Proc. Am. Phil. Soc.* 105:167–95.

Darlington, C. D. and K. Mather. 1949. *The Elements of Genetics*. London: Allen and Unwin.

Dobzhansky, Th. 1937. *Genetics and the Origin of Species.* 1st ed. New York: Columbia University Press.

———. 1951. *Genetics and the Origin of Species.* 3d ed. New York: Columbia University Press.

———. 1955. A review of some fundamental concepts and problems of population genetics. *Cold Spring Harbor Symp. Quant. Biol.* 20:1–15.

———. 1959. Evolution of genes and genes in evolution. *Cold Spring Harbor Symp. Quant. Biol.* 24:15–30.

Edwards, M. D., C. W. Stuber, and J. F. Wendel. 1987. Molecular-marker-facilitated investigations of quantitative-trait loci in maize. I. Numbers, genomic distribution and types of gene action. *Genetics* 116:113–25.

Fisher, R. A. 1930. *The Genetical Theory of Natural Selection.* Oxford: Clarendon Press.

Gould, S. J. and R. C. Lewontin. 1979. The spandrels of San Marco and the Panglossian paradigm: A critique of the adaptationist programme. *Proc. Royal Soc. London B.* 205:581–98.

Haldane, J. B. S. 1932. *The Causes of Evolution.* New York: Harper.

Helentjaris, R., M. Slocum, S. Wright, A. Shaeffer, and J. Nienhuis. 1986. Construction of genetic linkage maps in maize and tomato using restriction fragment length polymorphisms. *Theor. Appl. Genet.* 72:761–69.

Krimbas, C. B. and J. R. Powell, eds. 1992. *Drosophila Inversion Polymorphism.* London: CRC Press.

Lande, R. and R. Thompson. 1990. Efficiency of marker-assisted selection in the improvement of quantitative traits. *Genetics* 124:743–56.

Mather, K. 1949. *Biometrical Genetics.* 1st ed. London: Methuen.

———. 1953. The genetical structure of populations. *Symp. Soc. Exp. Biol.* 7:66–95.

Mather K. and Th. Dobzhansky. 1939. Morphological differences between the "races" of *Drosophila pseudoobscura. Am. Natur.* 73:5–25.

Paterson, A. H., E. S. Lander, J. D. Hewitt, S. Peterson, S. E. Lincoln, and S. D. Tanksley. 1988. Resolution of quantitative traits into Mendelian factors by using a complete linkage map of restriction fragment length polymorphisms. *Nature* 335:721–26.

Serebrovsky, A. S. 1925. Khromozomy i mekhanizm evoliutsii (Chromosomes and the mechanism of evolution). *Zhurnal eksperimental'noi biologii,* ser. B., 2 (1):49–75.

Shrimpton, A. F. and A. Robertson. 1988a. The isolation of polygenic factors controlling bristle score in *Drosophila melanogaster.* I. Allocation of third chromosome sternopleural bristle effects to chromosome sections. *Genetics* 118:437–43.

———. 1988b. The isolation of polygenic factors controlling bristle score in *Drosophila melanogaster.* II. Distribution of third chromosome bristle effects within chromosome sections. *Genetics* 118:445–59.

Sturtevant, A. H. and G. W. Beadle. 1936 The relations of inversions in the X chromosome of *Drosophila melanogaster* to crossing over and disjunction. *Genetics* 21:421–43.

Sumner, F. B. 1932. Genetic, distributional and evolutionary studies of the subspecies of deer mice (*Peromyscus*). *Bibliogr. Genetica* 9:1–116.

Templeton, A. R. 1980. The theory of speciation *via* the founder principle. *Genetics* 94:1011–38.

Thoday, J. M. and J. N. Thompson, Jr. 1976. The number of segregating genes implied by continuous variation. *Genetica* 46:335–44.

Wallace, B. 1968. *Topics in Population Genetics*. New York: Norton.

———. 1991. Coadaptation revisited. *J. Heredity* 82:89–95.

Weller, J. I., M. Soller, and T. Brody. 1988. Linkage analysis of quantitative traits in an interspecific cross of tomato (*Lycopersicon esculentum* X *Lycopersicon pimpinellifolium*) by means of genetic markers. *Genetics* 118:329–39.

Wright, S. 1932. The roles of mutation, inbreeding, crossbreeding and selection in evolution. *Proc. 6th Int. Congr. Genetics* 1:356–66.

———. 1977. *Evolution and the Genetics of Populations*. Vol. 3. *Experimental Results and Evolutionary Deductions*. Chicago: University of Chicago Press.

———. 1982. Character change, speciation and the higher taxa. *Evolution* 36:427–43.

Wright, T. R. F. 1963. The genetics of an esterase in *Drosophila melanogaster. Genetics* 48:787–801.

8

The Neo-Darwinian Legacy for Phylogenetics

Jeffrey R. Powell

There is little argument that one of Th. Dobzhansky's most significant and far-reaching contributions to biology was providing an objective definition of species and thus placing the concept of species on sounder footing. Beginning in 1935 and followed by two seminal treatments in 1937 (Dobzhansky 1935, 1937a, b) Dobzhansky put forward the essence of what was to become known as the "biological species concept" or BSC. While this concept was to undergo various minor modifications and embellishments, in fact, the idea that species boundaries are marked by *genetic* discontinuity due to some barrier to gene exchange (later to be known as "isolating mechanisms") has remained the essence and central idea of the BSC. It is safe to say that the BSC is still the dominant species concept and is applicable and useful in many contexts. This is not to imply that no significant advances have been made in species concepts in the last fifty years. The limitations of the BSC have been enumerated many times, and even the most ardent supporters of the BSC (Dobzhansky included) agree problems exist with it.

The theme of this paper is to evaluate how the BSC and related issues repeatedly discussed in Dobzhansky's writings created a mind-set that continues to influence our thinking about two major (but related) areas of concern in evolutionary biology today: systematics and conservation biology. I do not address the more mechanistic aspects of speciation such as the genetic basis of species formation.

I thank George Amato, Gisella Caccone, Hampton Carson, Juhnyong Kim, and Richard Lewontin for helpful comments that improved the presentation of this discussion. They should not, however, be held responsible for the sometimes idiosyncratic views expressed.

Systematics

Genetic Continuities and Discontinuities

As clearly stated in his 1937 edition of *Genetics and the Origin of Species*, Dobzhansky viewed the organic world as a series of discontinuous populations of organisms:

> If we assemble as many individuals living at a given time as we can, we notice at once that the observed variation does not form a single probability distribution or any other kind of continuous distribution. Instead, a multitude of separate, discrete, distributions are found. In other words, the living world is not a single array of individuals in which any two variants are connected by unbroken series of intergrades, but an array of more or less distinctly separate arrays, intermediates between which are absent or at least rare. Each array is a cluster of individuals, usually possessing some common characteristics and gravitating to a definite modal point in their variations. Small clusters are grouped together into larger secondary ones, these into still larger ones, and so on in an hierarchical order. (1937a:4)

A major argument of this influential first edition was that this commonly observed discontinuity has a genetic cause. This view of systematics as a *hierarchy of genetic relatedness* became more explicit in later editions of *Genetics and the Origin of Species* when this section of the book incorporated Sewall Wright's adaptive landscape imaging:

> Hence the living world is not a formless mass of randomly combining genes and traits, but a great array of families of related gene combinations, which are clustered on a large but finite number of adaptive peaks. Each living species may be thought of as occupying one of the available peaks in the field of gene combinations. The adaptive valleys are deserted and empty. Furthermore, the adaptive peaks and valleys are not interspersed at random. "Adjacent" adaptive peaks are arranged in groups, which may be likened to mountain ranges in which the separate pinnacles are divided by relatively shallow notches. (Dobzhansky 1970:26–27)[1]

---

[1] I one time wrote to S. Wright asking him if he though his shifting balance theory and adaptive landscape model could or should be applied to systematics or classification above the species level as Dobzhansky had done. He replied some-

How these ideas applied to systematics (or "classification" in those days) was most explicitly stated in the first edition:

The classification thus arrived at is to some extent an artificial one because it remains for the investigator to choose, within limits, which cluster is to be designated a genus, family, or order. But the same classification is *natural* so far as it reflects the *objectively ascertainable discontinuity of variation*, and the dividing lines between species, genera, and other categories are made to correspond to the gaps between the discrete clusters of living forms. (Dobzhansky 1937a:4, italics added)

This has a modern ring to it. Yet terms like *phylogeny* are not used by Dobzhansky, owing partly to the historical context, but perhaps owing also to a prejudice prevalent among geneticists at that time, a point I return to at the end of this paper. One sidelight of this view, and its evolution through successive editions of Dobzhansky's classic book, concerns the "meaning" of species vis-à-vis adaptation. Initially, in this first edition just quoted, Dobzhansky is noncommittal as to the causes of the genetic discontinuities. Later, as selection began to predominate in his thinking, Dobzhansky's allusions to Wright's adaptive landscape explicitly indicate that species represented adaptive units. Indeed, speciation was considered a mechanism of adaptation. This change in attitude is discussed further by Gould (1982), who describes the later versions of the synthesis as "hardened" to indicate an inordinate faith in the ubiquity of natural selection as the cause of all observations. However, the important point is that whatever the cause of species differences, whether speciation is an adaptive "mechanism" in some sense, or whether species formation is a random process, the major point of this paper remains valid: Dobzhansky's view that genetic discontinuities were a way of defining and recognizing levels of differentiation above the species level was an important contribution that forms the basis of a large part of modern-day systematics research.

---

what obliquely, "You are correct in supposing that my shifting balance theory was designed primarily for intraspecific evolution but I tacitly assumed that it applied to evolution in general" (S. Wright to J. R. Powell, February 18, 1981). Nowhere in his single-spaced, two page letter did Wright hint that he felt Dobzhansky had abused his concepts in this higher level extension. This is contrary to some authors (e.g., Eldredge and Cracraft 1980) who claim Wright never intended his model to apply to interspecific levels of evolution.

Microevolution, Macroevolution, and Interspecific Genetics

In retrospect, it has become common knowledge that one of the major achievements of the synthetic or neo-Darwinian theory of evolution was to show that the process of evolution was continuous, i.e., that the processes at work molding genetic variation within species (mutation, selection, drift, migration) could also account for evolution at higher levels such as species formation. "Speciation as a stage in evolutionary divergence," the title of an important Dobzhansky paper in 1940, emphasized this continuum view of evolution. The train of reasoning went that if the formation of species was a continuation of population processes, then all of macroevolution would fall into place under the same conceptual framework. Speciation was the crucial step, and all higher taxa were simply quantitative enhancement of species differences.

But a careful reading of early contributions to the synthetic theory would show that this "common knowledge" was far from clearly evident. Again, from the 1937 *Genetics and the Origin of Species*:

> Experience seems to show, however, that there is no way toward an understanding of the mechanisms of macro-evolutionary changes, which require time on a geological scale, other than through a full comprehension of the micro-evolutionary processes observable within the span of a human lifetime and often controlled by man's will. For this reason we are compelled at the present level of knowledge *reluctantly* to put a sign of equality between the mechanisms of macro- and micro-evolution, and, proceeding on this assumption, to push our investigations as far ahead as this working hypothesis will permit. (Dobzhansky 1937a:12 italics added)

It is clear from this that instead of being totally a philosophical or conceptual linking of micro-evolution and macroevolution, Dobzhansky's concerns were more practical. One simply could not study mechanisms of evolution on a macroevolutionary scale; methodological considerations dictated that one must work on the microevolutionary (intraspecific) scale. The only recourse was to assume what was true of mechanisms within a species also applied to evolution above the species level. What was this methodological roadblock? Simply that, by definition, one could perform experimental genetic studies only within a species. In classical genetics one needs to be able to cross organisms to understand their genetic differences and the mechanisms responsible for the transmission of traits. How could a geneticist study genetics above the species level when the very definition of a species stated that they could not interbreed? This emphasizes, of course, that "mechanisms of evolution" are essentially a question of genetic phenomena, a view not universally accepted.

This apparent roadblock is somewhat overstated as, indeed, Dobzhansky himself was a pioneer in studying interspecific genetic differences. Species can be found that form fertile hybrids of at least one sex (as in the famous case of *Drosophila pseudoobscura* and *D. persimilis*), which allows one to do classical segregation studies of genetic differences between species. But there is a clear limit to this approach in trying to understand higher level genetic differences: relatively few examples can be found of organisms that lend themselves easily to breeding studies, and (at least with animals) the level at which one can obtain fertile hybrids rarely extends beyond closely related sibling species. Using such classical approaches to try to quantify genetic differentiation among higher taxa (the "mountain ranges" of the metaphor) is clearly not possible.

This is precisely where the beauty and extreme usefulness of the molecular genetic approaches come in. One no longer needs to perform genetic crosses between taxa to obtain a quantitative estimate of genetic differences. This has opened up a whole area of genetics of interspecies comparison on a detailed level never before possible. A consequence is that what was the ending boundary for the geneticist—the species boundary—has been broken down.

The Geneticist as Phylogeneticist

What is it that the geneticist can contribute to systematics? Two subdisciplines within genetics have combined to form the basis for the empirical and theoretical molecular phylogenetics: molecular genetics and population genetics. On the empirical side, molecular genetics concepts and methodologies have proven a significant boon to gathering the kind of data needed to do genetically based systematics. Of prime importance is that molecular methods allow one to quantify genetic differentiation between any two taxa regardless of ability to either cross them or breed them. I will not go into methodological details of molecular systematics, for several recent treatises cover the subject in detail (e.g., Hillis and Moritz 1990; Hewitt, Johnston, and Young 1991; Miyamoto and Cracraft 1991). Suffice it to state, especially with the most recent DNA technology involving the polymerase chain reaction (PCR), that obtaining the DNA sequence of selected regions of the genome of species is becoming routine. The amount of tissue needed is minuscule, so that, for example, a hair or feather of a rare species is sufficient for considerable DNA analysis. A major stumbling block is simply having the resources to perform all the studies needed for a complete systematic survey; methodology is not the problem.

Given, then, that it is possible to gather comparative molecular data, usually direct DNA sequence data, on almost all taxa, what other assumptions are needed to use the data for systematic purposes? Obviously, going back to the adaptive landscape imagery, we would base our systematics on the genetic relatedness of taxa: the degree of genetic affinities would define which mountain range a species belonged to and then the mountain ranges could be grouped. Conceptually, then, given enough DNA sequence data, one could place all of classification or systematics on an objectively defined, genetically derived basis. Presumably, the discontinuities observed at the phenotypic level are reflected in (indeed due to) discontinuity at the DNA level, only at the DNA level it is much easier to objectively quantify the level of discontinuity to reach a hierarchical classificatory system (e.g., Powell 1991).

Given a classificatory system based on genetic relatedness, what further assumptions need to be made to take the next conceptual jump to phylogenetics? That is, why might we believe that the relative genetic relatedness of two taxa reflects their relative times since divergence? This is where the work in population genetics comes to bear. While one can rarely draw absolute generalities, a reasonably safe one is that virtually all of population genetics theory and observation confirm that when populations become isolated (no longer exchange genes) they genetically diverge monotonically with time. This is true whether the driving mechanism is mutation, genetic drift, selection, or some combination of these forces. Therefore, if we have a systematic classification based on quantitative genetic relatedness, it should to a very large degree also be a historical classification: taxa more closely related in the classification shared a common ancestor more recently than taxa more distant in the classification. This obviously arrives at a school of systematics often called evolutionary systematics wherein it is proposed that classification reflect genealogy. Clearly molecular systematics and evolutionary systematics share a common conceptual basis. (I have purposely avoided the question of taxonomy or naming of taxa and assigning various ranks, an issue not directly relevant to this analysis.)

The statements above are not meant to imply that molecular phylogenetics is predicated on a molecular clock. What it relies on is that isolated populations (taxa) genetically diverge *monotonically* with time, not that they diverge *linearly* with time. Deviations from clocklike behavior do not invalidate the molecular approach, but they can certainly complicate it. In fact this is only one of many complicating factors in molecular systematics, and the facile description above does not do justice to the intricacies of the data and problems encountered in extracting the phylogenetic information. In fact, the collection of data is no longer the limiting factor in molecular systematics. Rather, the development of methodologies to extract the maximum amount of phylogenetic information and to recognize misleading information is what is most needed at present. The development of such methodologies is also dependent upon as yet poorly understood mechanisms of molecular evolution. Thus, rather than conclude that all the problems of systematics and phylogenetics will be solved simply by filling in the missing sequences of A, T, C, and G, much is yet to be done to exploit this data base to extract as accurately as possible the historical information as well as to examine patterns for gaining an understanding of mechanisms. It is a vibrant and moving field in its infancy. Undoubtedly, as molecular systematics develops, we will be gaining insights into a whole array of biological phenomena from the highest level of relationships of major lineages to details of molecular mechanisms.

Thus using molecular data to reconstruct phylogenetic relationships, can be seen as a natural outgrowth of the neo-Darwinian synthesis. The biological species concept was the first systematic level to be defined strictly from a genetic standpoint. It is now possible, following from this intellectual legacy, to begin to place higher levels of relationship into genetic context. The prospects are exciting. Geneticists are making important contributions to these larger questions of evolution, while at the same time the use of the comparative method developed by systematists is providing insights into genetic mechanisms. Thus the fields of molecular systematics and evolution provide unifying territory for the biological sciences, which have historically been experiencing more and more fractionation.

Alternative Views

It is appropriate to mention the more recent challenges to the neo-Darwinian approach with regard to equating microevolutionary and macroevolutionary mechanisms. The most serious questions have been promulgated primarily by paleontologists who have claimed that the patterns of evolution in the long-term record are incompatible with population genetics theory and that something very different is happening on the macroevolutionary *versus* microevolutionary scale (e.g., Eldredge and Gould 1972). Punctuated equilibrium in particular was claimed to be at odds with population genetics; there followed a renewed interest in Gold-schmidt's "hopeful monsters" to explain sudden bursts of morphological changes associated with speciation events. Whether the punctuated *pattern* in the fossil record is correct or not is one problem; that the *mechanisms* of population genetics can account for such patterns is another. While the problem of interpreting patterns remains unresolved, most agree now, including some of the originators, that nothing in punctuated equilibrium theory negates the principles of population genetics. Mayr (1988) has dealt with this in some detail and amply documents that neo-Darwinism is compatible with punctuational patterns of evolution.

Nevertheless, it is interesting to consider the implications of such thinking on phylogenetic studies based on DNA sequences. It could be disastrous. If the majority of cladogenic events involved one or a very few genes, it is almost certain that molecular phylogeneticists do not routinely use these genes in their studies. How could one, in fact identify such genes? And what kind of changes would one expect to see that indicate speciation events? Would not such a limited part of the genome also be subject to considerable homoplasy? One could, of course, argue that it is just such parts of the genome one should not study to understand historical relationships. The "less important" parts of the genome evolve in ways more indicative of history—owing perhaps to drift. But how would we recognize the differences? Fortunately, the evidence that speciation is caused by a few "important" genes is not compelling.

On the other hand, one could interpret punctuated equilibrium to indicate a sudden burst of genetic change across much of the genome at speciation events. Rohlf et al. (1990) have analyzed the implications of such models for phylogenetic estimation. They conclude that if both lineages of the split have increased changes associated with the cladogenic event, their speciational model, it has little influence on the accuracy of phylogeny estimations. On the other hand, if only one lineage was associated with the sudden burst of changes, then phylogeny estimation is seriously less accurate. This latter model, their punctuational model, is the type implicated by the hopeful-monster-type scenarios, for it would be highly unlikely that both daughter lineages would have experienced at the same time such rare drastic events. So only if the most extreme punctuate model involving Goldschmidtian-type drastic changes in a single offshoot is the way evolution occurs, would this be a significant challenge to molecular phylogenetics.

One final issue deserves comment in this section: whether the cladistic approach used by so many molecular systematists is really in the same neo-Darwinian tradition as some other approaches to phylogenetics. In passages above, I have purposely used the ambiguous words (perhaps "weasel words") genetic *affinities* and genetic *relatedness*. This was intended to imply that one can define affinities or relatedness in various ways. One could use overall similarities and/or differences and arrive at what has become known as the phenetic approach. Or one could use only synapomorphies to define affinities as ardently propounded by the cladistic approach. Or one could use some other methodology to define relatedness in order to maximize the probability of extracting historically accurate information while avoiding misleading information. It seems less important in the present context whether one holds to one particular approach than whether one agrees that genetic relatedness is a legitimate basis for systematics and phylogenetics.

Some strict cladists might argue this point, saying that the cladistic approach has nothing to do with population genetics and could be used whether or not one believed, or was even aware of, a single population genetics equation. This may be true, but the reliance on genetic characters to perform cladistic analysis implies a faith in genetic data as indicating historical relationships. The dynamics of the variants being analyzed obey (to the best of our knowledge) the rules of population genetics. So whether the molecular cladist knows or admits population genetics is the basis of the analysis is akin to the automobile mechanic who knows nothing about Newtonian physics or thermodynamics.

Drosophila Phylogenies

Given that *Drosophila* were the favorite research organism throughout much of Dobzhansky's life, it is only appropriate to illustrate advances in phylogenetics by using molecular data collected on *Drosophila*. What follows is a small excerpt from a longer review containing much more detail, other examples, and documentation of the conclusions about to be drawn (Powell and DeSalle, in press). As indicated in one of the quotes above, the initial treatments of the neo-Darwinian synthesis were primarily concerned with microevolutionary phenomena and especially with mechanisms of evolutionary change at the expense of historical concerns that involve issues above the species level (macroevolution). In addition to presenting some phylogenetic analysis of *Drosophila*, I also present an example of how such analysis can be used to study mechanisms of evolution. Thus the methodological dictum implied by Dobzhansky's statement, that if one wants to study mechanisms of evolution one must work on the microevolutionary scale, is no longer strictly true.

Figure 8.1 presents a phylogeny of several groups of *Drosophila* most used by geneticists. This tree is based on a large amount of data but primarily molecular data including DNA-DNA hybridization and DNA sequence data. The topology of the tree is very well supported and we have considerable confidence in it; the relative branch lengths here are arbitrary and should not be taken as indications of quantitative relationships.

One question we asked of this phylogeny concerns the evolution of *Drosophila* karyotypes. *Drosophila* have what are called six elements or chromosome arms, five rods and a dot. These elements are recognized as being homologous among species and have been designated as letters (Muller

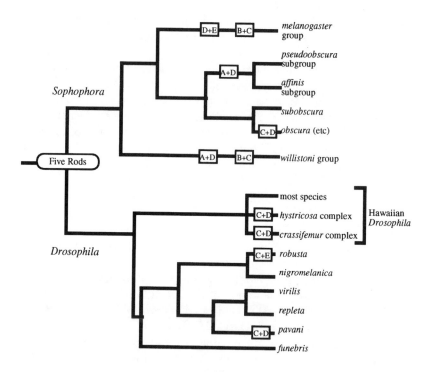

FIGURE 8.1.

Phylogeny of selected species of *Drosophila* for which chromosomal arm homologies have been established; data are in table 8.1. The tree topology was constructed based on molecular evidence, totally independently of the chromosome data. The most parsimonious evolution of karyotype configurations is shown on the assumption that five rods is the ancestral state. From Powell and DeSalle (in press), which should be consulted for details.

1940; Sturtevant and Novitski 1941). It has long been dogma among *Drosophila* evolutionists that the predominant direction of change in karyotypes among species is the fusion of arms whereas fissions are much less frequent (e.g., Patterson and Stone 1952). Ignoring the small dot, table 8.1 shows the configuration of the five rods in species for which homology of the five rod elements has been established. (The dot is ignored here, for it has not been well characterized and is sometimes missing as a dot, presumably incorporated into one of the rods.) Assuming the five-rod condition is the ancestral state, one most parsimonious explanation of karyotype evolution is noted in figure 8.1. Of most interest is that this explanation requires only fusions of arms. Some parallel changes are noted but no reversals. (A second equally most parsimonious scheme requires one fission.) Thus we see that the dogma of fusions predominating over fissions is consistent with the phylogenetic analysis. In fact this analysis lends support to the notion that fusions of centromeres are much more common than fissions. This is an example of how higher level (i.e., macroevolutionary) analysis can be used to study mechanisms of evolution.

TABLE 8.1

Karyotypes of Species and Groups of Drosophila *as Indicated by the Five Rod Elements*[*]; *"+" Indicates Fusion of Two Arms*[†]

Species or Group	Configuration of Elements				
melanogaster group	A	B+C	D+E		
pseudoobscura subgroup (excluding *miranda*)	A+D	B	C	E	
affinis subgroup	A+D	B	C	E	
subobscura	A	B	C	D	E
obscura	A	B	C+D	E	
willistoni group	A+D	B+C	E		
robusta	A	B	C+E	D	
nigromelanica	A	B	C	D	E
virilis	A	B	C	D	E
repleta	A	B	C	D	E
pavani	A	B	C+D	E	
funebris	A	B	C	D	E
Hawaiian *Drosophila*					
Most species	A	B	C	D	E
Crassifemur complex	A	B	C+D	E	
hystricosa complex	A	B	C+D	E	

[*]Defined by Muller (1940) and Sturtevant and Novitski (1941).
[†]Data from Patterson and Stone (1952), Ashburner (1989), and Carson and Yoon (1982).
NOTE: All species that have been studied for homology of chromosome arms and for which we have good phylogenetic data are included.

Conservation of Biodiversity

Issues surrounding the conservation of species and biodiversity have recently been raised in the consciousness of many biologists and nonbiologists alike. Such issues are closely related to problems in phylogenetics (e.g., Avise 1989; Eldredge 1992), and it is of interest to examine how the neo-Darwinian view continues to influence our views in this realm. Of prime importance in conservation efforts is to define the unit that should be conserved. Generally laws are written in which the units protected by legislation are named species or subspecies. Thus it has become incumbent upon the evolutionary biologist interested in conservation to have an unambiguous, defensible species concept that may, in fact, need to be defended in a court of law (Geist 1992). Many academic biologists, myself included, have spent countless hours debating species and subspecies concepts, most often with no conclusion. While such issues are the grist that makes academic life interesting, we should recognize that there are increasing pressure and need for coming to some kind of consensus on such issues. They are no longer the sole property of esoteric graduate seminars.

As with molecular phylogenetics, I would argue that much of what is being done in this realm stems directly (whether consciously or not) from the neo-Darwinian legacy. The recognition of what units should be conserved is often defined on a genetic basis, the whole nascent field of conservation genetics. The recognition of genetic discontinuities that represent significantly novel histories of populations is often used to decide whether the populations should be given special consideration. Whether one assigns subspecies or species names to the genetically discontinuous units is a moot point in many contexts. Ryder (1986) introduced the useful term *evolutionary significant unit* (ESU). This avoids the often contentious issues of species or subspecies concepts. But concepts like the ESU are direct outgrowths of neo-Darwinian thinking, that the important units are recognizable by genetic criteria. Populational genetic differences arise either through selected changes that indicate the populations represent different adaptive solutions or largely through genetic drift, in which case the genetic information is providing information about the relative times populations have been isolated from one another, and thus the units represent unique histories of varying lengths. Often the accurate assessment of the causes of observed genetic discontinuities is very difficult as is the judgment of the relative importance of qualitatively and quantitatively different degrees of genetic divergence. As with similar questions in phylogenetics, the collecting of the data is just the beginning, and good judgment and all kinds of other knowledge must be brought to bear on any given management decision based on genetic analysis. Few generalities concerning the "level" of genetic difference that warrants designation as an ESU worthy of special consideration are likely to emerge. Rather, each situation needs to be evaluated on its own. The genetic differentiation indicative of biologically significant individuality will differ among taxa. For example, in our own work, we have found species of closely related *Drosophila* that can even form interspecific hybrids that have more genetic (DNA) differences than different families of birds or mammals (Caccone and Powell 1990).

But the discussion above leaves begging the initial question: Can biologists agree upon subspecies and species concepts that can provide the basis for legislation? Can genetics provide such a definition other than the often reworked and reworded BSC? Can the application of molecular approaches provide the objective data to define legally binding species concepts? I do not have answers to these questions. Rather I suggest that genetics, and particularly molecular evolutionary genetics, will continue to provide guidelines if not definitions. Using genetic criteria to aid in these efforts is a natural extension of neo-Darwinian thinking.

Epilogue

Whereas I have tried to show that the rise of molecular phylogenetics stems directly from a neo-Darwinian view of evolution starting with the BSC, it is a curious fact that the major architects of the BSC rarely addressed species concepts in a phylogenetic framework. Today, some species concepts are explicitly phylogenetic (e.g., Cracraft 1987; for an explicitly molecular phylogenetic concept of considerable interest, see Avise and Ball 1991). Why did Dobzhansky in particular shy away from such considerations? He was, after all, initially a systematist working on ladybird beetles so that he certainly knew the field of systematics—at least as it existed in the first half of this century.

I think the answer lies in the prevailing attitudes toward evolution promulgated to a large degree by geneticists. The major dominant intellect in genetics for much of the first third of the century was T. H. Morgan. Morgan had a strong interest in evolution and even published three books on evolution from his perspective as a founder of modern genetics (1903, 1925, 1932). Curiously, however, Morgan did not even believe in the reality of species and considered them "scholastic distinctions." As a staunch mechanistic reductionist (he was the founder of the division of biology at Cal. Tech.), his view of evolution was far from populational, much less systematic or phylogenetic (see Powell 1987). The whole idea that one could ever know with any certainty the history of relationships of organisms would be scoffed at in such an atmosphere.

While Dobzhansky greatly admired and respected Morgan, he came from such a different background (taxonomy and the nascent population genetics of Russian origin) that he disagreed with Morgan's view of evolution. In promoting the reality of species, Dobzhansky clearly broke with the then genetics orthodoxy. However, he evidently remained highly skeptical of ever objectively, rigorously understanding phylogenetic relationships. When he realized that he could use the overlapping inversions of the third chromosome of *D. pseudoobscura* to infer the phylogeny of the inversions, he was astonished. He wrote to M. Demerec on February 17, 1936:

> Sturtevant and myself are spending the whole time studying the inversions in the third chromosome in geographical strains of *pseudoobscura*. We are constructing *phylogenies* of these strains, believe it or not. This is the first time in my life that I believe in constructing phylogenies, and I have to eat some of my previous words in this connection. (Quoted in Provine 1981)

To my knowledge this is the only phylogeny that Dobzhansky ever published. Of course, the faith in an inversion phylogeny is based on a major assumption: they are monophyletic in origin. Thus in modern terminology, they are completely ordered characters with no reversals, so that phylogeny reconstruction is trivial. The closest thing to a species phylogeny occurs in Spassky et al. (1971) for the *Drosophila willistoni* group, but this is in fact a set of overlapping circles indicating the "relationships" of the species; this could be interpreted as a phylogeny. So considering his skepticism in phylogenies (in general), one is not surprised that Dobzhansky did not discuss his species concept in a phylogenetic context.

REFERENCES

Ashburner, M. 1989. *Drosophila: A Laboratory Handbook*. Cold Spring Harbor, N.Y.: Cold Spring Harbor Press.

Avise, J. C. 1989. A role for molecular genetics in the recognition and conservation of endangered species. *Trends Ecol. Evol.* 4:279–81.

Avise, J. C. and R. M. Ball, Jr. 1991. Principles of genealogical concordance in species concepts and biological taxonomy. In D. Futuyma and J. Antonovics, eds., *Oxford Surveys in Evolutionary Biology*. Vol. 7, pp. 45–67. Oxford/New York: Oxford University Press.

Caccone, A. and J. R. Powell. 1990. Extreme rates and heterogeneity in insect DNA evolution. *J. Mol. Evol.* 30:273–80.

Carson, H. L. and J. S. Yoon. 1982. Genetics and evolution of Hawaiian *Drosophila*. In M. Ashburner, H. L. Carson, and J. N. Thompson, eds., *Genetics and Biology of Drosophila*. Vol. 3b, pp. 298–344, New York: Academic Press.

Cracraft, J. 1987. Species concepts and the ontology of evolution. *Biol. Phil.* 2:63–80.

Dobzhansky, Th. 1935. A critique of the species concept. *Phil. Sci.* 2:344–55.

———. 1937a. *Genetics and the Origin of Species.* New York: Columbia University Press.

———. 1937b. Genetic nature of species differences. *Am. Natur.* 71:404–20.

———. 1940. Speciation as a stage in evolutionary divergence. *Am. Natur.* 74:312–21.

———. 1970. *Genetics of the Evolutionary Process.* New York: Columbia University Press.

Eldredge, N., ed. 1992. *Systematics, Ecology, and the Biodiversity Crisis.* New York: Columbia University Press.

Eldredge, N. and J. Cracraft. 1980. *Phylogenetics and the Evolutionary Process.* New York: Columbia University Press.

Eldredge, N. and S. J. Gould. 1972. Punctuated equilibria: An alternative to phyletic gradualism. In T. J. M. Schopf, ed., *Models in Paleobiology*, pp. 305–32. San Francisco: Freeman, Cooper.

Geist, V. 1992. Endangered species and the law. *Nature* 357:274–76.

Gould, S. J. 1982. Introduction, In Theodosius Dobzhansky. *Genetics and the Origin of Species*, pp. xvii–xli, reprint of 1937 edition. New York: Columbia University Press.

Hewitt, G. M., A. W. B. Johnston, and J. P. W. Young, eds. 1991. *Molecular Techniques in Taxonomy.* Berlin: Springer-Verlag.

Hillis, D. M. and C. Moritz, eds. 1990. *Molecular Systematics.* Sunderland, Mass.: Sinauer.

Mayr, E. 1988. *Towards a New Philosophy of Biology.* Cambridge: Harvard University Press.

Miyamoto, M. and J. Cracraft, eds. 1991. *Phylogenetic Analysis of DNA Sequences.* Oxford: Oxford University Press.

Morgan, T. H. 1903. *Evolution and Adaptation.* New York: Macmillan.

———. 1925. *Evolution and Genetics.* Princeton: Princeton University Press.

———. 1932. *The Scientific Basis of Evolution.* New York: Norton.

Muller, H. J. 1940. Bearings of the *Drosophila* work on systematics. In Huxley, J., ed., *The New Systematics*, pp. 185–268. Oxford: Oxford University Press.

Patterson, J. T. and W. S. Stone. 1952. *Evolution in the Genus Drosophila.* New York: Macmillan.

Powell, J. R. 1987. "In the air"—Theodosius Dobzhansky's *Genetics and the Origin of Species. Genetics* 117:363–66.

————. 1991. Molecular approaches to systematics and phylogeny reconstruction. *Boll. Zool.* (Italy) 58:295–98.

Powell, J. R. and R. DeSalle (in press) *Drosophila* molecular phylogenies and their uses. *Evol. Biol.*

Provine, W. 1981. Origins of the genetics of natural population series. In R. C. Lewontin, J. A. Moore, W. B. Provine, et al., eds., *Dobzhansky's Genetics of Natural Populations I–XLIII*, pp. 1–76. New York: Columbia University Press.

Rohlf, F. J., W. S. Chang, R. R. Sokal, and J. Kim. 1990. Accuracy of estimated phylogenies: Effects of tree topology and evolutionary model. *Evolution* 44:1671–84.

Ryder, O. 1986. Species conservation and systematics: The dilemma of subspecies. *Trends Ecol. Evol.* 1:9–10.

Spassky, B., R. C. Richmond, S. Perez-Salas, O. Pavlovsky, C. A. Mourao, A. S. Hunter, H. Hoenigsberg, Th. Dobzhansky, and F. J. Ayala. 1971. Geography of the sibling species related to *Drosophila willistoni*, and the semispecies of the *Drosophila paulistorum* complex. *Evolution* 25:129–43.

Sturtevant A. H. and E. Novitski. 1941. The homologies of the chromosome elements in the genus *Drosophila*. *Genetics* 26:515–41.

9

Theodosius Dobzhansky—
a Theoretician Without Tools

Richard C. Lewontin

The field of population genetics is unique in biology in the degree to which experimental and theoretical aspects are so tightly intertwined. The current fad for molecular population genetics, for example, is entirely motivated by the possibility of using molecular data to estimate the selective and population structure parameters of the synthetic stochastic theory of population genetics elaborated by the Wrightian enterprise. The intimate relation between mathematical theory and experimental and natural historical observation is a continuation of a mode of science that we owe in large part to the work of Dobzhansky. This mode, more than any specific problem or theory, is the real continuing significance and influence of his work.

Theodosius Dobzhansky is thought of as one of the founders of experimental population genetics and especially as the person who brought together laboratory experimentation and observations of the genetics and behavior of natural populations in an attempt to explain the dynamics of genetic variation in nature. In a field that spans the modes of scientific work from the abstract probabilistic theory of Wright and Kimura to the ecological field studies of Cain and Heed, Dobzhansky is always placed firmly at the experimental end. I argue in this paper for a very different view of Dobzhansky. I claim that he was, in fact, a theoretician whose entire intellectual program was theoretical and conceptual but that, lacking the abstract mathematical tools that are the stock-in-trade of the conventional theoretician, he used the only tool at his disposal, the manipulation of living organisms. He was, in this view, an experimentalist *faute de mieux*, for whom organisms were a kind of analogue computer, rather than the subjects of interest in themselves. I bolster this heterodox view by a sketchy analysis of Dobzhansky's empirical work and by a clear look at his theory of *coadaptation*.

To anyone who was familiar on a day-to-day basis with Dobzhansky, the claim that he was a theoretician can seem only perverse. He repeatedly told his students that "science is 99% perspiration and only 1% inspiration" and that "statistics is a way of making bad data look good." He had a repertoire of stories of the follies of statisticians and theoreticians, such as the claim that C. I. Bliss once definitively misidentified a Live Oak in Pasadena as an orange tree on the grounds that 99.5% of all trees in California are orange trees. Dobzhansky himself was innumerate, or at least he appeared to be. He could not carry out and claimed not to understand the simplest algebraic manipulations and was unable to use the tables of the standardized normal deviate to draw a normal curve with a known mean and statistical deviation. He would on occasion demand from a theoretical colleague a formula that could be put to some use in manipulating data, but he never asked how the formula was arrived at.

This picture of Dobzhansky, familiar to his students and postdoctoral fellows, comes as a surprise to those who know his work only from publications. No other large corpus of empirical biological work is as deeply dependent upon and intertwined with mathematical theory as is Dobzhansky's "Genetics of Natural Populations" (GNP) series (see Lewontin et al. 1981). These papers use a variety of theoretical formulations of Sewall Wright and employ a number of techniques of estimation of the parameters of Wright's models, including sophisticated simultaneous least squares and regression methods. Moreover, the papers are characterized by considerable statistical analysis, including nonparameter methods and analysis of variance at a time when most biologists, even evolutionary biologists, were strangers to statistics. It is not surprising that the body of Dobzhansky's work on population genetics would give the reader the impression of someone very familiar with and at home with the most advanced theoretical biology of the time.

A closer examination of the papers and a knowledge of Dobzhansky's actual practice reveal the actual situation. Dobzhansky collaborated constantly with theoreticians. Four of the GNP series were coauthored with Sewall Wright; two with H. Levene, who also appears as coauthor on three other papers not in the series; and Dobzhansky's theoretically sophisticated graduate student and colleague Wyatt Anderson is coauthor on five of the GNP series and on two papers outside the series and is sole author on

GNP 41. Beyond formal coauthorship, virtually every paper in the GNP series acknowledges the theoretical and statistical advice and work of W. Anderson, H. Levene, R. C. Lewontin, S. Wright, and B. Wallace, all of whom designed and carried out statistical tests, wrote computer programs to estimate parameters, and even designed the experimental protocols according to theoretical and statistical principles.

What is most important, however, is not Dobzhansky's dependence upon statisticians and theoreticians for the mechanical details of data collection and analysis but the central role that theory played in the formulation of Dobzhansky's problematic, not only in general but also in specific papers. An excellent example of the hidden hand of theory at both levels is in GNP 6, "Microgeographic races in *Linanthus parryae*" (Epling and Dobzhansky 1942). Ostensibly a collaboration between an experimental genetical evolutionist and an evolutionary botanist to describe the variation in flower color along a geographical transect, the intellectual program of the paper arose directly out of Wright's theory of stationary distribution of gene frequencies under random genetic drift. Wright is acknowledged for help in the statistical analysis, but the real importance of Wright can be seen in figure 4 of the paper, showing "the frequency of samples containing different proportions of blue-flowered plants." Such a frequency distribution makes no sense at all outside the context of Wright's stationary distribution theory. It is precisely the *negation* of the then unusual treatment of such transect data that would have sought an ecotone or cline related to the external environment. Indeed, it is unclear why anyone would even want to *make* the observations in the absence of Wright's system of explanation. Note that Wright quite independently undertook a reanalysis of these data as part of his own program (Wright 1943, 1978).

The paper on *Linanthus parryae* leads us to the most general consideration of the structure of Dobzhansky's intellectual program. The functions of empirical work in biology are diverse, quite aside from those observations meant simply to augment the catalogue of diversity of living and dead organisms. A major mode for analytical and reductionist biology, characteristic of nearly all physiological, molecular, and developmental biology, is the elucidation of the material mechanisms that mediate biological phenomena, the description of the gears and levers of their articulations, such as, for example, the determination of the structure of DNA by Watson and Crick or the determination of the DNA decoding mechanisms by Khorana, Brenner, and Crick.

At the other end of the scale of description are experiments meant simply to rough out the description of the phenomena themselves: what happens if one sees the world through inverting prisms, inserts a bit of dorsal lip of a blastopore into the hind end of a gastrula, removes all members of a competing species from a community, and so forth? Then there are the classic experimental tests of hypotheses, a great deal rarer than older textbooks on the philosophy of science lead us to believe, but occasionally appearing in their idealized form like Meselson and Stahl's test of the semiconservative model of DNA replication by isotope-labeled precursors. Dobzhansky's work belongs to none of these modes. Rather, it is of two sorts: the experimental measurement of parameters of an already articulated theory and the illustrative demonstration of what were, for Dobzhansky, unquestioned general truths about nature through concrete examples.

One of the remarkable features of Dobzhansky's empirical work was that although he spent a large part of his career working on the selective importance of new mutations and inversion polymorphisms, he devoted no attention to the elucidation of the ecological and physiological variables that underlay the balanced polymorphisms in which he fervently believed! With the exception of an aborted attempt later in his experimental career to link changes in the frequency of chromosomal inversion in *Drosophila pseudoobscura* to insecticide use (Anderson et al. 1968, GNP 39), and the earlier suggestion (with no test of the hypothesis) that several years of drought might have been the cause of inversion frequency changes in *Drosophila pseudoobscura* and *D. persimilis* over a four-year period, Dobzhansky evinced no interest whatsoever, at least in his active scientific work, in finding out what ecological and physiological factors were responsible for the fitness differences inferred in nature. Note that the paper on insecticides is the only one of the GNP series with the word *test* in the title. When general principles are involved, the papers are never characterized as "tests" but sometimes as "proofs." The problem for Dobzhansky was to demonstrate selection in action, not to elucidate its mechanisms. In this sense, Dobzhansky's work on selection was purely formal and curiously unbiological for someone who thought of himself as a naturalist who spent much of his life outdoors and who enjoyed an immersion in the diversity of nature.

The formality of Dobzhansky's work on the inversion polymorphisms is seen in the two paths he took to demonstrating selection. Having shown that the repeated cycles of inversion frequency change in California could not be explained by migration of flies from one altitude to another as the seasons changed, his first demonstration that selection was occurring was to show that selection would change inversion frequencies in the laboratory. These experiments (Wright and Dobzhansky 1946, GNP 12), although entitled "Experimental *reproduction* of some of the changes *caused by natural selection* in certain populations of *Drosophila pseudoobscura*" (my emphasis), *reproduced* nothing. They were not intended to actually mimic the natural ecological situation, a hopeless goal in any case, or to isolate the causes of the selective differences, or to reproduce the seasonal cyclicity or characteristic frequencies of any of the inversions in any natural populations. Rather, they demonstrated that it was possible, in a highly crowded laboratory population, at a constant temperature of 25°C, neither of which conditions was "natural," to observe repeatable changes in inversion frequencies and eventual stable equilibrium of intermediate frequencies and to demonstrate that these changes were the result of differential fitness of genotypes with heterozygotes having the highest fitness. Note that a failure to observe these changes would *not* have been taken as evidence against natural selection in nature. In fact, no changes were observed in cages kept at 16°C, an observation that was folded into the explanatory scheme as an example of how fitnesses of genotypes are sensitive to environment.

The second route of demonstration of natural selection was to show that, *in nature*, there was a consistent excess of heterozygous adults above the expected Hardy-Weinberg proportions, a demonstration that, in the words of Dobzhansky and Levene (1948), was "Proof of the operation of natural selection in wild populations of *Drosophila pseudoobscura*." Again, the demonstration depends in no way on any ecological or physiological observations but entirely on the deviation of observed from theoretical frequencies. And again, the failure to find a significant deviation from Hardy-Weinberg frequencies could not have been taken as disproof of the operation of natural selection since a variety of selective mechanisms including frequency-dependent selection (a favorite of Sewall Wright) and fertility differentials would not necessarily be detectable in this way (see Prout 1965). Both the laboratory cages and the genotypic frequencies from nature *illustrated* what Dobzhansky and Wright had already concluded from their observations of frequency patterns in nature.

Over and over again, Dobzhansky performed experiments to exemplify general principles without attempting to cash them out in detail for specific cases. A number of experiments showed how fitness of genotypes depended on environment (Dobzhansky and Spassky 1944, GNP XI; Dobzhansky et al. 1955, GNP 23; Dobzhansky and Levene, 1955, GNP 24), but the environments chosen for these demonstrations, three different temperatures and three different species of yeast food not natural to the flies, were not meant to explain in material detail the fate of genes in nature; they were simply different environments chosen to illustrate the general principle that each genotype had its unique norm of reaction, that one could not predict how well a genotype would be in the environment from its performance in another environment, and that heterozygotes were generally less susceptible to environmental fluctuations than homozygotes. There was never any question whether these propositions were true in general. The empirical problem was to illustrate them in practice. The "experiments" were demonstrations.

The other chief preoccupation of Dobzhansky's empirical work was the estimation of parameters such as mutation rates, migration rates, the size of the "panmictic unit" or, more generally, the hierarchical structure of F-statistics, in order to fit the observations of natural genetic variation into the analytic apparatus created by Sewall Wright in his general theory of evolution in Mendelian populations (Wright 1931). It is important to understand that Wright's "theory" is untestable in the classic sense, because it is not a "theory that" but a "theory of," that is, a synthesis of all the relevant forces operating on gene frequencies in populations. The parameters of that theory, especially the parameters of breeding structure, are not estimated autonomously and then put into the structure. Rather, the observations of frequency patterns are themselves the data for a tauto-logical estimation of the parameters, given the theoretical apparatus. Indeed, we do not even know what we mean by "effective breeding size" in a natural population except as a dual representation of the actual degree of coancestry among individuals. Many of Dobzhansky's measurements of these parameters cannot be understood except as the filling-in of quantita-tive values in an unquestioned theoretical apparatus.

Dobzhansky's relations with theoreticians were rather like those of an architect building a skyscraper to structural engineers. He understood intuitively what the theory entailed but depended upon the practitioners to provide the technical expertise to turn the intuitive theory into a structure that would stand up. Dobzhansky depended upon them absolutely (and therefore, perhaps, resented them, too) for the detailed planning and analysis of the experiments. There is, for example, a plaintive letter from Dobzhansky to Wright, dated May 14, 1941, in which he begs Wright to tell him what the correct experimental plan is to be for the summer work at Mather in which dispersal of *D. pseudoobscura* is to be measured. More-over, to Wright he owed his original understanding of population structure and the interaction between drift and selection. (The suggestion has been made that, in fact, A. H. Sturtevant first showed Dobzhansky the impor-tance of Wright's theory for work on natural populations. See Provine's essay in Lewontin et al. 1981 and letters from Dobzhansky to Wright in the early 1930s from Pasadena in the Wright-Dobzhansky correspondence. Whether or not Sturtevant's role was crucial, there is no doubt that Dobzhansky developed an extraordinarily subtle understanding of Wright's theory, despite his own lack of mathematical facility.)

However, Dobzhansky never depended upon theoreticians for the conceptualization of the experiments and never allowed them to influence his interpretation of the results. This last point is material to the claims that Dobzhansky was a "theoretician without tools." Dobzhansky knew at all times what the experiments were intended to demonstrate and what the conclusion was from the results. Indeed, if my claim is correct that the experiments were largely *demonstrations* rather than explorations, these conclusions were already in existence *before* the experiments were done. The role of the theoretician was to use his expertise to draw the rigorous quantitative connection between the data and the conclusions that Dobzhansky had already decided upon. Any real independence on the part of the theoretician to draw other conclusions from the data was not tolerated, and this applied even to Sewall Wright, whom Dobzhansky believed to be "one of the two greatest living geneticists." (The other was Muller.)

Two incidents illustrate the point. In 1951–52, A. R. Cordeiro, working at Columbia, carried out a series of experiments, planned by Dobzhansky, on chromosomal heterozygotes and homozygotes in *D. willistoni*, designed to demonstrate the fitness superiority of heterozygotes (Cordeiro 1952). Howard Levene was asked to carry out the statistical analysis of a fairly complex experimental design, and his first analysis showed the expected superiority of heterozygotes and their lower variation in fitness. Shortly thereafter, Levene reworked the analysis and reported to Dobzhansky that his second, better analysis failed to show the results expected. Dobzhansky was exceedingly annoyed and said the paper could not be published. He pressured Levene into yet another analysis, which now showed significant results. These were immediately incorporated into the paper, and without further ado, the paper was published.

Yet more extraordinary were the circumstances of publication of GNP 37 (Pavlovsky and Dobzhansky 1966). The population cage experiments were analyzed by W. Anderson, who was originally one of the authors of the paper, and the conclusion was the conventional one of fitness superiority of inversion heterozygotes. Wright was a reviewer of the paper and carried out his own analysis, concluding that there was actually frequency-dependent fitness rather than unconditional superiority of heterozygotes. On reading Wright's review, Anderson concurred, but Dobzhansky refused to change his conclusions. Anderson was dropped from authorship, although the results of his original calculations were used in the paper, and Dobzhansky simply reiterated his already predetermined conclusion. (See discussion of this incident in Lewontin et al. 1981:807–8 for details.)

It should not be concluded that Dobzhansky was incapable of rejecting a long-held theory in the face of overwhelming evidence. On the contrary, he twice abandoned strongly held theoretical commitments. The first, in 1941, was the change from his original belief that the inversions of *D. pseudoobscura* were entirely subject to random genetic drift to his unshakable conviction after that time that the inversion frequencies were determined by natural selection (see Lewontin et al. 1981:303). The second was his abandonment in 1953 of the theory of coadaptation, in favor of the belief in the unconditional superiority of allelic heterozygotes. I now turn my attention to this later change.

By the time of the appearance of the third edition of *Genetics and the Origin of Species* in 1951, Dobzhansky had formed a well-articulated theoretical view. This view included the following elements:

1. The fitnesses of genotypes, like other aspects of their phenotype, were contingent on environment and were generally unpredictable from one environment to another.

2. There is no "wild-type" genotype because all populations of sexually reproducing organisms are characterized by a large amount of genetic variation, so that the "typical" individual is necessarily a heterozygote at many loci.

3. Homozygotes *from natural populations* were less fit than heterozygotes *from natural populations* in most environments, but there are environments in which particular homozygotes may be superior. Homozygotes were "narrow specialists" while heterozygotes were "broad generalists" in fitness.

4. In particular, inversion heterozygotes had higher fitness than inversion homozygotes *in the natural populations from which those chromosomes were taken.*

5. The superiority of heterozygotes over homozygotes is itself a consequence of natural selection and is not some intrinsic property of heterozygosity per se. That is, populations will accumulate, merely as a mechanical consequence of the stable equilibrium of gene frequencies caused by fitness overdominance, a set of alleles at a locus that show overdominance in fitness when

combined with each other. Mutations that do not show overdominance with other alleles will be either fixed or lost by natural
selection and drift. This is the theory of *coadaptation*.

A long-known result of simple population genetic theory was that if the
fitnesses of homozygotes *AA* and *aa* were only $(1-s)$ and $(1-t)$ while the
heterozygote had relative fitness 1.0, then a stable equilibrium of allele frequency was produced with $p(A) = t/s+t$. If, on the other hand, the
homozygotes were more fit than heterozygotes, there would be fixation of
allele frequencies at 1 or 0. Dobzhansky reasoned that every population
was constantly undergoing mutations and that those mutant alleles that
had the heterotic property would accumulate in the population. But since
heterosis is a *relational* property between two alleles, then which alleles
were accumulated would depend on which other alleles were present at a
particular time in the life of the population.

This historical contingency in conjunction with the random events
associated with finite population size and finite population lifetime would
mean that the particular ensemble of heterotically maintained alleles in a
population would be different from the ensemble in a different population.
There was, therefore, no reason to suppose that a random allele from population A would be heterotic in combination with a random allele from
population B. This contingency would be even greater because of epistatic
effects between loci. Since fitness of a genotype depended on environment,
it also depended on the *genetic* environment at other loci. Whether or not
a heterozygote $A_i A_j$ was heterotic would depend, not only on the identity
of $A_i A_j$, but also on the alleles at other loci that were historically present.
Thus, Dobzhansky pictured a process of selective *coadaptation* taking place
during the history of a population, the coadaptation of alleles at a locus to
each other, and of the alleles at one locus to the alleles at other loci. These
two elements of coadaptation are parallel to Mather's notions of *relational
balance* and *internal* balance in artificially selected populations. The alleles
at loci along a haploid chromosome would be internally balanced to form
a superior linked unit, and this unit would, in turn, be relationally balanced
with other haplotypes in the population.

Dobzhansky's theory of coadaptation included the inversion heterozygotes and homozygotes. The fitness properties of inversion types, according to Dobzhansky, were the result, not of the inversion breaks themselves,
but of the allelic content of the inverted chromosomes locked up in nonre-
combining blocks. It followed, then, that an inversion heterozygote, say,
ST/CH, would be superior in fitness to the homozygotes ST/ST and

CH/CH only if the ST and CH chromosomes were taken from the same natural population where the allelic contents of the two gene pools had become coadapted by selection. If ST and CH were taken from two different populations, the alleles would have no selective history relative to each other and there would be no heterosis. GNP 19 (1950) was to be a demonstration of this truth. As the title "Origin of heterosis through natural selection in populations of *Drosophila pseudoobscura*, suggests, the result was as expected.[1] Whereas several previous experiments (Wright and Dobzhansky 1946; Dobzhansky 1947) had shown a superior fitness of inversion heterozygotes, when inversions from Mather, California; Piñon Flats, California; and Chihuahua, Mexico, were crossed with one another, the heterozygotes were always either intermediate or even inferior in fitness to chromosomal homozygotes. This observation held without any exception for all combinations of different inversions, ST, AR, and CH, both for larval viability and in net fitnesses in population cage experiments. ST chromosomes from Piñon Flats had no selective experience of AR chromosomes from Mather, so no coadaptation had occurred. This experiment is a classic in the "demonstration" mode and was a particular triumph for Dobzhansky who had wagered and won a dime, framed and displayed in his office, against the contrary prediction of H. J. Muller.

The next step in the demonstration of the universality of coadaptation was to carry it to chromosomes not involved in inversion polymorphism. By 1951, it was universally agreed that if chromosomes are sampled from a natural population and made homozygous, their carriers would generally be of lower viability and fertility than random reconstituted heterozygotes from those same chromosomes. No experiment had ever given a different result. (See, for example, GNP 8, Dobzhansky, Holz, and Spassky 1942; GNP 11, Dobzhansky and Spassky 1944; and later experiments such as GNP 22, Dobzhansky and Spassky 1954.) The theory of coadaptation predicted, however, that if heterozygotes are made between chromosomes from geographically distant populations, no heterosis would be seen. No such experiment had been done, but Dobzhansky was certain of its outcome and assigned the demonstration to M. Vetukhiv, an exiled Ukrainian geneticist who had a fellowship to work at Columbia.

---

[1]After completing the present paper, I became acquainted with Jane Maienschein's very fruitful concept of "epistemic styles" (Maienschein 1981). What has been illustrated here is an epistemic style that was of great influence on the future of population genetics.

The experimental design was meant to eliminate the effect of homozygosity for rare recessive deleterious genes that is always the result of mating isogenic chromosomal homozygotes. That is, the previous experiments, cited above, always contrasted totally homozygous chromosomes with random heterozygotes. But a total homozygote would, on any theory, show inbreeding depression because it is almost certain that any chromosome in a natural population carries a few rare recessive deleterious genes. The Vetukhiv experiment, then, did not make chromosomal homozygotes but contrasted the fitness components of crosses between chromosomes taken from the same geographical population (*intrapopulation* crosses) with crosses between chromosomes from different populations (*interpopulation* crosses). These crosses were then carried to the F_2 to show the effects of recombination of genes from different populations, breaking up the internally balanced haplotype. Several sets of experiments were carried out, measuring viability (Vetukhiv 1953, 1954), one measuring fecundity of females (Vetukhiv 1956), and one measuring longevity (Vetukhiv 1957). A fourth set of a more complex experimental design was carried out by D. Brncic (1954).

The results of the Vetukhiv experiments produced a major intellectual crisis for Dobzhansky. Instead of the interpopulational hybrids being intermediate or lower in fitness than intrapopulational crosses, heterosis was observed in all components of fitness for all five of the populations involved. There did appear to be fitness breakdown in the F_2 recombinant generation, as predicted from coadaptational theory, but the clear heterosis in F_1s between chromosomes that had never experienced joint selective histories in nature was a fatal blow to coadaptation. There was some skepticism in the laboratory at Columbia about the experimental design because the parental (intrapopulation), F_1 (interpopulation), and F_1 crosses were not made and measured simultaneously. This was partly remedied by carrying out a new set of experiments in which parental and F_1 crosses were simultaneous and a separate set of parental and F_2 crosses were made simultaneously. The results were the same. Coadaptation was dead and in its place was the extraordinary result that any heterozygote was likely to be superior in fitness to its constituent homozygotes, irrespective of joint selective history and eliminating the effect of rare deleterious mutants.

Dobzhansky's reaction was to encourage more experiments by Vetukhiv and Brncic and to convert his own theoretical view from coadaptation to a radical belief in the fitness superiority of heterozygotes per se. From 1955 onward, only five years after the triumphant outcome of the wager with Muller, the theory of coadaptation disappears from the central dogma of the Dobzhansky school. In its place are an equally assured and unquestioned adherence to a belief in the intrinsic superiority of heterozygotes and a long sequence of demonstration experiments in support of this view (for example: Dobzhansky and Levene 1955, GNP 24; Spassky et al. 1960, GNP 29; Dobzhansky, Krimbas, and Krimbas 1960, GNP 30; Wallace and Dobzhansky 1962; Dobzhansky, Spassky, and Tidwell 1963, GNP 32). (Typical of a demonstration experiment, the paper by Wallace and Dobzhansky 1962 is entitled "Experimental *proof* of balanced genetic loads in *Drosophila*" (emphasis added). It is reminiscent of GNP 17, "Proof of the operation of natural selection in wild populations of *D. pseudoobscura*. A rhetorical analysis of Dobzhansky's paper titles is beyond this essay but would be well worthwhile.) The subsequent history of the struggle between the "balanced" and "classical" schools over the importance of heterosis and overdominance is also not in the scope of the present paper (but see Lewontin 1974 for one version).

What is the explanation of Dobzhansky's conversion from one theoretical dogma to another? Certainly the Vetukhiv experiments were a potent force. We might say that while Dobzhansky was ready to discount *alternative* explanations of phenomena that did not agree with his theoretical prior position, even when those who disagreed with him were of the highest intellectual level in his pantheon (S. Wright and H. Muller, for example), he was not prepared to sweep observed *contradictions* under the rug. But the distribution between alternative and contradictory is not so clear. All scientists must be prepared to explain away apparently contradictory observations in the face of well-grounded and strongly held belief. Dobzhansky was not an exception, and he explained away contradictory results from other laboratories as experimental artifacts. Given my claim that Dobzhansky was essentially a theoretical scientist, engaged, not in experimental tests, but in experimental demonstration, it is more likely that Dobzhansky was already in the process of a *theoretical* shift to which the Vetukhiv experiment gave a final push.

That theoretical shift was a result of the influence of M. Lerner and his developing ideas of "genetic homeostasis." Lerner had a powerful personal influence on Dobzhansky who felt very close to him. When *Genetic Homeostasis* (1954) appeared, Dobzhansky took it up immediately and it became central to laboratory discussion and to the training of graduate students. Lerner's view was that heterozygotes have the property of greater developmental and physiological buffering against disturbance of the adapted phenotype and that this greater buffering, an *intrinsic* property of heterozygosis, lay at the base of the higher fitness of heterozygotes. Such a theory was consonant with and unified all the elements of Dobzhansky's theoretical apparatus, except for the theory of coadaptation (see above), and at the same time predicted Vetukhiv's observations. Its theoretical power must have been irresistible to Dobzhansky. Especially, it provided a general explanation of the central core of Dobzhansky's observation of nature, the genetic heterogeneity among individuals that Dobzhansky regarded as the critical fact of biology.

To make the theory of intrinsic heterozygous superiority the central theoretical commitment of his view, Dobzhansky had to give up a general theory of the operation of natural selection, one for which he had offered a final triumphant demonstration only five years before. It also involved putting out of consideration the observations on which the theory of coadaptation was based and which were, in turn, contradictory to the new theory. So Dobzhansky's science changed, not by the progressive reduction of contradictions between observations and theory, but by the choice of a theoretical apparatus on the basis of quite other considerations that are yet to be fully understood.

REFERENCES

Anderson, W. W., C. Oshima, T. Watanabe, Th. Dobzhansky, and O. Pavlovsky. 1968. Genetics of Natural Populations. XXXIX: A test of the possible influence of two insecticides on the chromosomal polymorphism in *Drosophila pseudoobscura*. *Genetics* 58:423–34.

Brncic, D. 1954. Heterosis and the integration of the genotype in geographic populations of *Drosophila pseudoobscura*. *Genetics* 39:77–88.

Cordeiro, A. R. 1952. Experiments on the effects in heterozygous condition of second chromosome from natural populations of *Drosophila willistoni*. *Proc. Natl. Acad. Sci.* 38:471–78.

Dobzhansky, Th. 1947. Genetics of Natural Populations. XIV: A response of certain gene arrangements in the third chromosome of *Drosophila pseudoobscura* to natural selection. *Genetics* 32:142–60.

———. 1948. Genetics of Natural Populations. XVIII: Experiments on chromosomes of *Drosophila pseudoobscura* from different geographical regions. *Genetics* 33:588–602.

———. 1950. Genetics of Natural Populations. XIX: Origin of heterosis through natural selection in populations of *Drosophila pseudoobscura*. *Genetics* 35:288–302.

Dobzhansky, Th., A. M. Holz, and B. Spassky. 1942. Genetics of Natural Populations. VIII: Concealed variability in the second and fourth chromosomes of *Drosophila pseudoobscura* and its bearing on the problem of heterosis. *Genetics* 27:463–90.

Dobzhansky, Th., C. Krimbas, and M. G. Krimbas. 1960. Genetics of Natural Populations. XXX: Is the genetic load in *Drosophila pseudoobscura* a mutational or balanced load? *Genetics* 45:741–53.

Dobzhansky, Th. and H. Levene. 1948. Genetics of Natural Populations. XVII: Proof of the operation of natural selection in wild populations of *Drosophila pseudoobscura*. *Genetics* 33:537–47.

———. 1955. Genetics of Natural Populations. XXIV: Developmental homeostasis in natural populations of *Drosophila pseudoobscura*. *Genetics* 40:797–808.

Dobzhansky, Th., O. Pavlovsky, B. Spassky, and N. Spassky. 1955. Genetics of Natural Populations. XXIII: Biological role of deleterious recessives in populations of *Drosophila pseudoobscura*. *Genetics* 40:781–96.

Dobzhansky, Th. and B. Spassky. 1944. Genetics of Natural Populations. XI: Manifestation of genetic variants in *Drosophila pseudoobscura* in different environments. *Genetics* 29:270–90.

———. 1954. Genetics of Natural Populations. XXII: A comparison of the concealed variability in *Drosophila prosaltans* with that in other species. *Genetics* 39:472–87.

Dobzhansky, Th., B. Spassky, and T. Tidwell. 1963. Genetics of Natural Populations. XXXII: Inbreeding and the mutational and balanced loads in natural populations of *Drosophila pseudoobscura*. *Genetics* 48:361–73.

Epling, C. and Th. Dobzhansky. 1942. Genetics of Natural Populations. VI: Microgeographic races of *Linanthus parryae*. *Genetics* 27:317–32.

Lerner, I. M. 1954. *Genetic Homeostasis*. New York: Wiley.

Lewontin, R. C. 1974. *The Genetic Basis of Evolutionary Change*. New York: Columbia University Press.

Lewontin, R. C., J. A. Moore, W. B. Provine, and B. Wallace. 1981. *Dobzhansky's Genetics of Natural Populations I–XLIII*. New York: Columbia University Press.

Maienschein, J. 1981. Epistemic styles in German and American embryology. *Science in Context* 4:407–27.

Pavlovsky, O. and Th. Dobzhansky. 1966. Genetics of Natural Populations. XXXVII: The coadapted system of chromosomal variants in a population of *Drosophila pseudoobscura*. *Genetics* 53:843–54.

Prout, T. 1965. The estimation of fitness from genotype frequencies. *Evolution* 19:546–52.

Spassky, B., N. Spassky, O. Pavlovsky, M. G. Krimbas, C. Krimbas, and Th. Dobzhansky. 1960. Genetics of Natural Populations. XXIX: The magnitude of the genetic load in populations of *Drosophila pseudoobscura*. *Genetics* 45:723–40.

Vetukhiv, M. 1953. Viability of hybrids between local populations of *Drosophila pseudoobscura*. *Proc. Natl. Acad. Sci.* 39:30–34.

———. 1954. Integration of the genotype in local populations of three species of *Drosophila*. *Evolution* 8:241–51.

———. 1956. Fecundity of hybrids between geographic populations of *Drosophila pseudoobscura*. *Evolution* 10:139–46.

———. 1957. Longevity of hybrids between geographic populations of *Drosophila pseudoobscura*. *Evolution* 11:348–60.

Wallace, B. and Th. Dobzhansky. 1962. Experimental proof of balanced loads in *Drosophila*. *Genetics* 47:1027–42.

Wright, S. 1931. Evolution in Mendelian populations. *Genetics* 16:97–159.

———. 1943. An analysis of local variability of flower color in *Linanthus parryae*. *Genetics* 28:139–56.

———. 1978. *Evolution and Genetics of Populations*. Vol. 4, pp. 194–223. Chicago: University of Chicago Press.

Wright, S. and Th. Dobzhansky. 1946. Genetics of Natural Populations. XII: Experimental reproduction of some of the changes caused by natural selection in certain populations of *Drosophila pseudoobscura*. *Genetics* 31:125–56.

PART THREE

Chromosomal Polymorphisms

10

Convergent Linkage Disequilibrium in Disparate Populations of *Drosophila robusta*

Max Levitan and William J. Etges

The title page of Levitan's paper (1970) carries the footnote, "Dedicated to a beloved teacher, Prof. Th. Dobzhansky, on his 70th birthday." Unfortunately, it was not possible to dedicate a similar work to him on the next major milestone, but the influence of this genius on us has continued, nevertheless, and probably on everyone in the world who studies the part played by genetics in evolution.

When M. L., the one of us who was his direct student, looks back, he realizes that this influence never consisted to any great extent of direct transmission of facts in lecture sessions, as is generally considered to be the role of the college teacher. Indeed, there was only one course with him: a seminar in which Dobzhansky shared the leadership with Dr. L. C. Dunn and almost all the "lecturing" was done by the students. And when it came time to choose a thesis problem, Dobzhansky did not, as was done by so many others in his position, dictate one; rather he asked, "Levitin (I believe he never learned to pronounce my name the way I do, perhaps because he thought I might be

We thank D. White for assisting in the collection of Mt. Magazine and Mill Creek flies, A. Badyaev for chauffeuring W. J. E. around the back roads of the Ozarks during the Fane Creek collection, and the department of biological sciences, University of Arkansas, for technical support. M. L.'s work was supported by the Anatomy Research Fund of the Mt. Sinai School of Medicine. We are also grateful to Professor H. L. Carson and the American Philosophical Society for permission to study the protocols of the Steelville data.

related to a great Russian painter with a similar patronymic), what would you like to do?" When the reply was, "I would like to do something similar to what you are doing, something that could combine fieldwork and laboratory studies," he did not assign *D. pseudoobscura* or one of its relatives, but without hesitation suggested that I might find interesting an eastern species, *D. robusta*, whose chromosomes H. L. Carson and H. D. Stalker were about to describe in the first volume of *Evolution* (Carson and Stalker 1947).

Thus he could not even teach me anything about this species, which he had probably never seen, and whose chromosomes he could not recognize. Yet he was the quintessential teacher, for he showed the way, he inspired by his example, in addition, of course, to his masterful ability to synthesize and organize the pieces of knowledge from many species and many sources into hypotheses and theories that continue to guide evolutionary study even when his physical presence is gone.

It is fitting, therefore, that the species he suggested should be the subject of this paper. Described by Sturtevant (1916, 1921), it is one of the more common species of the genus in the deciduous forest of North America east of the Rocky Mountains and north of 28° N latitude. Its haploid set of four consists of two metacentric chromosomes of nearly equal size, the X- and second chromosomes, a nearly metacentric third chromosome, and an acrocentric dot chromosome. Its most common gene arrangements were first described by Carson and Stalker (1947), and Carson (1958) and Levitan (1992) have described those that have been found subsequently in natural populations.

Levitan (reviewed by Levitan, 1982, 1992) found that the gene arrangements of *Drosophila robusta* are often present in natural populations in linkage disequilibrium. The data of this report show that deviations from linkage equilibrium of the arrangements of these chromosomes exist in populations of the Ozark and Ouachita Mountain regions, and many of these associations take forms characteristic of far distant populations rather than those of nearby areas.

Materials and Methods

Adult *D. robusta* were collected in the Ozark Mountain region in (1) June through July 1989 and 1992 in a wooded residential area in Fayetteville, Arkansas (elevation 1400 ft; Etges 1991); (2) June 1989 on the north side of Mount Magazine in the southern district of the Ozark National Forest (elevation 1375 ft); and (3) late July 1992 in the Ozark National Forest along Fane Creek in the White Rock Wildlife Management Area (elevation 680 ft). A population from the Mill Creek Recreation Area in the Ouachita National Forest (elevation 880 ft) was collected in mid-July 1990. All populations were sampled by sweep-netting over aged, fermented bananas in buckets wired to trees located near flowing streams. Individual females were placed in vials and allowed to oviposit until they exhausted all stored sperm. Adults were shipped to one of us (M. L.) for analysis by the methods described by Levitan (1955), except for the Mill Creek sample, which was karyotyped by the second author (W. J. E.).

The Steelville collections were originally described by Carson (1958). The arrangement frequencies shown there were derived primarily from "egg sample" analyses, that is, from larval progeny of collected females that had been inseminated in the wild; in cases where more than one larva was reported from the same female, complete karyotypes could be inferred, especially for the X-chromosome. This was supplemented by data from analysis of some collected adults, in several ways: (1) complete analysis, seeing the salivaries from at least six larval offspring from crosses to flies of known constitution; and (2) crosses of collected males and females, leading in most cases to inferences of their linkage karyotypes but, in the case of the autosomes, not being able to determine which karyotype belonged to the male and which to the female. The arrangement combinations in these data were derived by one of us (M. L.) from the archival records of these collections, courtesy of Dr. H. L. Carson and the Library of the American Philosophical Society in Philadelphia, Pennsylvania. It was not possible to assign combination frequencies to thirty "egg sample" larvae that were doubly heterozygous in the X-chromosome and eighteen that were doubly heterozygous in the second chromosome.

Statistical significance of the data is based on log-likelihood, or G-, tests (also known as 2I tests), for these give a closer approximation to the theoretical chi-square distribution than the traditional chi-square tests (Sokal and Rohlf 1981; Lewontin 1992). For an even better fit, the correction of Williams (1976): $1 + (a+1)/6nv$, where a = number of items of data, n = the total data in the sample, and v = degrees of freedom, was applied to all G-test results with more than one degree of freedom. For data with one degree of freedom, such as 2×2 tables, where the departures from continuity with the chi-square distribution are likely to be more severe, especially if the expected values in some cells are small, a more stringent Williams correction was applied (see Sokal and Rohlf 1981: 737). The data with more than one degree of freedom also often contain cells with small frequencies, but a number of writers (e.g., Cochran 1954; Lewontin and Felsenstein 1965; and Sokal and Rohlf 1981) have shown that drastic adjustments, such as Yates's, tend to be overcorrections even for conventional chi-square tests and would certainly be so for G-tests that were adjusted with Williams's correction.

To save space, the text and tables will often use the shorthand notation for the arrangements introduced by Carson (1953): The Standard arrangement of each arm is dubbed "S." Each of the other arrangements is referred to by the Arabic numeral after the hyphen in its name. A fly with karyotype 2L/2L-3, 2R/2R-1, for example, would be S/3,S/1 in this notation. Depending on the linkage combination of the arrangements, it is also either *SS/31* or *S1/3S*.

Results

Except for the smallest sample—Mt. Magazine, where n = 22 and one of the arrangements (XR-1) is absent in the karyotyped adults (but was present in the egg samples from 1 individual), all the Ozark and Ouachita region data deviate in a highly significant way from the numbers expected on the assumption of linkage equilibrium (table 10.1). If the Mt. Magazine sample is combined with the larger one from Fane Creek, with which it is statistically homogeneous, the deviations for the combined sample, too, are highly significant.

Table 10.2 examines the same data in more detail. Part A shows that in every population, except again the small Mt. Magazine sample, the combination of XL-2 and XR-2 appears more frequently, and the combinations of XL-2 with XR and XR-1 less frequently, than would be expected if the linkages were random. The deviations are highly significant at Steelville, in both Fayetteville samples, and in the Mill Creek sample, but they are not statistically significant at Fane Creek or in the combined Fane Creek and Mt. Magazine samples. Similar results are obtained if the linkage of XL-2 and XR-1 is omitted from the calculations. At Steelville, for example, the frequencies of XR and XR-2 are nearly equal, about 25:28. At equilibrium one would expect 14.2 of the 30 instances of XL-2 to be linked to XR, 15.8 to be linked to XR-2. Instead the ratio is 0 XL-2.XR:30 XL-2.XR-2, a highly significant deviation.

Part B of table 10.2 compares how often XR-1 is combined with the left-arm arrangements and the number expected if there were no association. In the absence of association, the numbers would reflect the relative frequencies of XL, XL-1, and XL-2 in each sample. This cannot be tested in the small Mt. Magazine adult sample, where XR-1 is absent. All the other data differ significantly at the 1 percent level from random expectation, in that

TABLE 10.1.

X-Chromosome Gene Arrangements in Adult Samples of Ozark and Ouachita Region Natural Populations of Drosophila robusta

	Fayetteville 1989	Fayetteville 1992	Fane Creek (FC)	Mt. Magazine (MM)	FC & MM	Mill Creek	Steelville
			Gene Arrangements				
XL	77.45	53.57	16.38	9.09	15.22	26.62	77.88
XL-1	17.65	34.82	53.45	40.91	51.45	5.46	15.35
XL-2	4.90	11.61	30.17	50.00	33.33	67.92	6.77
XR	30.39	18.75	1.72	9.09	2.90	11.95	25.28
XR-1	50.98	31.25	3.45	—*	2.90	3.07	46.73
XR-2	18.63	50.00	94.83	90.91	94.20	84.98	27.99
N	102	112	116	22	138	293	443

					Left-Right Combinations									
	a	e	a	e	a	e	a	e	a	e	a	e	a	e
S S	25	24.0	18	11.3	1	0.3	1	0.2	2	0.6	35	9.3	99	75.4
S 1	50	40.3	33	18.7	4	0.7	0	0.0	4	0.6	9	2.4	179	139.4
S 2	4	14.7	9	30.0	14	18.0	1	1.8	15	19.8	34	66.3	37	83.5
1 S	6	5.5	3	7.3	1	1.1	0	0.8	1	2.1	0	1.9	1	14.9
1 1	2	9.2	2	12.2	0	2.1	0	0.0	0	2.1	0	0.5	8	27.5
1 2	10	3.4	34	19.5	61	58.8	9	8.2	70	66.9	16	13.5	29	16.5
2 S	0	1.5	0	2.4	0	0.6	1	1.0	1	1.3	0	23.8	0	6.5
2 1	0	2.5	0	4.1	0	1.2	0	0.0	0	1.3	0	6.1	0	12.1
2 2	5	0.9	13	6.5	35	33.2	10	10.0	45	43.3	199	169.1	30	7.2
Totals	102	102.0	112	112.0	116	116.0	22	22.0	138	138.0	293	293.0	383	383.0
G-tests:	45.684		36.821		16.604		3.284		19.004		139.566		174.856	
df	4		4		4		2		4		4		4	
P	<<0.01		<<0.01		<0.01		>0.15		<0.01		<<<0.01		<<<0.01	

NOTE: Combinations are given in numbers, arrangements in percent; a = actual, e = expected on the assumption of random equilibrium; G-tests measure possible deviation of these data from linkage equilibrium.
*None in adult sample; present, however, in egg sample from one individual.

XL.XR-1 is more frequent, and *XL-1.XR-1* and *XL-2.XR-1* are less frequent, than expected. The association holds true even if *XL-2.XR-1*, which has never been found in nature, is eliminated from consideration. At Steelville, for example, the ratio of XL:XL-1 is about 5:1, but the ratio of *XL.XR-1:XL-1*. XR-1 is about 22:1.

Part C examines the data if the combinations involving XL-2 and those involving XR-1 are removed, leaving only the combinations of XL and XL-1 with XR and XR-2. This part tests, therefore, for a possible "S,1:S,2 coupling-repulsion association." In every Ozark and Ouachita region sample the coupling combinations, *XL.XR* and *XL-1.XR-2*, are more frequent

TABLE 10.2.

Tests for the Three Kinds of X-Chromosome Linkage
Disequilibria in the Table 10.1 Data

	Fayetteville 1989		Fayetteville 1992		Fane Creek (FC)		Mt. Magazine (MM)		FC & MM		Mill Creek		Steelville	
A. The XL-2:XR-2 Association														
	a	e	a	e	a	e	a	e	a	e	a	e	a	e
2 S	0	1.5	0	2.4	0	0.6	1	1.0	1	1.3	0	23.8	0	7.6
2 1	0	2.5	0	4.1	0	1.2	0	0.0	0	1.3	0	6.1	0	14.0
2 2	5	0.9	13	6.5	35	33.2	10	10.0	45	43.3	199	169.1	30	8.4
Totals	5	4.0	13	13.0	35	35.0	11	11.0	46	45.9	199	199.0	30	30.0
G-tests:	15.131		17.143		3.627		0.000		2.899		64.369		62.491	
df	2		2		2		2		2		2		2	
P	<<0.01		<<0.01		>0.15		1		>0.20		<<0.01		<<0.01	
B. The XL:XR-1 Association														
	a	e	a	e	a	e	a	e	a	e	a	e	a	e
S 1	50	40.3	33	18.7	4	0.7	0		4	0.6	9	2.4	179	145.6
1 1	2	9.2	2	12.2	0	2.1	0		0	2.0	0	0.5	8	28.7
2 1	0	2.5	0	4.1	0	1.2	0		0	1.3	0	6.1	0	12.7
Totals	52	52.0	35	35.0	4	4.0	0		4	3.9	9	9.0	187	187.0
G-tests:	15.260		29.688		11.952				12.418		20.724		43.770	
df	2		2		2				2		2		2	
P	<<0.01		<<<0.01		<0.01				<0.01		<<<0.01		<<<0.01	
C. The S,1:S,2 Coupling:Repulsion Association														
	a	e	a	e	a	e	a	e	a	e	a	e	a	e
S S	25	20.0	18	8.9	1	0.4	1	0.2	2	0.6	35	28.3	99	81.9
S 2	4	9.0	9	18.1	14	14.6	1	1.8	15	16.4	34	40.6	37	54.1
1 S	6	11.0	3	12.1	1	1.6	0	0.8	1	2.4	0	6.6		18.1
1 2	10	5.0	34	24.9	61	60.4	9	8.2	70	68.6	16	9.4	29	11.9
Totals	45	45.0	64	64.0	77	77.0	11	11.0	88	88.0	85	85.0	166	166.0
G-tests:	10.666		25.199		0.894		3.284		3.128		18.939		34.033	
df	1		1		1		1		1		1		1	
P	<0.01		<<<0.01		>0.30		>0.05		>0.05		<<<0.01		<<<0.01	

NOTE: The testing procedures are discussed in the text; a = actual, e= expected on the assumption of random equilibrium; G-tests measure possible deviation of these data from linkage equilibrium.

than would be expected if the combinations were random, and the repulsion forms, *XL.XR-2* and *XL-1.XR*, less frequent. At Fayetteville, Mill Creek, and Steelville, the deviations are highly significant, and marginally so $(0.05 < p < 0.1)$ in the Mt. Magazine and the combined Mt. Magazine and Fane Creek samples. The association is clearly absent, however, in the substantial Fane Creek data.

As a whole, the second chromosome combinations in the Fayetteville, Steelville, and Mill Creek samples do not deviate significantly from the

numbers expected on the assumption of linkage equilibrium (table 10.3).
Significant deviations are encountered only in the Fane Creek and Mt.
Magazine samples, especially in the statistically homogeneous combination of these two samples.

Table 10.4 shows a special portion of the second chromosome data of
table 10.3, namely, the relative numbers of (1) double heterozygotes containing the same left- and right-arm arrangements and (2) arrangement
combinations with the same left-arm arrangement in the double heterozygotes. Part A shows that in the Ozark samples the double heterozygotes
containing the combination 2L-1.2R-1 (*11*) are consistently more frequent
than the double heterozygotes containing 2L-1.2R (*1S*). The deviations
from equality reach statistical significance in the Fayetteville, all Ozarks,

TABLE 10.3.

*Chromosome 2-Gene Arrangements in Ozark and
Ouachita Natural Populations of* D. robusta

	Fayetteville 1989		Fayetteville 1992		Fane Creek (FC)		Mt. Magazine (MM)		FC & MM		Mill Creek		Steelville	
Gene Arrangements														
2L	44.50		41.01		46.15		50.00		46.90		25.81		69.13	
2L-1	38.50		46.76		37.36		34.09		36.73		49.87		22.18	
2L-2	14.50		5.76		6.04		11.36		7.08		23.56		6.65	
2L-3	2.50		6.47		10.44		4.55		9.29		0.75		2.03	
2R	85.50		83.45		74.73		79.55		75.66		81.95		77.26	
2R-1	14.50		16.55		25.27		20.45		24.34		18.05		22.74	
N	200		139		182		44		226		399		541	
Left-Right Combinations														
	a	e	a	e	a	e	a	e	a	e	a	e	a	e
SS	73	76.1	48	47.6	63	62.8	20	17.5	83	80.2	86	84.7	275	269.7
1 S	66	65.8	51	54.2	44	50.8	10	11.9	54	62.8	162	163.6	85	86.6
2 S	27	24.8	8	6.7	10	8.2	3	4.0	13	12.1	77	77.3	31	26.0
3 S	5	4.3	9	7.5	19	14.2	2	1.6	21	15.9	3	2.5	9	7.9
S 1	16	12.9	9	9.4	21	21.2	2	4.5	23	25.8	17	18.3	82	79.4
1 1	11	11.2	14	10.8	24	17.2	5	3.1	29	20.2	37	35.4	19	25.5
2 1	2	4.2	0	1.3	1	2.8	2	1.0	3	3.9	17	16.7	3	7.6
3 1	0	0.7	0	1.5	0	4.8	0	0.4	0	5.1	0	0.5	1	2.3
Totals	200	200.0	139	139.0	182	182.0	44	44.0	226	226.0	399	399.0	505	505.0
G-tests:		3.886		7.022		16.015		9.780		19.862		1.291		7.578
df		3		3		3		3		3		3		3
P		>0.25		>0.05		<0.01		<0.025		<<0.01		>0.50		>0.05

NOTE: Combinations are given in numbers, arrangements in percent; a = actual, e = expected on the assumption
of linkage equilibrium; G-tests measure possible deviation of these data from linkage equilibrium.

TABLE 10.4.

Double Heterozygotes (Part A) and the Included Gene Arrangements
(Part B) in the Data of Table 10.3.

	Fayetteville 1989	Fayetteville 1992	All Fayetteville	Fane Creek	Mt Magazine	Steelville	All Ozarks	Mill Creek	Grand Total
				Double Heterozygotes (Part A)					
SS/11	4	7	11	7	4	5	27	11	38
S1/1S	2	2	4*	5	1	4	14*	6	20*
SS/21	2	0	2	1	0	2	5	2	7
S1/2S	1	0	1	0	0	2	3	3	6
SS/31	0	0	0	0	0	0	0	0	0
S1/3S	0	1	1	0	0	1	2	0	2
1S/21	0	0	0	0	1	0	1	12	13
11/2S	2	2	4	3	0	0	7*	6	13
1S/31	0	0	0	0	0	0	0	0	0
11/3S	0	0	0	2	0	0	2	0	2
2S/31	0	0	0	0	0	0	0	0	0
21/3S	0	0	0	0	0	0	0	1	1
Totals	11	12	23	18	6	14	61	41	102
				The Included Combinations (Part B)					
S S	6	7	13	8	4	7	32	13	45
S 1	3	3	6	5	1	7	19	9	28*
1 S	2	2	4	5	2	4	15	18	33
1 1	6	9*	15**	12*	4	5	36**	17	53
2 S	3	2	5	3	0	2	10	9	19
2 1	2	0	2	1	1	2	6	15	21
3 S	0	1	1	2	0	1	4	1	5
3 1	0	0	0	0	0	0	0	0	0
Totals	22	24	46	36	12	28	122	82	204

NOTE: Asterisks denote the pairs of karyotypes with the same gene arrangements (part A) and the pairs of combinations with the same left-arm arrangements (part B) that deviate significantly from the equality that would be expected in each case on the assumption of linkage equilibrium.
*$p < 0.05$
**$p < 0.01$

and total data that compare the numbers of SS/11 and S1/1S. The consistency manifests itself in part B by the regular excess of *11* over *1S* in the double heterozygotes in the Ozark region. That inequality is significant at the 1% level in the total Fayetteville and all Ozarks samples and at the 5% level in the 1992 Fayetteville and Fane Creek samples. Here the Oauchita Mountain sample differs markedly from the Ozark data, primarily because there were 12 *1S/21:6 11/2S* at Mill Creek, whereas in the Ozarks the ratio of these karyotypes was 1:7. The opposite result, excess of the combinations containing 2R over the ones containing 2R-1, is found in the ratios of *SS:S1, 2S:21* (but not at Mill Creek) and *3S:31*, but only one of these inequalities achieves statistical significance, the 45 *SS*:28 *S1* in the total sample.

In chromosome 3, only 3R and 3R-1 occur in this area. The frequencies of the karyotypes present are shown in table 10.5. With the exception of the small Mt. Magazine sample, the numbers of arrangement heterozygotes exceed, and the numbers of homozygotes fall short of, Hardy-Weinberg expectations. The discrepancy is significant at the 5% level at Fane Creek. In the other cases, including the combined Fane Creek and Mt. Magazine sample, the data do not contradict the assumption of random mating equilibrium.

Only a few years after Painter (1934) introduced the giant chromosomes of the larval salivary glands as a useful tool in *Drosophila* genetics, Dobzhansky, in collaboration with Alfred H. Sturtevant, used them to demonstrate chromosomal polymorphism in natural populations (Sturtevant and

TABLE 10.5.

Chromosome 3-Gene Arrangements in Ozark and Ouachita Natural Populations of D. robusta

	Fayetteville 1989	Fayetteville 1992	All Fayetteville	Fane Creek (FC)	Mt Magazine (MM)	FC & MM	Mill Creek	Steelville
Gene Arrangements								
3R	41.00	43.75	42.15	15.08	11.36	14.35	1.75	60.51
3R-1	59.00	56.25	57.85	85.00	88.64	85.65	98.25	39.49
N	200	144	344	179	44	223	399	547
Karyotypes								
	a e	a e	a e	a e	a e	a e	a e	a e
3R/3R	13 16.8	12 13.8	25 30.6	0 2.0	1 0.3	1 2.3	0 0.1	89 93
3R/3R-1	56 48.4	39 35.4	95 83.9	27 22.8	3 4.4	30 27.3	7 6.8	129 121
3R-1/3R-1	31 34.8	21 22.8	52 57.6	62 64.2	18 17.3	80 81.4	192 192.1	37 39
Totals	100	72	172	89	22	111	199	255
G-tests:	2.466	0.732	2.842	4.735	1.450	1.203	0.125	0.615
df	1	1	1	1	1	1	1	1
P	>0.10	>0.30	>0.05	<0.05	>0.20	>0.20	>0.50	>0.30

NOTE: Karyotypes are given in numbers, arrangements in percent; a = actual, e = expected on the assumption of Hardy-Weinberg equilibrium; G-tests measure possible deviation of these data from that equilibrium.

Dobzhansky 1936a). In a seminal publication Dobzhansky and Sturtevant (1938) showed that multiple inversions occurring on the same chromosome fall into three categories: (1) "independent," that is, separated inversions with recombination between them—dubbed a "linkage plexus" by Levitan (1958a); (2) "included," that is, having both breaks of one inversion between the breaks of a larger one; and (3) "overlapping," that is, having one break of an inversion fall *between* the breaks of a second one and one break *outside* the breaks of the second one. Although some of his earliest work on natural populations of *Drosophila* concerned what later proved to be an example of the first category, namely, the inversions of the SR ("sex ratio") complex of *D. pseudoobscura* (Sturtevant and Dobzhansky 1936b; Dobzhansky 1944), most of the studies by Dobzhansky and his students and collaborators in this period concerned inversions of the third category (reviewed by Powell 1992). The work with *D. robusta*, members of the *melanica* group, and many other species (see reviews by Levitan 1982 and Levitan and Fukatami 1993) has demonstrated, on the other hand, that multiple inversions of the first category also constitute major adaptations of Diptera to their environment. This report reinforces previous data that *D. robusta* populations especially develop complexes of the first category in its two largest chromosomes, and it shows that these complexes are often similar to those in geographically distant locations.

The association of arrangements XL-2 and XR-2, with concomitant virtual absence of any XL-2s linked to XL, occurs in every locality in which XL-2 exists (reviewed by Levitan 1992), and the Ozark and Ouachita populations described here maintain this pattern (table 10.2, part A). One instance of XL-2 linked to XR was found where more than 36 were expected under linkage equilibrium. The discrepancy of XL-2s linked to XR-1 is even more extreme. Indeed, to date, not a single instance of such a linkage has been found where the two arrangements coexist. In these data more than 27 such combinations would have been expected on the assumption of linkage equilibrium. It is interesting that the association of these two arrangements, like that of the SR complex of *D. pseudoobscura*, is maintained in this extreme fashion despite the large area of intervening euchromatin available for crossing-over. In both the *D. robusta* and *D. pseudoobscura* cases, recombination between X-chromosome inversions is very rare (Wallace 1948; Carson 1953; Levitan 1958b; Powell 1992).

In contrast, the nonrandom combinations of arrangement XR-1 with arrangements of the left arm observed in the Ozark and Ouachita populations (table 10.2, part B) are far from universal. Indeed, the presence of this disequilibrium in Fayetteville, Arkansas, and Steelville, Missouri, marks its only instances in places with high frequencies of XR-1. The disequilibrium is strongest in southeastern Pennsylvania; nearby Princeton, New Jersey; and northeastern New Jersey, where XR-1 is quite rare. The XR-1 present is almost exclusively linked to XL, even though XL-1 is the most common left-arm arrangement in many of these localities (Levitan 1973).

The disequilibrium is absent, however, in substantial samples from Kentucky, Ohio, northern Indiana, southern Michigan, southern Minnesota, Iowa, east-central Missouri, and eastern Nebraska, where XR-1 attains some of its highest frequencies in the presence of both XL and XL-1 (as may be gleaned from the data in Levitan 1992). The disequilibrium is also absent in the areas to the east of the Ozark region and to the south of Pennsylvania, where XR-1 (and, in low elevations, also XL-1) are nearly or completely absent. The only place the disequilibrium has been encountered between the northeast and the areas of this report has been in the Bloomington, Indiana, area. There XR-1 is much rarer than expected on the basis of its typical tendency to increase from east to west: 209 XR-1 of 1388 X-chromosomes, about 15%, in contrast to frequencies of more than 50% in Dayton, Lima, and Columbus, Ohio, to the east of it (Levitan 1992); 185 of the 209 are linked to XL, 24 to XL-1, whereas on a random basis 168.1 would be expected to be *XL.XR-1*, 40.9 *XL-1.XR-1*, a highly significant deviation.

Levitan (1961) noted that the X-chromosome S,1;S,2 coupling-repulsion association may take two forms: one in which the "coupling" combinations (*SS* and *12*) are in excess of the numbers expected under equilibrium and one in which the "repulsion" combinations (*S2* and *1S*) are in excess. Highly significant associations of the first form characterize populations near Blacksburg, Virginia, in the Allegheny plateau; Lexington, Kentucky; southeastern Pennsylvania; northeastern New Jersey; and many midwestern localities (including the Bloomington area of Indiana), whereas the second form is characteristic of all studied elevations of the Great Smoky Mountains of Tennessee, the Blue Ridge mountains in North Carolina and Georgia, and the Allegheny Mountains in western Virginia and eastern West Virginia (reviewed by Levitan 1992; see also Etges 1984). It is interesting that the Ozark and Ouachita populations take the same form of this association as the populations of southeastern Pennsylvania, northeastern New Jersey, and the Bloomington, Indiana, area, just as the XR-1.XR association does.

Levitan (1955, 1958a, 1964) and Levitan and Scheffer (1993) found that the major features of second chromosome associations in *D. robusta* were the relative numbers of (1) double heterozygotes containing the same left-

and right-arm arrangements and (2) arrangement combinations with the same left-arm arrangement in the double heterozygotes. Equilibrium theory predicts equalities in each of these categories if the populations are in linkage equilibrium (Levitan 1964). Since the expectation of equalities is independent of the underlying gene or karyotype frequencies, the numbers in different samples may be added, and the sums for these categories are also expected to be equalities.

As noted earlier, the Ozark populations contain a number of significant departures from the random equilibrium equalities in the second chromosome data (table 10.4). Compilations of such data (Levitan 1964, 1982) have shown that the deviations from linkage equilibrium are not of the same intensity in all populations. Additional data published by Levitan (1992) strengthen the previous findings of very large deviations from equilibrium expectations in the areas with high frequencies of 2L-3, especially in the higher elevations of the South and in western localities of the North. In these populations heterozygotes combining 2L-3 and 2R with *2L.2R-1* or *2L-1.2R* (*S1/3S* or *11/3S* in the shorthand notation) tend to be significantly more abundant than the heterozygotes with the same arrangements in the opposite configurations, that is, *SS/31* or *1S/31*. The corollary result is that *3S* is usually much more frequent than *31* in all forms of the double heterozygotes, accompanied by an excess of *S1* or *11* (or both) over *SS* or *1S*, respectively. The effect is much stronger in males than in females. Although the female data are often in the same direction, these are less likely to be statistically significant.

In areas where 2L-3 is rarer, such as the Ozark and Ouachita regions, *3S* is again usually more frequent that *31*, but the numbers are usually so small that they do not reach statistical significance; an exception is the cumulative total for the southern areas with 2L-3 less than 25%. Instead these areas tend to emphasize excesses of *2L-1.2R-1* (*11*) over *2L-1.2R* (*1S*), or, less often, *2L.2R-1* (*S1*) over *2L.2R* (*SS*). With respect to these arrangements, it is interesting that the Ozark region populations resemble more closely those of the southeastern lowlands (mainly Raleigh, North Carolina) than the results from nearby northcentral Missouri (St. Louis area), on the same side of the Mississippi River, where, contrary to pattern, *1S* is significantly more frequent than *11*. The two localities also differ in that in the St. Louis area *S1* is, as usual, significantly more frequent than *SS*, whereas in the Ozark-Ouachita region, as in the southeastern lowlands (and the females of southern Indiana), the opposite is true.

Although based on a limited number of populations, a pattern of geographic variation in inversion polymorphism in the Arkansas-Missouri highlands was apparent. Extending from the Ozark plateau (Steelville and Fayetteville) down through the lowland transition (Fane Creek) to the Arkansas River valley, across to the Mt. Magazine escarpment, and farther south to the more isolated Ouachita highlands, the most western extension of the Appalachian chain, *D. robusta* populations exhibit considerable genetic differentiation across this complex landscape. Ozark plateau populations share similar inversion frequencies, particularly for the X-chromosome (table 10.1), although the matter may be complicated by evidence of year-to-year temporal variation in X-chromosome inversion frequencies (Etges 1991 and unpublished data; note also the difference between the two Fayetteville samples in table 10.1). Lowland Ozark populations on the north side of the Arkansas River are considerably distinct from the plateau populations, as shown by the large X-chromosome gene arrangement differences between Fayetteville and the two sites that have been sampled in the White Rock area, Fane Creek (this report, table 10.1) and Shores Lake (Etges 1991), located 10 km away. The similarity of the latter two populations to the Mt. Magazine population was surprising, given that they are separated from Mt. Magazine by the Arkansas River valley, more than 60 km.

No other data are known from the Ouachitas, but the Mill Creek sample suggests that XL-2 and the *XL-2.XR-2* combination reach their highest frequencies here in the entire range. The previous high in substantial samples was the 53.7% at Emory, Georgia (Levitan 1992). This calls into question previous hypotheses about the location of the epicenter of the supposed "radiate" distribution of XL-2 (Carson 1958; Levitan 1992).

Second chromosome frequencies are much more indifferent to these geographical differences than the X-chromosome arrangements. This might suggest considerable gene flow if it were not for the X-chromosome variation. Such a large regional effect may be explained by the comparatively high frequencies of 2L-1 and 2L-2, "warm weather" arrangements that increase in frequency at lower latitudes and altitudes (Levitan 1992). These arrangements are associated with shorter egg-to-adult development times, particularly under warm temperature conditions (Etges 1989). This suggests a mechanism for maintaining the uniformity in second chromosome frequencies in these populations, where the summers are typically warm and dry. However, the nature of the selective forces shaping the extensive variation in X-chromosome arrangements and in the patterns of linkage disequilibrium in both chromosomes remain to be determined.

One of Dobzhansky's major accomplishments was to demonstrate the selectional basis of the overlapping inversion complexes of *D. pseudoobscura* and its relatives. The data of this paper further strengthen the hypothesis

118   *Chromosomal Polymorphism*

that linkage complexes, too, have a selectional basis. The fact that some populations are at equilibrium for the linkage combinations that are in disequilibrium in other localities negates, for example, the possibility that the disequilibria represent merely a stage in the historical process toward equilibrium, that these populations just have not had enough time to accomplish sufficient recombination to reach the random proportions of equilibrium theory. This is further enhanced by the findings that several localities have developed (or marshaled selectional forces to *maintain*) the same disequilibrium even though the intervening populations have either developed a different disequilibrium of the same arrangements (Levitan 1961) or are at linkage equilibrium for them.

REFERENCES

Carson, H. L. 1953. The effect of inversions on crossing over in *Drosophila robusta*. *Genetics* 38:168–86.
———. 1958. The population genetics of *Drosophila robusta*. *Adv. Genet.* 9:1–40.
Carson, H. L. and H. D. Stalker. 1947. Gene arrangements in natural populations of *Drosophila robusta* Sturtevant. *Evolution* 1:113–33.
Cochran, W. G. 1954. Some methods for strengthening the common χ^2 test. *Biometrics* 10:417–51.
Dobzhansky, Th. 1944. Chromosomal races in *Drosophila pseudoobscura* and *Drosophila persimilis*. In Th. Dobzhansky and C. Epling, eds., *Contributions to the Genetics, Taxonomy, and Ecology of Drosophila pseudoobscura and Its Relatives*. Washington, D.C.: Carnegie Inst. Wash. Publ. 554:47–144.
Dobzhansky, Th. and A. H. Sturtevant. 1938. Inversions in the chromosomes of *Drosophila pseudoobscura*. *Genetics* 23:28–64.
Etges, W. J. 1984. Genetic structure and change in natural populations of *Drosophila robusta:* Systematic inversion and inversion association frequency shifts in the Great Smoky Mountains. *Evolution* 38:675–88.
———. 1989. Chromosomal influences on life-history variation along an altitudinal transect in *Drosophila robusta*. *Am. Natur.* 133:83–110.
———. 1991. Seasonal variation among gene arrangements in *Drosophila robusta*. *Drosophila Info. Serv.* 70:65–66.
Levitan, M. 1955. Studies of linkage in populations. I: Associations of second chromosome inversions in *Drosophila robusta*. *Evolution* 9:62–74.
———. 1958a. Non-random associations of inversions. *Cold Spring Harbor Symp. Quant. Biol.* 23:251–68.
———. 1958b. Studies of linkage in populations. II: Recombination between linked inversions of *D. robusta*. *Genetics* 43:620–33.
——— 1961. Proof of an adaptive linkage association. *Science* 134:1617–19.

———. 1964. Geographic and sex factors in an autosomal linkage disequilibrium: An exercise in basic research. *J. Am. Med. Women's Assn.* 19:370–79.
———. 1970. Chromosomal breaks with attachments to the nucleolus. *Chromosoma* 31:452–61.
———. 1973. Studies of linkage in populations. VII: Temporal variation and X-chromosomal linkage disequilibriums. *Evolution* 27:476–85.
———. 1982. The *robusta* and *melanica* groups. In M. Ashburner, H. L. Carson, and J. N. Thompson, Jr., eds., *The Genetics and Biology of Drosophila*, 3b:141–92. London: Academic Press.
———. 1992. Chromosomal variation in *Drosophila robusta* Sturtevant. In C. B. Krimbas and J. R. Powell, eds., *Drosophila Inversion Polymorphism*, pp. 221–338. Boca Raton, Fla.: CRC Press.
Levitan, M. and A. Fukatami (in press). 1993. Studies of linkage in populations. XIII: Unibrachial and dibrachial inversion disequilibria in *Drosophila lutescens* and *Drosophila rubida. Evol. Biol.*
Levitan, M. and S. J. Scheffer. 1993. Studies of linkage in populations. X. Altitude and autosomal gene arrangements in *Drosophila robusta. Genet. Res. Camb.* 61:9–20.
Lewontin, R. C. 1992 (private communication).
Lewontin, R. C. and J. Felsenstein. 1965. The robustness of homogeneity tests in 2 x N tables. *Biometrics* 21:19–33.
Painter, T. S. 1934. A new method for the study of chromosome aberrations and the plotting of chromosome maps in *Drosophila melanogaster. Genetics* 19:175–88.
Powell, J. R. 1992. Inversion polymorphism in *Drosophila pseudoobscura* and *Drosophila persimilis.* In C. B. Krimbas and J. R. Powell, eds., *Drosophila Inversion Polymorphism*, pp. 73–126. Boca Raton, Fla.: CRC Press.
Sokal, R. R. and F. J. Rohlf. 1981. *Biometry: The Principles and Practice of Statistics in Biological Research.* 2d ed. New York: W. H. Freeman.
Sturtevant, A. H. 1916. Notes on North American Drosophilidae with descriptions of twenty-three new species. *Ann. Entomol. Soc. Am.* 9:323–43.
———. 1921. *The North American Species of Drosophila.* Washington, D.C.: Carnegie Inst. Wash. Publ. 301.
Sturtevant, A. H. and Th. Dobzhansky. 1936a. Inversions in the third chromosome of wild races of *Drosophila pseudoobscura*, and their use in the study of the history of the species. *Proc. Natl. Acad. Sci.* 22:448–50.
———. 1936b. Geographical distribution and cytology of "sex ratio" in *Drosophila pseudoobscura. Genetics* 21:473–90.
Wallace, B. 1948. Studies on "sex-ratio" in *Drosophila pseudoobscura. Evolution* 2:189–217.
Williams, D. A. 1976. Improved likelihood ratio tests for complete contingency tables. *Biometrika* 63:33–37.

11

Studies on Mexican Populations
of *Drosophila pseudoobscura*

*Louis Levine, Olga Olvera, Jeffrey R. Powell, Robert F. Rockwell,
Mariá Esther de la Rosa, Víctor M. Salceda, Wyatt W. Anderson,
and Judith Guzmán*

Our study of Mexican populations of *Drosophila pseudoobscura* had its origin during the Thirteenth International Congress of Genetics, which took place in Berkeley, California, in 1973. On one of the days of the congress, Theodosius Dobzhansky (Dodik), Jeffrey R. Powell, and Wyatt W. Anderson approached me (Levine) and asked that I arrange a meeting with Dr. Alfonso L. de Garay who was then Director of the Department of Radiobiology and Genetics at the National Institute of Nuclear Research of Mexico. I had been associated with the genetics group at the institute since 1965 and had participated in a number of their projects. Dodik, Jeff, and Wyatt were interested in studying the Mexican populations of *D. pseudoobscura* and needed both a base of operations and the collaboration of a group of local geneticists. I was delighted to arrange the requested meeting, and Dr. de Garay recognized immediately the worthiness of the project and agreed to have his laboratory participate in it. My own inclusion in the study was a result of an unexpected withdrawal of Wyatt from the project, which was followed by Dodik's "invitation" or, more accurately stated, "command" that I join the investigation.

Most unfortunately, about eighteen months after the study began, Dobzhansky died, a tragic loss for all of us. Jeff, who had been principal investigator of the project since its inception, found that Wyatt was able to join the group and invited him to do so. In 1980, because of outside commitments, Jeff and Wyatt withdrew from active participation in the study and Robert F. Rockwell joined us. The Mexican component of our group (Judith Guzmán, Olga Olvera, María Esther de la Rosa, and Víctor M. Salceda) has had the good fortune of a long, pleasant, and rewarding association with one another and Dr. de Garay over a period of time preceding the establishment of this project and continuing to the present. They are to be envied. The data reported in this paper were gathered at various times, covering the different periods of active participation of the authors.

Chromosomal Polymorphism

Dobzhansky's interest in the *D. pseudoobscura* of Mexico was sparked many years before the Berkeley meeting. In 1935 he went on a collecting trip to Colorado, Arizona, New Mexico, and Mexico. As related in Provine (1981), Dobzhansky was extremely excited by his studies of the third chromosome gene arrangements of the Mexican populations sampled. On January 26, 1936, he wrote to his friend and colleague Milislav Demerec as follows:

> Sturtevant and myself are gone crazy with the geography of inversions in *pseudoobscura*, and working on this whole days—he with crosses and myself with the microscope. . . . As to our inversions, Mexico seems to be an inexhaustible source of them, and I am beginning to regret that last year only relatively few Mexican strains were collected.

Two years later he returned to collect in Mexico, and, in addition, in Guatemala.

A paper on the gene arrangements of the Mexican and Guatemalan populations was published as paper 4 of the famous series "Genetics of Natural Populations" (Dobzhansky 1939a). At that time the total number of gene arrangements recorded was 13, and of this number, 9 were found in Mexico. In 1944 Dobzhansky and Epling published their classic work on the chromosomal polymorphism of *D. pseudoobscura*. As shown in figure 11.1, the number of third chromosome gene arrangements had risen to 16, which included the inversion labeled "hypothetical" (HY). This gene arrangement has not been found in any population but has to be postulated as the intermediate link between the Standard (ST) and Santa Cruz (SC) inversions. Of the 15 observed gene arrangements, 10 (67 percent) occurred in Mexico. The study of the species's third chromosome has continued both in the United States and in Mexico. In 1985, Olvera et al. published an update on its inversion polymorphism. As shown in figure 11.2,

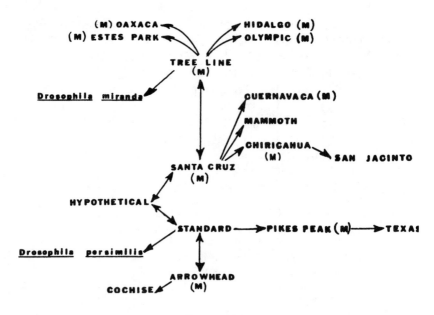

FIGURE 11.1.

The phylogeny of gene arrangements in the third chromosome of *D. pseudoobscura* as of 1944. Those found in Mexico are indicated by (M).

(From Olvera et al. 1985. Copyright © 1985 by the American Genetic Association.)

at that time, 40 gene arrangements had been described, including HY. Of the 39 observed inversions, 26 (67%) have been found to occur in Mexico.

More recently a report by Anderson et al. (1991) presents the frequencies of the various third chromosome gene arrangements found in the United States and Canadian *D. pseudoobscura* populations, covering the period 1936–1983. In the paper, it is stated that a number of new gene arrangements have been observed over the time period, but no description of breakpoints or of positions in the chromosomal phylogenetic tree of the species is given. Until the breakpoints of the "undescribed" inversions are determined, it will not be possible to know which are really "new" gene arrangements. On the basis of the information currently available in the literature, one can state with confidence only that of the 39 observed gene arrangements, 22 (56%) are found in the United States.

Whether the above-noted differences in degree of chromosomal polymorphism reflect differences in age of the southern (Mexican) and north-

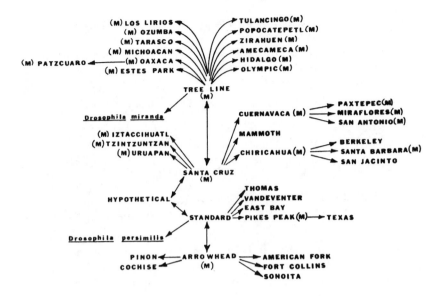

FIGURE 11.2.

The phylogeny of gene arrangements in the third chromosome of *D. pseudoobscura* as of 1985. Those found in Mexico are indicated by (M).

(From Olvera et al. 1985. Copyright © 1985 by the American Genetic Association.)

ern (western United States) populations is as yet unknown. It is hoped that further research will provide information on this point. At this time we can do no better than to take note of Dobzhansky's cautious statement on this question (Dobzhansky et al. 1975):

We feel that it would be premature to try to decide whether the southern or the northern populations are phylogenetically more ancient in their chromosomal composition. Since the populations of central Mexico lack the chromosomes of the Standard phylad, but the northern ones have both the Standard and the Santa Cruz derivatives, the North would seem to have a kind of priority. The related species, *D. miranda* and *D. persimilis*, are also northern and are not known to occur in Mexico. Yet the Mexican populations are not recently arisen ones, as shown by the presence of endemic chromosomes and by the high degree of polymorphism.

Geographic Gradients of Inversion Frequencies

In his many investigations of third chromosome gene arrangements from neighboring populations of *D. pseudoobscura*, Dobzhansky noted that the frequencies of the various gene arrangements often varied in a definite pattern. He characterized the situation as follows (Dobzhansky and Epling 1944):

> The study of geographic gradients has shown that, with some exceptions, the composition of populations becomes more and more different as the distance separating them in space increases. The gradients are steeper in some regions, gentler in others, but the rule generally holds.

What has been widely regarded as his classic study of geographic gradients (clines) of inversion frequencies involves a series of collections made just north of the United States–Mexican border. His findings along this west-east transect are shown in figure 11.3. The data demonstrate: (1) an inversion high in frequency in the west, which decreases as one moves to the east (Standard); (2) an inversion with the reverse frequency pattern (Pikes Peak); (3) an inversion high in frequency in the center of the distribution, which decreases both to the west and to the east (Arrowhead); and (4) the exceptions (Chiricahua and Tree Line), which show either no clines or very limited clines, and then only over certain regions of the transect.

It was of interest to investigate whether a similar situation exists along a west-east transect across Mexico. Such a study was made, and the data obtained are shown in figure 11.4 (Guzmán et al. 1993). In order to make comparisons along the more closely located lines of longitude, only the four most easterly collections made by Dobzhansky were considered. Clinal variation along the transect approximately 1200 km (750 miles) south of the United States–Mexican border is similar to that found just north of the border. However, different chromosome arrangements are involved in the two regions. For example, Santa Cruz in Mexico has a pattern strikingly similar to that of Arrowhead in the United States. Not quite as dramatic are the parallel gradients shown by Cuernavaca and Pikes Peak. The Tree Line gene arrangement has a similar frequency pattern both in northern and southern populations but is somewhat higher in frequency in all localities in Mexico. Lastly, Estes Park in Mexico and Standard in the United States do not exhibit any easily characterized frequency gradients over the common segment of their distribution.

Although not all transects demonstrate inversion frequency gradients, and Dobzhansky was quick to admit this, the existence of parallel gradients from west to east across Mexico and a comparable region in the United

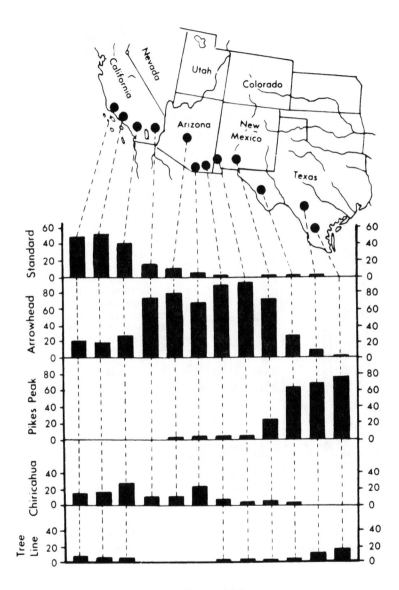

FIGURE 11.3.

West-east distribution of third chromosome gene arrangements of *D. pseudoobscura* along the United States–Mexico border. Heights of columns symbolize relative frequencies in percent of inversions. Circles indicate geographic origin of population samples. From Dobzhansky and Epling 1944.

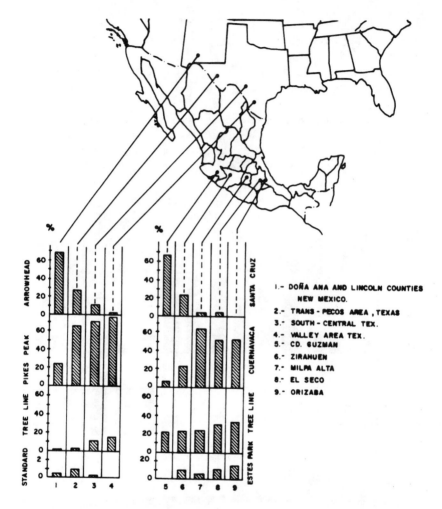

FIGURE 11.4.

West-east distribution of third chromosome gene arrangements of *D. pseudoobscura* across Central Mexico and the corresponding transect along the United States–Mexico border. For details, see figure 11.3. From Guzman et al. 1993.

States may be characteristic of this region. Support for this possibility is found in a second west-east study made by Dobzhansky at a latitude approximately 480 km (300 miles) north of the border, as shown in figure 11.5 (Dobzhansky and Epling 1944). The pattern of inversion frequencies observed was very similar to that found at the border, thus complementing the earlier study.

Dobzhansky would have found the recently obtained data from Mexico "very interesting" and would no doubt have urged or rather "ordered" that the populations along a transect 480 km south of the United States–Mexican border be investigated.

Microgeographic Genetic Differentiation

Dobzhansky was not only instrumental in demonstrating the existence of gradual directional changes in inversion frequencies of populations spread over a wide area of the distribution of D. *pseudoobscura*, he was also the one to establish the existence of microgeographic genetic variation when nearby populations are sampled. The initial study involved the populations of eleven neighboring mountain ranges, some of which were as close as 32 km whereas others were as far as 96 km from each other (Dobzhansky and Queal 1938). Only three ranges yielded samples whose differences were not significant statistically; the others were all different. This study was followed by an investigation by Koller (1939) of the populations found in six canyons within one of the mountain ranges. The distances between canyons were 6–10 km. The differences in the frequencies of the same gene arrangements in the various localities were found to be statistically significant.

Quite clearly it was of interest to determine what was the smallest distance between populations that would permit genetic differentiation to develop. This information is of special importance for a flying form like *Drosophila*. Dobzhansky (1939b) reported on samples taken from three populations in Texas that were 1–2.5 km apart. Although the collections were small, the data obtained strongly suggested a significant heterogeneity among the samples examined.

A further examination of microgeographic genetic differentiation was conducted by Taylor and Powell (1977), using the sibling species D. *persimilis*. In their investigation, flies were collected from traps spaced 40–50 meters apart in an ecologically diverse area consisting of two regions of dry open woods, two of moist dense woods, and one meadow. The traps were spread over distances of 100–500 meters within a given region. A chi-

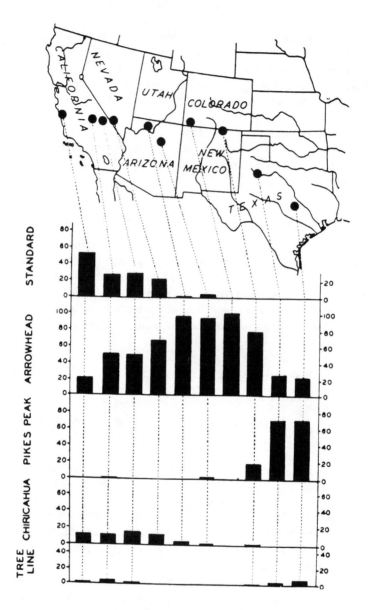

FIGURE 11.5.

West-east distribution of third chromosome gene arrangements of *D. pseudoobscura* about 480 km north of the United States–Mexico border. For details, see figure 11.3. From Dobzhansky and Epling 1944.

square test for heterogeneity of the inversion frequencies from the five regions showed the populations varied in their genetic composition from habitat to habitat.

It was of interest to us to determine whether genetic variation could be detected within the populations at some of our collecting sites. One locality chosen was Amecameca, shown in figure 11.6, which is located about 50 km southeast of Mexico City. This site is a nursery for *Pinus moctezuma* and has an open grassy understory. Flies were collected from 27 traps, arranged as shown by the numbers in figure 11.7a. The second locality is called Tulancingo, also shown in figure 11.6, which is located about 100 km northeast of Mexico City. Our collecting site there is an open area at the edge of an oak forest. Flies were collected from 35 traps, arranged as shown by the numbers in figure 11.7B. Both collection areas are about 30 meters by 50 meters in size, and the average distance between traps was about 3.5 meters.

Tables 11.1 and 11.2 show the results obtained at Amecameca and Tulancingo respectively. In both collections, when performing the chi-square test for heterogeneity, we considered only two classes of gene arrangements: CU (Cuernavaca) and all others combined. This was necessary because sample sizes were not large enough to consider each gene

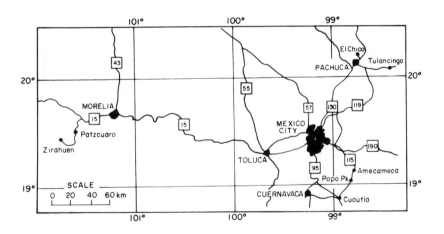

FIGURE 11.6.

Locations of collection sites (Zirahuen, Amecameca, and Tulancingo) of *D. pseudoobscura* in Central Mexico.

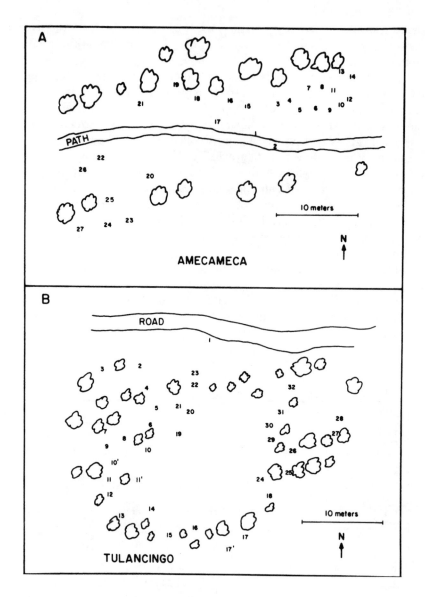

FIGURE 11.7.
Maps of Amecameca and Tulancingo collection sites with numbers indicating positions of
baits; trees are also indicated.

TABLE 11.1.

Chromosomal Frequencies at Amecameca

Traps	CU	TL	EP	SC	Others
1–2	16	7	—	1	—
3–6	28	12	—	—	—
7	73	38	3	—	—
8	38	22	1	1	—
9–10	13	10	1	—	—
11–14	27	3	—	—	—
15	32	19	2	—	1
16–19	11	4	1	—	—
20	15	15	—	—	—
21	18	8	1	—	1
22	60	30	2	—	—
23	17	1	—	—	—
24	19	4	—	1	—
25–27	9	6	1	1	—

$\chi^2 = 23.85$ with 13 d.f.; prob. = 0.033

NOTE: All gene arrangements other than CU were combined into a single class and adjacent traps were grouped until the numbers were large enough to permit a chi-square test for homogeneity.

arrangement separately. Likewise, not all traps yielded sufficient sample size, and therefore, as shown in the tables, the data from adjacent traps were sometimes combined. Despite the lumping of categories, which diminishes the power of the test, we obtained a statistically significant heterogeneity chi-square for the population at Amecameca. This was not the case for the population at Tulancingo.

We cannot at this time ascribe the difference in data from the two populations to any particular factor or complex of factors. However, these results do demonstrate that microgeographic genetic differentiation can exist over quite small distances. We can only conclude as did Koller (1939):

Local genetic diversification can arise in a population occupying a continuously habitable territory. How small the elementary breeding units are in such a territory remains to be determined.

Seasonal Changes in Inversion Frequency

One of the characteristics of Dobzhansky was that he advanced his ideas with fervor. When, however, he discovered that an idea was incorrect, he would change his position and pursue the new approach with equal vigor.

TABLE 11.2.

Chromosome Frequencies at Tulancingo

Trap	CU	TL	EP	HI	SC	OL	Others
1	65	58	5	3	2	—	—
2	146	139	8	2	1	9	2
3	24	22	—	2	—	1	—
4	18	13	—	—	—	1	—
5+6	29	26	—	1	—	—	—
7	50	34	1	1	—	1	—
8	32	42	1	1	1	1	—
9	64	55	5	3	1	3	—
10	14	11	1	—	—	—	—
10′	26	45	1	3	—	3	—
11	38	44	3	—	—	1	—
11′	8	7	—	—	—	1	—
12	14	15	—	—	—	—	1
13	7	7	—	1	1	—	—
14	21	11	—	—	—	—	—
15	28	29	2	—	—	1	—
16	54	56	2	—	—	—	—
17	5	5	—	—	—	—	—
17′	20	20	1	—	1	2	—
18	38	40	1	—	—	3	—
19+20	4	7	1	—	—	—	—
21	34	33	2	—	—	—	—
22	7	6	—	—	—	1	—
23	19	18	1	—	—	1	—
24	12	12	—	—	—	—	—
25	4	4	—	—	—	—	—
26	9	14	—	—	—	1	—
27	14	13	—	—	1	—	—
28	9	16	—	1	—	—	—
29	23	23	—	—	—	—	—
30	2	11	1	—	—	—	—
31	5	9	—	—	—	—	—
32	8	13	—	1	—	—	—

χ^2 = 30.06 with 32 d.f.; prob. = 0.56

Chi-square test for homogeneity was performed as in table 11.1

This was the situation that occurred with relation to the question of whether inversions did or did not have adaptive value for their carriers. As always, Dobzhansky (1947) explained the situation best.

Because the inversion homozygotes and heterozygotes are indistinguishable in external appearance, there was no reason to suppose that inversions are other than adaptively neutral characters. Not until Dobzhansky (1943 in *D. pseudoobscura*) and Dubinin and Tiniakov

(1945, 1946 in D. *funebris*) found that populations which live in different habitats often differ in the relative frequencies of their gene arrangements, and that the composition of a single population may vary appreciably from season to season, was it realized that carriers of different gene arrangements may be favored or discriminated against by different environments.

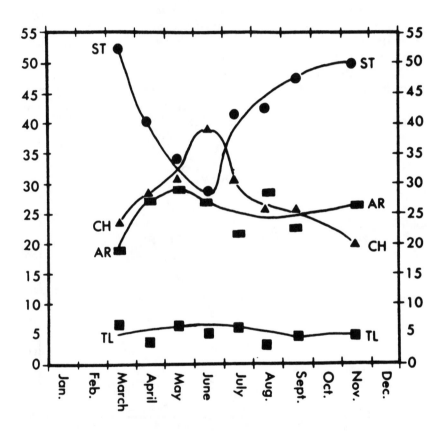

FIGURE 11.8.

Monthly frequencies in percent of third chromosome gene arrangements in the D. *pseudoobscura* population at Piñon Flats. Redrawn from Dobzhansky 1947.

The most striking example of seasonal variation in chromosome frequencies occurs in the *D. pseudoobscura* population of Piñon Flats, California (fig. 11.8). Flies come to the traps only from about March to November, presumably being rendered inactive during the winter months by the extremely low temperatures. As can be seen in figure 11.8, at Piñon Flats, Standard (ST) is relatively high in frequency in both the early spring and late fall collections, falling to a relatively low frequency in midsummer. Chiricahua (CH) follows a reverse pattern while the frequencies of both Arrowhead (AR) and Tree Line (TL) change relatively little over the seasons. Similar seasonal patterns of changes in frequencies of gene arrangements have been noted in various populations. However, there are a number of localities in which no seasonal shifts in inversion frequencies have been found (see detailed discussion in Powell 1992).

Although we have collected over the years at three locations, Zirahuen, Tulancingo, and Amecameca (fig. 11.6), our most nearly complete data on seasonal changes come from Amecameca. Because of the relatively mild weather conditions prevailing at this locality, collections can be made

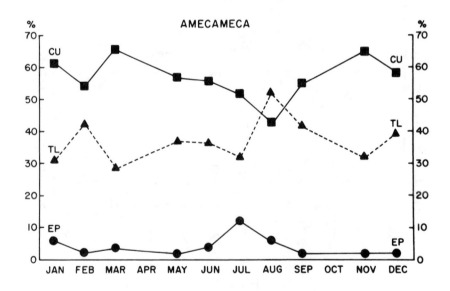

FIGURE 11.9.
Monthly frequencies in percent of third chromosome gene arrangements in the *D. pseudoobscura* population at Amecameca.

throughout the year. There are three gene arrangements that, depending on the collection, account for 95–99% of the inversions recorded (fig. 11.9). These are Cuernavaca (CU), Tree Line (TL), and Estes Park (EP). CU and TL are the predominant gene arrangements, with CU having the higher frequency except in the month of August, when TL becomes the most frequent inversion in the population. August occurs toward the latter part of the warmer (April–October), wetter (May–October) period of the year at Amecameca, and we may be recording the effects of a change in adaptive values of these two gene arrangements for their carriers under those particular environmental conditions.

It is of interest to compare Dobzhansky's and our results on seasonal changes. TL is the only gene arrangement found in both localities. In Piñon Flats, TL is in very low frequency and shows no response to seasonal changes. In Amecameca, however, TL is a major component of the chromosomal polymorphism of the population and does show a change in frequency during the midsummer period, becoming the most frequent inversion in the population. This is, of course, the same season of the year in which ST and CH show sharp changes in their frequencies in the Piñon Flats population. For populations that exhibit seasonal changes in gene arrangements, it may be that the environmental factors during the summer period are especially important.

A further word on TL is in order. The difference in role of TL in Piñon Flats and Amecameca is, most probably, the result of allele differences in the TL inversion from the two localities. It has been found that the TL gene arrangement from different populations, as is also true of the other inversions, carries different alleles for a number of loci (Lewontin 1974).

Long-Term Changes in Inversion Frequency

One of the purposes of long-term genetic studies of natural populations is to determine whether any directional (evolutionary) changes are occurring. Where seasonal variation takes place, as we find in inversion frequencies of many *D. pseudoobscura* populations, it is necessary to restrict one's examination of the data to those gathered in comparable periods of time. As was mentioned earlier, our most nearly complete data come from the population at Amecameca. In table 11.3 are listed the frequencies of the three most common gene arrangements recorded in the months of January, May, and November, at seven-, eight-, and nine-year intervals, respectively. A chi-square test of the raw data for each of these months yielded the following p values: >0.05, <0.01, >0.05. We do not know why there have

been significant changes in CU and TL frequencies during the month of
May over this period of time or whether in recent years there may have
been a reversal of this situation. It is clearly of great interest to sample the
population again in the near future.

When we examine the numerous sets of data collected by Dobzhansky
for any indication of long-term changes, it is the population at Piñon Flats
that offers the greatest amount of information (Dobzhansky 1971). In table
11.4 are listed the frequencies of the three most common gene arrangements recorded in the months of March, April, and May at a thirteen-,
followed by a ten-year, interval. There were highly significant increases in
the frequency of ST and "other" gene arrangements and concomitant
decreases in the frequencies of AR and CH in all three months for each
time period (chi -square in each case has a p value of <0.001).

A comparison of the situation at Amecameca and Piñon Flats reveals
that although there were significant changes in inversion frequencies in the
D. pseudoobscura populations at both localities, the increase in TL percentage at Amecameca was about 50%, whereas the increases in ST percentages at Piñon Flats were as much as 100% of the previously recorded val-

TABLE 11.3.

Percentage Frequencies of Inversions at Amecameca (n = Sample Size)

Month and Year	CU	TL	EP	Others	n
Jan. 1977	59	31	8	2	973
Jan. 1984	64	29	4	3	335
May 1975	62	30	3	5	232
May 1983	49	46	2	3	184
Nov. 1976	66	31	2	1	619
Nov. 1985	61	36	1	2	176

TABLE 11.4.

Percentage Frequencies of Inversions at Piñon Flats (n = Sample Size)

Month and Year	ST	AR	CH	Others	n
March 1940	45	20	30	5	386
March 1953	60	19	5	16	412
March 1963	80	6	3	11	114
April 1940	35	28	34	3	176
April 1953	48	23	14	15	588
April 1963	76	11	3	10	300
May 1940	28	27	40	5	202
May 1953	42	21	15	22	342
May 1953	63	14	5	18	190

ues. As mentioned earlier, the weather conditions at the two locations are quite different, and the more extreme situation at Piñon Flats may account for the sharper long-term changes in inversion frequencies at that location.

A final comment on long-term studies is warranted. Anderson, et al. (1991), reporting on "Four decades of inversion polymorphism in *Drosophila pseudoobscura*," stated, "There is only one pronounced trend over time: the increase in frequency of the Tree Line inversion in Pacific coast populations." This conclusion results from examining the combined over-all yearly data of all the North American populations. That approach cannot detect long-term changes in individual populations, especially those that show seasonal variation in inversion frequencies.

Discussion

The inversion polymorphism seen in *D. pseudoobscura* is probably the most intensively studied example of chromosomal polymorphism to date. Yet investigations of natural populations continue to discover new gene arrangements that not only indicate the complexity of this phylogeny but also imply its antiquity. These studies also cause us to anticipate the finding of an increased role that linked blocks of genes have played in evolution.

Our study of geographic gradients of inversion frequencies has revealed a situation in Mexico that parallels the one discovered earlier in the United States. Further work, not only in Mexico, but also in the United States is certainly warranted if a broad picture of the importance of different gene arrangements in various natural populations is to be achieved. Such research will, it is hoped, lead to investigations of the environmental factors that determine inversion frequencies in particular populations.

Studies on microgeographic genetic variation, in effect, attempt to answer the question, "How small is micro?" This is not a facetious question, especially for a flying form; rather, it raises the problem of the existence of breeding units. With succeeding investigations, it has become clear that the distance over which distinct breeding units of *D. pseudoobscura* may form can be quite small and the lower limit may not yet have been discovered. What also remain to be determined are the environmental and behavioral factors that control the size of the area in which a distinct breeding unit may form in a given locality.

Seasonal changes in inversion frequency have been important historically in providing proof of the selective value of linked blocks of genes in natural populations. However, there appears to be no rule that predicts whether seasonal changes will occur in a given population or, if they do occur, how drastic they will be. On the basis of the research that has been done, apparently each situation may very well be unique.

For the geneticist interested in studying the evolutionary process in natural populations of *D. pseudoobscura*, long-term changes in inversion frequency offer an opportunity to monitor the actual course of events. Although, in time, any given evolutionary change will be reflected by the entire species, the initial genetic alteration will occur and become established in one or more relatively small populations. With about forty years of data available from a number of sites, we would urge that certain populations of the species be earmarked for year-round monthly sampling at five- or ten-year intervals. Certainly, the population at Piñon Flats should be designated one such "evolutionary resource," and we look forward to continuing our studies of the Mexican populations.

Conclusion

We have reviewed our work on the Mexican populations of *D. pseudoobscura* and have indicated how central Theodosius Dobzhansky was and remains in our investigations. The evolutionary process he studied so intensively reveals its complexity with each new piece of research and continues to attract and excite geneticists. If we have provided information that sparks further discussion or have indicated some areas of research that hold promise for a better understanding of the process, we will have done what we believe Dobzhansky would have wanted us to do.

We also want to express our gratitude to Dr. A. L. de Garay not only for making the project possible but also for continually supporting us in our work. Although he has consistently refused coauthorship on any of our papers, he has had an integral role in the planning of our investigations. We can conclude this review in no better way than to quote the acknowledgment that has been in each of our papers since 1975:

> This paper is dedicated to Dr. Alfonso L. de Garay and to the memory of Th. Dobzhansky, initiators and guiding spirits of this program.

REFERENCES

Anderson, W. W., J. Arnold, D. G. Baldwin, A. T. Beckenbach, C. J. Brown, S. H. Bryant, J. A. Coyne, L. G. Harshman, W. B. Heed, D. E. Jeffery, L. B. Klaczko, B. C. Moore, J. M. Porter, J. R. Powell, T. Prout, S. W. Schaeffer, J.

C. Stephens, C. E. Taylor, M. E. Turner, G. O. Williams, and J. A. Moore. 1991. Four decades of inversion polymorphism in *Drosophila pseudoobscura*. *Proc. Natl. Acad. Sci.* 88:10367–71.

Dobzhansky, Th. 1939a. Genetics of natural populations. IV: Mexican and Guatemalan populations of *Drosophila pseudoobscura*. *Genetics* 24:391–412.

———. 1939b. Microgeographic variation in *Drosophila pseudoobscura*. *Proc. Natl. Acad. Sci.* 25:311–14.

———. 1947. Adaptive changes induced by natural selection in wild populations of *Drosophila*. *Evolution* 1:1–16.

———. 1971. Evolutionary oscillations in *Drosophila pseudoobscura*. In R. Creed, ed., *Ecological Genetics and Evolution*, pp. 109–33. London: Oxford University Press.

Dobzhansky, Th. and C. Epling. 1944. Contributions to the genetics, taxonomy, and ecology of *Drosophila pseudoobscura* and its relatives. Carnegie Inst. Wash. Publ. 554:1–183.

Dobzhansky, Th., R. Felix, J. Guzman, L. Levine, O. Olvera, J. R. Powell, M. E. de la Rosa, and V. M. Salceda. 1975. Population genetics of Mexican *Drosophila*. I: Chromosomal variation in natural populations of *Drosophila pseudoobscura* from Central Mexico. *J. Heredity* 66:203–6.

Dobzhansky, Th. and M. L. Queal. 1938. Genetics of natural populations. I: Chromosomal variation in populations of *Drosophila pseudoobscura* inhabiting isolated mountain ranges. *Genetics* 23:239–51.

Guzman, J., O. Olvera, M. E. de la Rosa, and V. M. Salceda. 1993. Population genetics of Mexican *Drosophila*. IX: East-west distribution of inversion polymorphism in *Drosophila pseudoobscura*. *Southwestern Natur.* 38:52–87.

Koller, P. C. 1939. Genetics of natural populations. III: Gene arrangements in populations of *Drosophila pseudoobscura* from contiguous localities. *Genetics* 24:22–33.

Lewontin, R. C. 1974. *The Genetic Basis of Evolutionary Change*. New York: Columbia University Press.

Olvera, O., R. F. Rockwell, M. E. de la Rosa, M. I. Gaso, F. Gonzalez, J. Guzman, and L. Levine. 1985. Chromosomal and behavioral studies of Mexican *Drosophila*. III: Inversion polymorphism of *D. pseudoobscura*. *J. Heredity* 76:258–62.

Powell, J. R. 1992. Inversion polymorphisms in *Drosophila pseudoobscura* and *Drosophila persimilis*. In C. B. Krimbas and J. R. Powell, eds., *Drosophila Inversion Polymorphism*, pp. 73–126. Boca Raton, Fla.: CRC Press.

Provine, W. B. 1981. Origins of the Genetics of Natural Populations series. In R. C. Lewontin, J. A. Moore, W. B. Provine, and B. Wallace, eds., *Dobzhansky's Genetics of Natural Populations I–XLIII*, pp. 1–85. New York: Columbia University Press.

Taylor, C. E. and J. R. Powell. 1977. Microgeographic differentiation of chromosomal and enzyme polymorphisms in *Drosophila persimilis*. *Genetics* 85:681–95.

12

Population Genetics of *Drosophila mediopunctata*

Louis Bernard Klaczko

Following a good old tradition, common among lower middle-class Brazilian students interested in science, I went to medical school. Of course, I was the worst Physician of my class. In spite of that, during this period I worked with various scientists interested in basic genetics. After graduation I got a master's degree in genetics and was introduced to some of the papers of Dobzhansky. Those on chromosome inversions impressed me the most (e.g., Dobzhansky 1943; Dobzhansky and Pavlovsky 1957; Wright and Dobzhansky 1946). Chromosome inversions are of great artistic beauty; they can be used to address fundamental questions and are a simple and cheap technique—important in Brazil, since inexpensive methods can ensure that work can be done even with small grants.

Looking for a good education, I applied for a Ph.D. with Jeffrey Powell at Yale University. He accepted me, apparently not without wondering what an M.D. wanted to do with *Drosophila* population genetics. In the years that followed, Jeff revealed himself to be a very good friend, supportive in all regards (especially, when my salary from Brazil would arrive up to

A large portion of the results mentioned were obtained by my students: Blanche C. Bitner-Mathé, Antonio B. Carvalho, Alexandre A. Peixoto, Helder Marques, Paulo G. Martins, and Claudiney F. Moraes. Sergio F. Reis helped me by providing computer facilities. I thank Eliane J. A. Klaczko and two anonymous reviewers for valuable suggestions in the manuscript. During various parts of this work we received financial support from: CAPES, CNPq, FAPESP, FINEP, FUJB, and Universidade Federal do Rio de Janeiro—where I was a faculty member during part of the time covered.

six months late). But, most importantly, Jeff did not think that I was crazy for dreaming of keeping alive Dobzhanky's tradition of research in Brazil—a dream of forming new generations of intellectually independent—though nonxenophobic—students interested in evolution. With Jeff, instead of merely learning the latest fashionable techniques, I learned Dobzhansky's way of working and the basic concepts and issues of population genetics. Using *D. pseudoobscura's* third chromosome inversions, in the field and in the laboratory, I worked with classical techniques and methods that could certainly be applied in Brazil—something Jeff was clearly concerned with.

I never met Dobzhansky. But once, at Mather, on a field trip with Jeff and C. Taylor (cf. Klaczko, Taylor, and Powell 1986), we received the visit of Olga Pavlovsky, who was brought by T. Prout and M. Green. Knowing her life was close to an end, she came all the way from New York to give her last farewell to her beloved "boss." Jeff led us, showing the way, to the place in the woods where Dobzhansky was resting with his wife. There, from a tape recorder, brought by this great lady, we silently listened to Mozart's *Requiem*. I think this was the moment I was really introduced to Dobzhansky.

Returning to Brazil in 1984, I found the country in a state of political euphoria. The clear signs of the end of the dictatorship, so anxiously awaited, were finally present. However, there was a very serious economic recession. To make things even worse a heavy bureaucracy was installed, and in science, priority was given to medical or applied research (an old trend in Brazilian politics). This, by the way, may explain why some of Dobzhansky's Brazilian students were driven out of the laboratory into scientific politics or moved away from *Drosophila* to work with other organisms, often with economic or medical interest. At this moment, to start a new laboratory working on *Drosophila* population genetics seemed an impossible task.

I realized that an important part of Dobzhansky's legacy is to have transformed some species into experimental models for evolutionary genetics. This is clearly the case in the United States with *D. pseudoobscura* and *D. persimilis*, which are still intensively studied. In Brazil, Dobzhansky and his local students developed *D. willistoni* and *D. paulistorum* into paradigms for studies of adaptation and speciation. Thus, I decided to start over, to transform another species into an experimental model for evolutionary genetics with emphasis on adaptation and coadaptation, my major interests.

I intended to work intensively with one species, studying, simultaneously, various aspects of its biology. I hoped that each aspect would shed light on the others, revealing new problems and new phenomena. I hoped also to develop original methods and to obtain results of general interest

(Klaczko 1990). Moreover, this would give me the opportunity to form students in an integrated research, where collaboration is important but where each one is trained in an independent aspect of the project.

After a few preliminary collections I decided to work with *Drosophila mediopunctata* (Dobzhansky and Pavan 1943). This species belongs to the *tripunctata* group of the subgenus *Drosophila* and has a wide geographical distribution—it has been found in many parts of Brazil and in El Salvador (Val, Vilela, and Marques 1981). Its life history is quite distinct from other species of *Drosophila* commonly used in genetics—most of them of the subgenus *Sophophora*—such as *D. melanogaster* or *D. pseudoobscura*. The maturation time of *D. mediopunctata* is long—it starts laying eggs by the seventh day after hatching—and its fecundity is relatively poor, but it has an extended longevity. I hoped that this peculiar biology could prove interesting by revealing facets not seen with the other species commonly studied.

The strategy I initially decided to use was to concentrate all field work at one locality (Parque Nacional do Itatiaia, States of Rio de Janeiro and Minas Gerais). Following Dobzhansky's paradigm at Mount San Jacinto (Dobzhansky 1943), I would be looking for altitudinal clines and temporal cyclic variations in the inversion frequencies. To minimize costs the flies collected were also to be simultaneously used for the other aspects under study. I also opted for using simple and inexpensive techniques, at least for the initial phase of my work.

Since the very beginning of my work, I have been aware of the necessity of having available genetic markers. *D. mediopunctata* has five pairs of acrocentric chromosomes and a pair of dots. Now, we have obtained fifteen visible mutants marking all chromosomes, except the dot (Marques et al. 1991). They were obtained by inbreeding flies from the field, by x-ray and EMS (ethyl methane sulfonate) treatments, and some were spontaneous mutations that appeared in laboratory cultures.

In this paper I present a short review of the most important results I obtained with my undergraduate and graduate students. I discuss the second chromosome inversions, the sex ratio trait, and morphological variation in this species, although we have also been studying other aspects of the biology of this fly.

Chromosome Inversions

After analyzing thousands of chromosomes directly from the field—from Itatiaia and from other localities—we have found inversions in chromosomes X, II, and IV. There are two gene arrangements in chromosome IV

(*St* and *1*). Chromosome X shows three inversions derived from the *St* gene arrangement—*1* and *2*—that do not overlap with each other, and *3*, which overlaps with both *1* and *2*.

The second chromosome is the most polymorphic. It may be divided didactically into two nonoverlapping regions: distal and proximal. In the distal region we have found 9 gene arrangements (*DA, DI, DS, DP*, etc.). In the proximal region there are also 9 different gene arrangements (*PAO, PA8, PBO, PCO, PC1*, etc.), which may be divided into 3 phylads (*PA, PB*, and *PC*).

After some training, inversions of the distal region can be easily recognized in routine preparations, both in heterokaryotypic and homokaryotypic individuals. However, inversions of the proximal region can be recognized with confidence in routine preparation only in heterokaryotes. Actually, this situation occurs in a number of species of *Drosophila*.

In studies using the "egg sample method"—each collected female, already inseminated, has one larva of its F_1 examined—it is common to analyze only the frequency of heterozygotes and not the "allele" frequencies. This is unfortunate because in a two-allele locus a heterozygosity of, say, 32% may be due either to a frequency of *A* equal to 80% or to 20%, and one cannot decide which of the two alternatives is correct. However, one should mention that with the "male method" this problem does not exist. Here each collected male is crossed to females of a known homokaryotypic strain and its karyotype inferred by the analysis of a number of F_1 larvae. Only one type of homozygote can show up in the offspring (with the same known genotype of the mother) and the various heterozygotes can be recognized.

At first we felt very discouraged with this problem since the "egg sample method" may be valuable, particularly when the number of animals collected is small. But eventually we realized that when a locus has 3 alleles, there are 3 different heterozygotes plus one class with all homozygotes lumped. This makes 3 degrees of freedom for the karyotype frequency data, and there are only 2 independent parameters to be estimated (the frequencies of 2 of the alleles). This means that, at least theoretically, each allele's frequency can be estimated. We then developed a maximum likelihood method to estimate the allele frequencies for the case when we can distinguish the heterozygotes but not the homozygotes and there are more than 2 alleles at the locus.

We applied this method to our data on the proximal inversions by using the "egg sample method," lumping the various inversions per phylad—in practice only inversions *PAO, PBO*, and *PCO* are common. We estimated their frequencies and compared them with the frequencies estimated by the "male method," finding no difference. We also applied, with success, this method to published data on overlapping inversions of other species, including *D. willistoni* and *D. nebulosa* (Klaczko, Otto, and Peixoto 1990; a program is available upon request).

Since in the second chromosome there are 9 inversions in the proximal region and another 9 in the distal one, it is possible to have 81 different combinations (haplotypes). We examined 2130 chromosomes directly from the field by using the "male method" and discovered that there is a very strong linkage disequilibrium. Of the eighty-one possible combinations, only 31 appeared, and 4 of them represented more than 87% of the chromosomes sampled. Their frequencies and the values of D' (D/D_{max}) are, respectively: *DA-PAO:* 47.7%, 0.976; *DI-PBO:* 19.8%, 0.857; *DP-PCO:* 7.2%, 0.975; and *DS-PCO:* 12.6%, 0.948 (a complete list can be found in Peixoto and Klaczko 1991). This means that *DA* is almost always associated with *PAO* in the haplotype *DA-PAO, etc.* This linkage disequilibrium has two consequences: a theoretical and a practical one.

On the theoretical side one may ask why does such a linkage disequilibrium exist? It may be the consequence of natural selection, which has favored harmonious associations between distal and proximal inversions, that is, a result of coadaptation. However, if the recombination rate between the two regions is small and selection favors one (or more) of the "alleles" in *one* of the regions, linkage disequilibrium may ensue as a result of a "hitchhiking" effect. And it may persist in the population for a long time, depending on the recombination rate.

To rule out this last hypothesis (or even a drift effect) we tried to estimate the recombination rate between the distal and proximal regions. First we analyzed directly the polytene chromosomes of the F_1 of double heterozygous females, crossed to appropriate males. We examined 3,600 gametes obtaining only 1 recombinant (Martins and Klaczko 1990). Another approach used was to analyze the progeny of double heterozygous females for visible mutations (*Delta* and *Antennapedia*) marking distal and proximal inversions. Among 14,081 individuals of the progeny, not a single recombinant was observed. Thus, the recombination rate is so small that we have no grounds to state that the linkage disequilibrium is due to coadaptation.

The practical side of the strong linkage disequilibrium found is that we can use the inversions of the distal region as markers of the haplotypes. Since they are easily recognized we may work solely with them, simplifying our language (and statistical treatments). Thus instead of talking about *DA-PAO* or *DS-PCO* we just mention *DA* or *DS*.

From 1986 to 1988 we carried out at Itatiaia 9 collections and in each occasion we collected flies at various altitudes. We observed that the frequency of *DA* tended to be higher in the colder months (varying from 52.8% to 54.5%) than in the warmer months (from 43.8% to 48.2%), while the opposite happened with the pooled frequencies of *DS* and *DP* (colder months varying from 15.6% to 21.1%; warmer months: from 19.3% to 27.6%). For *DI* no pattern was evident. We then tried to correlate these inversions frequencies with the altitude of the point of collection. For *DA* we found a positive and significant correlation ($r = 0.87$, 6 d.f., $p < 0.01$); for *DS* and *DP* the correlations were negative and significant ($r = -0.86$, $p < 0.01$ and $r = -0.77$, $p < 0.05$, respectively), while for *DI* no significant correlation was detected (Klaczko 1990). This strongly suggests that this polymorphism is under some form of selection. However, we have not yet been able to characterize which components of fitness are being affected.

Sex Ratio

In one of our first field trips we brought to the laboratory a female that produced a progeny of 22 females but no males. This was the first clue for the presence of a "sex ratio" phenomenon in *D. mediopunctata*. Sex ratio is the production of progenies with excess of females by affected males. It is a particular case of meiotic drive affecting the X-chromosome. It is known in several species of *Drosophila*, and it is usually, but not always, associated with inversions on the X-chromosome (James and Jaenike 1990).

Our first approach was to analyze the progenies of the males collected at Itatiaia for the inversion chromosome study. Each male had its karyotype determined (including the X-chromosome inversions) by the "male method" (see above). All we had to do, then, was to sex and count the emerging progenies to see whether there was any difference in the sexual proportion for males with different X-chromosome karyotypes. We found that males carrying inversion *2* had on the average progenies with about 20% of males, while those with a *St* X-chromosome had approximately 47%. With a few confirmatory crosses in the laboratory we could demonstrate that the sex ratio trait does, in fact, occur in *D. mediopunctata* and is associated with X-chromosome inversion *2* (Carvalho, Peixoto, and Klaczko 1989). Actually the situation is a bit more complex, since in the X-chromosome there is also a strong, but not complete, linkage disequilibrium between inversions *1* and *2*. But for the present purposes we may ignore inversion *1*.

In *D. mediopunctata* the expression of "sex ratio" is very variable. There are males carrying inversion *2* on their X-chromosome (hereafter called *X:2*) producing progenies that vary from 0 to 65% males (fertile). Our next step was to examine the causes of the variability of this character.

First, we found that in the population of Itatiaia there is a complicated system of suppressors of sex ratio. There are Y-chromosome suppressors and nonsuppressors, but there are also suppressible and insuppressible *X:2* chromosomes. The progeny of a male carrying a suppressor Y and a suppressible *X:2* is normal, while males with all other combinations are "sex ratio."

To detect the possible existence of autosomal modifiers of sex ratio, we selected a homokaryotypic strain carrying X-chromosome inversion *2* for a high proportion of females in the progeny. After 5 generations of selection the males were producing "normal" progenies. Using a strain with visible mutants marking all autosomes (except the dot), we carried out a series of crosses to test the effect of substituting each of the chromosomes. We could then detect the presence of modifiers (suppressors) in all autosomes. Moreover, we could show that chromosome IV had the strongest effect, which might indicate the existence of a major gene (Carvalho and Klaczko 1993).

Finally, we also investigated the effect of age on the expression of sex ratio. We took autosomally suppressed, Y suppressed and unsuppressed *X:2* males and analyzed their progenies throughout their life span. They were crossed successively with different sets of females. We found that age did not affect the sexual proportion of the progenies of suppressed males. They were always "normal." For the unsuppressed males age had a marked effect. Although young males had progenies with excess of females, this was greatly accentuated in the older ones (Carvalho and Klaczko 1992).

Morphological Variation

For various reasons the study of morphology has been traditionally outside mainstream population genetics (Lewontin 1974). Yet, it is probably at this level where adaptations and coadaptations are most evident. I wanted to analyze the morphological variation of *D. mediopunctata*, separating its components due to size and shape, as an initial attempt to study coadaptation in this species. As in any other aspect of my work, the first step was to characterize the variation present in a natural population. For this purpose I had to define the material to be used, the measurements to be taken, and the method of analysis.

The females collected in the field for the chromosome inversion study—used for the "egg sample" method—and up to 3 of their daughters raised in the laboratory that were saved became our material. For each of them (in a total of more than 1,500 animals) we took 13 measurements, 2 from one of the legs and 11 measurements from one of the wings. These were taken between points of insertion of the wing veins, according to the "truss-box" method (see Bookstein et al. 1985). Wing measurements were in larger number, thus with greater weight, since this is essentially a two-dimensional organ, and I hoped to be able to reconstruct, eventually, any modifications of its shape.

To the data we applied a principal component analysis (PCA). This method consists essentially in an axis rotation of an n-dimensional system where each dimension is one of the original measurements. The rotation is done in such a way that its first axis, or first principal component (PC1) coincides with the direction of maximal variation in the original data set. The second component (PC2) is orthogonal to the PC1—that is noncorrelated—and coincides with the direction of the second largest variation in the data, and so on. We can visualize this process by thinking that each observed animal is initially represented by a point in an n-dimensional space of measurements. After the axis rotation, the points, although at the same place, have other values in a new system of variables. However, and most importantly, after rotation the total variance remains the same, but the new variables are uncorrelated.

Usually in morphological studies there is a positive correlation between PC1 and each of the original variables. For this reason PC1 is taken as a measure of a "latent" variable representing overall size. And since all other components are orthogonal to PC1 they are size-free and represent "shape."

Among other results, we found that the variance due to size (0.0137) in the flies coming from the field was about three fourths of the total variance (0.018). For the progeny of these flies, grown in the laboratory, the total variance (0.009) was reduced to half of their mother's variance. Moreover, this was due to a reduction in the size variance (0.0047), while the rest of the variation (shape) was the same for mothers and daughters (0.0043) (Klaczko 1992). Although PCA can partition the total variance into components due to size and shape, and also have further applications, it does not have a simple geometric interpretation. This is why we were trying to find a geometric model that could describe the wings of *Drosophila*.

The ellipse is one of the simplest geometric figures. An ellipse located in a plane defined by a system of Cartesian coordinates is described by the general equation: $AX^2 + BY^2 + CXY + DX + EY + F = 0$. Performing a rotation and translation of the coordinate axis, such that the major axis of the ellipse coincides with the abscissae, and its minor axis with the ordinates, the previous equation is simplified to the well-known formula: $x^2/a^2 + y^2/b^2 = 1$. Where a and b are the radii of its two director circles. The area of an ellipse is equal to $3.14159ab$. So, if one takes the geometric mean of a and b—$SI = (ab)^{1/2}$—one can have a shape-free measure of size. In fact, SI is equal to the radius of a circle with the same area of the ellipse. On the other hand, the ratio $SH = b/a$ is a size-free measure of an ellipse's shape. Actually, a number of other measures can also be used, but these are probably the simplest ones. The position of any point on the outline of an ellipse can be determined by the angle between the line joining it to the center and the major axis. This measure is also size-free, like the position of the hours in a watch.

We tried to adjust an ellipse to the outline of the wing of *D. mediopunctata*. Using photographs, we took the Cartesian coordinates of points along

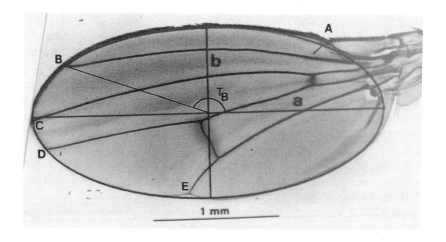

FIGURE 12.1

Wing of *Drosophila mediopunctata* and adjusted ellipse. The two director radii are respectively *a* and *b*. A, B, C, D, and E correspond, respectively, to the insertion points on the wing's border of the first, second, third, fourth, and fifth longitudinal veins. For each of these points the position is given by the angle between the line that joins it to the center of the ellipse and the major radius. For example, for point *B* the position is given by the angle *TB*.

TABLE 12.1

*Averages and Standard Errors of Parameters of the Ellipses
Adjusted to the Wings of Flies from the Field and Their
Daughters Grown in the Laboratory at 16.5°C and 20°C.*

	Flies			
	Field $n = 17$	16.5°C $n = 21$	20°C $n = 21$	F d.f. = 2
SI	338.62^A	369.48^B	341.23^A	18.81***
	5.72	3.63	2.70	
SH	0.4437	0.4453	0.4407	0.78^{ns}
	0.0018	0.0029	0.0033	
TA	0.5338	0.5426	0.5322	0.87^{ns}
	0.0048	0.0074	0.0053	
TB	2.7954^A	2.8237^B	2.8309^B	12.53***
	0.0046	0.0049	0.0056	
TC	3.1404	3.1374	3.1378	0.21^{ns}
	0.0026	0.0034	0.0038	
TD	3.3056	3.3104	3.3039	1.05^{ns}
	0.0044	0.0034	0.0026	
TE	4.3232^A	4.2810^{A,B}	4.2505^B	8.39***
	0.0137	0.0103	0.0129	

NOTE: The last column shows the F values from a one-way ANOVA (ns: $p>0.05$; ***: $p<0.001$). Averages with the same superscript are not significantly different by the Tukey test at the 0.05 level. *SI* is the size parameter; *SH* represents shape; *TA, TB, TC, TD,* and *TE,* the positions of points *A, B, C, D,* and *E*, respectively, expressed in radians (see text for details).

the internal edge of the wing. Then, we fit the general equation of the ellipse to these points, finding its reduced form. When we drew the ellipses and overlaid them on the original photographs, we were impressed by the nearly perfect fit (Klaczko and Bitner-Mathé 1990; see fig. 1). Hence, the ellipse's parameters can be used to describe the wing. Besides size (*SI*) and shape (*SH*), we can characterize the position of the insertion points of the veins (*A, B, C, D,* and *E*) by the angles they make with the major axis (*TA, TB, TC, TD,* and *TE,* see figure 12.1). We have also examined the wing of a large number of other species of *Drosophila* and *Zygothrica*, finding a good fit in all cases (Klaczko and Bitner-Mathé 1990).

We analyzed the wings of 17 females brought from the field and 42 of their daughters raised in the laboratory, half at 16.5°C and half at 20°C. Table 12.1 shows the averages of the ellipses's parameters and the positions of the veins' insertion points for the three groups, as well as the F values of a one-way analysis of variance (ANOVA). While no shape differences were found, there were clear size differences, the females from the field and their daughters raised at 20°C being smaller than those raised at 16.5°C. For the positions of the veins' insertion points we detected differences only for *TB* and *TE*.

Another advantage of the use of ellipses, to represent *Drosophila* wings, is that these results can be graphically summarized. Figure 12.2 shows the "average ellipses" depicting the "average wings" of the flies grown at 16.5°C (outer ellipse), and their mothers come from the field (inner ellipse). Inspection of the figure reveals at once the similarities and differences between the wings of the two groups. The "average ellipse" of the flies from 20°C (not shown) is virtually identical in size and shape to the one from the flies of the field (if they are drawn together, they fall each on top of the other). But the positions of the veins' insertion points of the 20°C flies are very close to the ones from 16.5°C flies.

For *TB* there were significant differences between the females from the field and their daughters raised at 16.5°C and 20°C but not between these

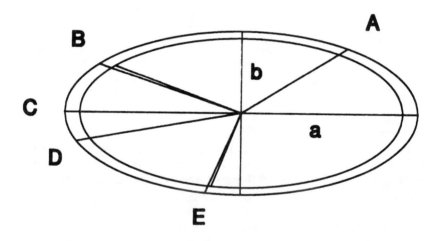

FIGURE 12.2

"Average ellipses" representing the "average wings" of the flies directly coming from the field (inner ellipse) and of the flies grown at 16°C (outer ellipse). Each ellipse was drawn by using the average values of the parameters of the ellipses adjusted to each of the wings of the two groups (shown in table 12.1). Points *A, B, C, D,* and *E* represent the position of the points of insertions of the veins (see figure 12.1 and text).

two groups (cf. table 12.1). This made us suspect that *TB* was not affected by temperature. We have carried out a preliminary experiment with *D. melanogaster* that showed that *TB* is influenced by density but not by temperature.

Examining our data we found that in the flies coming from the field we could fit a linear regression of *SI* on *TB*:$SI = -1576.817 + 685.22\ TB$ (the correlation was significant $r = 0.56$, $p = 0.02$). Since we know that size is influenced both by temperature and density, we tried to remove the effect of density in the females from the field by using the regression estimate. That is, we tried to estimate what would be the size of the flies from the field if they had grown at the same density of the laboratory conditions. Substituting the average of *TB* in the laboratory (2.827) in the equation, we found *SI* equal to 360.4. Assuming that the relationship between *SI* and temperature is linear, we interpolated this value to the averages of *SI* at the two temperatures in the laboratory (369.5 and 341.2), finding an estimated temperature of 17.6°C for the flies in the field. This agrees well with the average temperature of Itatiaia in the month of collection: 17.1°C (Klaczko 1992).

An important part of Dobzhanky's legacy—not always recognized—is to have developed some *Drosophila* species, notably *D. pseudoobscura* and *D. persimilis*, into models for evolutionary genetics. The main objective of my work is to transform *D. mediopunctata* into another experimental model for studies of adaptation and coadaptation.

Following Dobzhanky's paradigm, I started investigating, with my students, the polymorphism of chromosome inversions. I looked for altitudinal clines and cyclic variations in their frequencies. However, I also analyzed, simultaneously, other aspects of the biology of the flies, trying to discover new problems and propose original methods of analysis. So far, only simple and inexpensive techniques have been used.

For the chromosome inversion polymorphism, evidences were found that it is under selection, although I have not been able to characterize what components of fitness are affected. Additionally, we developed a method to estimate allele frequencies when only the heterozygotes can be recognized and there are more than two alleles—a situation that is fairly common with chromosome inversions and that Dobzhansky, himself, dealt with.

Luckily, as in *D. pseudoobscura*, there is in *D. mediopunctata* a "sex ratio" trait. In contrast, however, this trait is very variable in this latter species. Studying the causes of its variation, for the first time, it was possible to demonstrate the existence of modifiers (suppressors) of "sex ratio" in all major autosomes.

To analyze the morphological variation and look for a method to examine the variations in shape and size, I discovered that the wings of *Drosophila* can be described by ellipses. That is, if one tries to adjust an ellipse to the edge of a wing, the fit is nearly perfect. Thus, one can use ellipses to characterize the size and shape of wings, as well as the positions of their veins' insertion points. This is a first step to determine the causes— genetic and environmental—of morphological variation in the field and in the laboratory.

Dobzhansky obviously believed that an autochthonous school of population geneticists, in spite of all the problems and difficulties, could be maintained in Brazil. So do I.

REFERENCES

Bookstein, F. L., B. Chernoff, R. L. Elder, J. M. Humphries, G. R. Smith, and R. E. Strauss. 1985. *Morphometrics in Evolutionary Biology.* Philadelphia: Academy of Natural Sciences of Philadelphia.

Carvalho, A. B. and L. B. Klaczko. 1992. Age and sex-ratio expression in *Drosophila mediopunctata. Genetica* (Netherlands) 87:107–11.

———. 1993. Autosomal suppressors of sex-ratio in *Drosophila mediopunctata. Heredity* 71:546–51.

Carvalho, A. B., A. A. Peixoto, and L. B. Klaczko. 1989. "Sex-ratio" in *Drosophila mediopunctata. Heredity* 62:425–28.

Dobzhansky, Th. 1943. Genetics of natural populations. IX; Temporal changes in the composition of populations of *Drosophila pseudoobscura. Genetics* 28:162–86.

Dobzhansky, Th. and C. Pavan. 1943. Studies on Brazilian species of *Drosophila. Bol. Faculd. Fil. Ciên. Letr. S. Paulo* 36:7–72.

Dobzhansky, Th. and O. Pavlovsky. 1957. An experimental study of the interaction between genetic drift and natural selection. *Evolution* 2:311–19.

James, A. C. and J. Jaenike. 1990. "Sex-ratio" meiotic drive in *Drosophila testacea. Genetics* 126:651–56.

Klaczko, L. B. 1990. A Genética de Populações e a origem das adaptações. Atas do Encontro de Ecologia Evolutiva. *Publicação ACIESP* 69:8–17.

———. 1992. Métodos para a análise das variações de tamanho e forma em populações de drosófilas. *Braz. J. Genetics* 15 (Suppl. 1):269–73.

Klaczko, L. B. and B. C. Bitner-Mathé. 1990. On the edge of a wing. *Nature* 346:321.

Klaczko, L. B., P. A. Otto, and A. A. Peixoto. 1990. Allele frequency estimates when only heterozygotes can be recognized: Method of estimation and application in the case of chromosomal inversion polymorphisms in *Drosophila. Heredity* 64:263–70.

Klaczko, L. B., C. E. Taylor, and J. R. Powell. 1986. Genetic variation for dispersal by *Drosophila pseudoobscura* and *Drosophila persimilis*. *Genetics* 112:229–35.

Lewontin, R. C. 1974. *The Genetic Basis of Evolutionary Change*. New York: Columbia University Press.

Marques, H. V. S., A. B. Carvalho, C. A. Elias, and L. B. Klaczko. 1991. Mutants in *Drosophila mediopunctata*. *Drosophila Info. Serv.* 70:280.

Martins, P. G. R. and L. B. Klaczko. 1990. Recombinação entre inversões cromossômicas em *Drosophila mediopunctata*. Res. 36th Congr. Nac. Genética, Soc. Brasil. *Genética*, p. 162.

Peixoto, A. A. and L. B. Klaczko. 1991. Linkage disequilibrium analysis of chromosomal inversion polymorphisms of *Drosophila*. *Genetics* 129:773–77.

Val, F. C., C. R. Vilela, and M. D. Marques. 1981. Drosophilidae of neotropical region. In M. Ashburner, H. C. Carson, and J. N. Thompson, Jr., eds., *The Genetics and Biology of Drosophila*. Vol. 3a. London: Academic Press.

Wright, S. and Th. Dobzhansky. 1946. Genetics of natural populations. XII: Experimental reproduction of some of the changes caused by natural selection in certain populations of *Drosophila pseudoobscura*. *Genetics* 31:125–50.

13

Colonization of Chile by *Drosophila subobscura* and Its Consequences

Danko Brncic

I began to study the Chilean species of *Drosophila* after my stay at Columbia University in Dobzhansky's laboratory. There is no doubt that my interest was derived from Dobzhansky's paradigmatic idea about the role played by the ecological genetics research of *Drosophila* on evolutionary thought. This influence was double because before my stay in Columbia University in 1952–53, I had spent a year (1951) in the Old Institute of Biosciences of the University of São Paulo, Brazil, dedicated to learning about *Drosophila* genetics from the young and clever Dobzhansky students Crodowaldo Pavan and Antonio Brito da Cunha.

Continental Chile is situated at the Pacific coast of South America between parallels 17°32′S and 56°S. Biogeographically, it is strongly isolated from the rest of the neotropical zone in the north by the Chilean-Peruvian deserts and in the east by the Andes mountains. The Andean areas of Chile and Argentina, from the lake region in the south down to the southern tip of the continent, constitutes a biogeographic region that is separated from the rest of South America by the arid and cold barrier of the Patagonian "Pampa."

I thank Professors Anssi Saura, Francisco José Ayala, and Remigio López-Solis for their valuable comments and corrections of the manuscript. The work has been supported by Grant 91–0967 from Fondo Nacional de Ciencia (FONDECYT) Chile.

The isolation of Chile from the rest of the neotropical region has determined the endemic nature of its flora and fauna (Fuenzalida 1950). The Drosophilidae represent a good example of this situation. About 60% of the 35 species described (Brncic 1987) correspond to local species, practically restricted to Chile and at most to the neighboring Andean regions. The remaining species are cosmopolitan and widespread species. In the last forty years, we have accumulated a large amount of data about the genetic structure and ecology of some of the species that constitute the *Drosophila* communities in Chile. In recent years, substantial changes in the Chilean assemblage of *Drosophila* species have been observed. These changes are coincident with the degradation of the environment by increasing urbanization, industrialization, intensive farming, and other synanthropic factors. One of the most spectacular changes was the invasive colonization by the palearctic species *D. subobscura* (see below).

In this article I report and discuss some of the changes that have occurred in the local communities of *Drosophila* in the central region of Chile since the arrival of *D. subobscura* and the conditions of coexistence of that species with the local fauna. For some of the species, such as the endemic *D. pavani* and the cosmopolitan *D. immigrans*, which have been extensively studied in relation to chromosomal polymorphisms, I report and discuss what has happened with their polymorphisms.

The Colonization of *D. subobscura*

D. subobscura is a palearctic species whose distribution is described in the paper by Anssi Saura. In Chile, the species was collected for the first time in 1978 in the locality of Puerto Montt (lat. 41°28'S) (Brncic and Budnik 1980), a place in which routine collections of *Drosophila* were being performed almost every year since 1954. However, it is hard to know exactly when the species arrived in Chile, although it seems reasonable to infer that it could not have been much earlier than 1978. Since its first finding, the species underwent a rapid expansion in about two years. At present its distributional area in Chile extends from about La Serena in the north (lat. 29°55'S) to Punta Arenas in the Strait of Magellan in the south (lat. 53°10'S) (Brncic et al. 1981) and, in the neighboring Argentinean regions of Bariloche and Mendoza (Prevosti et al. 1989) and in Mar del Plata, at the Atlantic coast of Argentina (López 1985). In 1982, *D. subobscura* was described in North America in the Northwest Pacific near Post Townsend, Washington (Beckenbach and Prevosti 1986). Now, the species in North America has spread from British Columbia to 100 km northwest of Los Angeles, California, and to the eastern slope of Sierra Nevada (Prevosti et al. 1989).

Prevosti et al. (1983, 1985) have studied the genetics of the South and North American populations of *D. subobscura* by comparing the chromosomal and allozymic polymorphisms in various newly established localities in the New World, with those of the well-known European populations, in order to determine the possible origin of the colonizers.

The chromosomal polymorphism of *D. subobscura* has been extensively studied in the Old World. About 80 different gene arrangements related to the presence of paracentric chromosomal inversions have been discovered (Krimbas 1992). Large interpopulational differences exist in the number and frequencies of the gene arrangements. Some inversions are present, although with different frequencies, all over the distributional area of the species, and other arrangements are widespread without being ubiquitous; still others are restricted only to certain localities. Twenty different arrangements have been discovered in South America (Brncic et al. 1981, Prevosti et al. 1985, 1989). One of them, the E_{17} inversion in chromosome E, which does not exist in the Old World, has been found only three times in the first collections in Chile but in the later collections has not been detected. The remaining 19 arrangements are shared by the South and North American populations (Prevosti et al. 1989).

The presence of the same inversions in approximately the same frequencies in both regions is highly suggestive that the origin of the colonizers must be the same. According to Brncic et al. (1981) and Prevosti et al. (1985), the most probable origin of the colonizers is the western Mediterranean region of Europe. All, except one, of the arrangements present in the Americas are also present in that region of the Old World. However, the existence of the O_5 arrangement in chromosome O, which has never been found in the western Mediterranean regions, makes that assumption uncertain. Inversion O_5 has been reported in eastern Mediterranean regions and in northern Europe. Only the populations studied by Burla and Götz (1965) near Zurich in Switzerland, contain all the chromosomal arrangements reported in the Americas but in quite different frequencies.

The colonization of *D. subobscura* in the Americas illustrates the manner of expansion of cosmopolitan species that are passively transported by human agents to environments that are similar to the places of origin, many of them closely related to human activities. The individuals having the greatest chance of being transported are those from regions where the species is abundant and highly polymorphic, such as the western Mediterranean regions of Europe. The comparison of the chromosomal arrangement frequencies in the Americas and Europe shows that most of the inversions found in low frequencies in the possible original European populations are not represented in the Americas. This fact supports the hypothesis that the number of colonizers was rather low.

Brncic et al. (1981), on the basis of both the genetic composition of the most probable original populations in Europe and the gene arrangements found in the Americas, estimated a maximum probability of 10 to 15 individuals (20 to 30 chromosomes) as the number of colonizers arriving in Chile. This estimation is concordant with the one made by Mestres et al. (1990) on the basis of the frequency of the O_5 inversion. The existence of the same recessive lethal gene linked to that inversion both in South and North American populations is indicative of a common origin of both colonizers. At the present, there are no ways of knowing which part of America was colonized first and which one represents a second colonization from the first. The alternative that both regions were colonized independently by individuals coming from the same European region seems to be highly improbable (Ayala, Serra, and Prevosti 1989).

The analysis of enzyme polymorphisms by means of gel electrophoresis confirms the idea of a common origin of the colonizations of *D. subobscura* in South and North America (Prevosti et al. 1989). Both qualitative and quantitative data of enzyme polymorphism show that all the electrophoretic variants found in the New World are present in the western Mediterranean regions of Europe. But, like the inversion polymorphism picture, the alleles that in Europe are in low frequency are not represented in the Americas.

The study of mitochondrial DNA (Latorre, Moya, and Ayala 1986) has shown that the nucleotide polymorphism is greater for the European than for the American populations. Only 2 of the 8 haplotypes have been found in America. Rozas et al. (1990) confirmed and extended those data. For a total of 26 different nucleomorphs detected for the species, just the 2 most widely distributed were found in America.

Microdifferentiation Process in the Colonizing Populations

The colonizing process of *D. subobscura* in Chile has been very rapid over a north-south expansion of about 3000 km, from the dry and semiarid areas in the north as far as the wet and cold forest in the south. It is easily collected over fermented banana baits in a variety of environments such as gardens, orchards, and other synanthropic places as well as in some wild habitats. Its relative abundance in *Drosophila* collections in late winter and spring could reach more than 90 percent (Brncic et al. 1981; Benado and Brncic 1994). The great adaptive capability of *D. subobscura* to different environments and the rapid genetic differentiation of the American populations of *D. subobscura*, compared with the ancestral European ones, constituted strong indications that the local populations of the species could also undergo a relatively rapid process of genetic microdifferentiation.

Prevosti et al. (1985) found incipient clines in the frequencies of the chromosomal arrangements in the Chilean populations of *D. subobscura* that follow the same latitudinal direction as in the Old World. The same authors (Prevosti et al. 1989) found that in the Pacific region of North America there is also a north-south gradient in the frequency of some of the inversions, resembling the latitudinal clines observed in the Old World. The existence of well-defined latitudinal clines resembling the ones observed in Europe supports the hypothesis of the adaptive importance of the chromosomal polymorphism and the rapid effect of selective forces. Brncic and Budnik (1987a, b) studied the chromosomal polymorphism of different populations of *D. subobscura* in an altitudinal gradient in the pre-Andean zone of central Chile, from sea level to an elevation of 1900 m. In these studies, they demonstrated significant differences in the frequencies of some of the chromosomal arrangements between populations. However, the observed process of microdifferentiation does not seem to be correlated with the altitudinal gradient.

Interpopulational differences were also detected by comparing European and Chilean stocks with regard to well-known quantitative characteristics, such as wing size, egg-to-adult developmental time, and preadult viability (Budnik, Cifuentes, and Brncic 1991). Highly significant differences were observed between the stocks irrespective of the continent from which the flies were derived. In Europe, interpopulational differences in wing measures in *D. subobscura* have been correlated with latitude (Prevosti 1955; McFarquhar and Robertson 1963; Misra and Reeve 1964; Pfriem 1983). But, according to Budnik, Cifuentes, and Brncic (1991), no clines related to latitude are found between the Chilean stocks.

Another argument for the existence of an interpopulational differentiation process in *D. subobscura* in Chile emerged through the analysis of sexual isolation (Brncic and Budnik 1984). Males and females of Chilean stocks of *D. subobscura* exhibit a clear tendency toward homogamic mating. This tendency shows a relationship with the distance between the localities from which the respective stocks were obtained. Nevertheless, when the Chilean flies are confronted with European stocks, the ethological isolation is observed only in some cases but not in others, depending on the geographic origin of the stocks. Ochando, Rey, and Alarcon (1991) observed in Chilean stocks a north-south cline for mating speed in the females.

Local Community Studies

A real evaluation of the colonizing process in Chile and the impact of the introduction of an alien species on the already existing assemblage of species can be understood only through an analysis of the local community structure. The outcome of local processes, like the acquisition of new adaptations by changes in the genetic makeup of the populations and the conditions of coexistence with the existing species, will depend on the available resources and other ecological characteristics existing in that particular place.

The distributional range of the species depends always on the ecology of the individual species. Hence, it is expected that a species will be distributed in all the sites in which there exist ecological conditions like the ones in which it originates. The synanthropic species are good examples of that. Prevosti et al. (1989) and Ayala, Serra, and Prevosti (1989) argue that the success of the colonization of *D. subobscura* in two regions of the New World, the Pacific Coast of North America, and the Pacific Coast and southern cone of South America is a result of the Mediterranean climate of those regions, which is similar to the corresponding one in the Old World. Nevertheless, the outcome of a microdifferentiation process and its impact on the *Drosophila* community in the various sites depends mostly on local factors.

Analysis of a Local Community in Central Chile

As is usual in most regions in which the phenology of drosophilid communities is studied, the assemblage of species varies from month to month and from season to season (Dobzhansky and Pavan 1950, Lumme et al. 1979; Brncic, Budnik, and Guiñez 1985; Burla and Bachli 1991; Benado and Brncic 1994). The analysis of these fluctuations is fundamental for understanding population growth, patterns of distribution, migration trends, and species interactions. Furthermore, all the abiotic and biotic factors associated with the succession of seasons are basic for delimiting the pool of species that could occupy a specific area or habitat. Through the study of the interactions of such factors, the community structure can be understood (Durrson and Travis 1991).

The analysis of monthly collections of drosophilids made between August 1984 and December 1991 in La Florida, a district on the outskirts of Santiago (Chile) has given us certain insights into seasonality and species turnover of that community. The place in which collections were made was a small orchard, where most vegetations are ornamental plants, a few old trees (plums, apricots, figs, mulberries, medlar trees, vines), and some native plants. Forty years ago, La Florida was basically a suburb with about 25,000 people. At that time, it was dedicated primarily to farming and to horticulture. Since then, the district has become more urbanized, and it is now a rather prosperous area of Santiago, with about 400,000 people.

Table 13.1 summarizes the collections made in La Florida between 1984 and 1991. A total of 65,998 flies and 16 species of drosophilids were sampled. Six species (*D. melanogaster, D. simulans, D. immigrans, D. subobscura, D. pavani,* and *Scaptomyza denticauda*) made up more than 95% of the total abundance. *D. pavani* and *S. denticauda* are native species; *D. subobscura,* as was mentioned before, is a recent palearctic migrant; the remaining 3 *Drosophila* are "cosmopolitan." (For further characteristics of the species, see Brncic, Budnik, and Guiñez 1985; Benado and Brncic 1994.) Relative abundances of the 6 most common species are depicted in figure 13.1.

TABLE 13.1.

Number of Flies Collected Since August 1984 to December 1991 in "La Florida" (Outskirts of Santiago)

Species	1984*	1985	1986	1987	1988	1989	1990	1991	Total
A) *D. busckii*	21		8	59	28	45	13	10	184
D. immigrans	292	146	403	638	555	1076	638	696	4444
D. funebris	3	2	11	17	5	24	14	26	102
D. hydei	14	10	15	62	212	106	129	30	578
D. melanogaster	227	1636	729	281	675	912	1337	1293	7090
D. repleta	29	13	11	8	32	24	21	47	185
D. simulans	168	4808	5325	5475	6291	5075	6475	5452	39,069
B) *D. buzzatii*		1		4		1	1	2	9
D. nigricruria		2		2		2			6
D. subobscura	929	1768	2035	1375	376	597	1308	919	9307
C) *D. araucana*			1						1
D. flavopilosa						1			1
D. pavani	64	147	106	69	56	126	223	176	967
D) *S. denticauda*	112	477	462	491	752	244	297	448	3283
S. melancholica	3	161	140	174	62	17	36	176	769
S. multispinosa	2						1		3
Total	1864	9171	9246	8655	9044	8250	10493	9275	65,998

NOTE: Species have been clustered in Cosmopolitan (A), Subcosmopolitan (B), endemic *Drosophila* (C), and endemic *Scaptomyza* (D).
*Only samples from August to December.

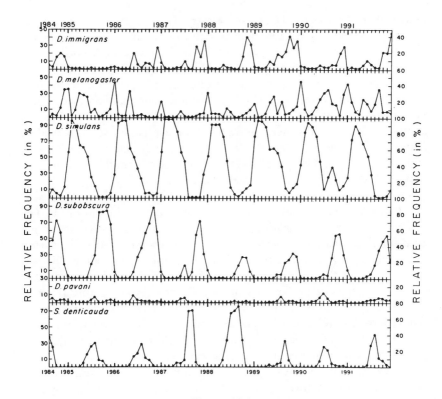

FIGURE 13.1.

Seasonal variation of the six most common species from La Florida (August 1984–December 1991).

(From Benado and Brncic 1994. With the kind permission of Blackwell Wissenschafts-Verlag GmlH, taken from Zeitschrift für Zoologische Systematik und Evolutionforschung.)

Autocorrelation analyses have indicated that the abundances display rather regular fluctuations (one-month and twelve-month cycles for all the species, except *D. melanogaster.*

Our data indicate that population numbers fluctuated rather regularly over the eight-year span of the study. Interestingly, seasonal changes in La Florida (Santiago) do not coincide with changes elsewhere. Thus, in Mediterranean Europe (David and Tsacas 1980), and in Australia (Parsons and Stanley 1981), *D. melanogaster* shows a single population peak at the beginning of summer. However, in Saporo, Japan, Watanabe (1979) reports that *D. melanogaster*, *D. simulans*, and *D. immigrans* are very scarce in spring and summer and that they flush in autumn. In Chile, *D. subobscura* shows a single peak in spring, yet in Europe there is an additional one in autumn (Shorrocks 1975; Begon 1976; David and Tsacas 1980).

As a rule, cosmopolitan species and subcosmopolitan ones, like *D. subobscura*, are more tolerant of climatic changes than endemic species are. However, racial differences with regard to cold and desiccation tolerance, as well as for several other external factors, are common in *Drosophila* (Ogaki and Nakashima-Tanaka 1966; David and Bocquet 1975; Parsons and Stanley 1981). Thus, it is likely that these seasonal differences represent local adaptations to environmental changes.

Nevertheless, other factors, like interspecific competition, may also affect population abundance, at least in some species. In La Florida, two of the most common *Drosophila* species, *D. simulans* and *D. subobscura*, are bred simultaneously from fermenting fruits of several species of orchard trees, and there is evidence that their larvae compete strongly within the rotting pulp of the fruits (Brncic and Budnik 1987b). The species also modify their rates of development and viabilities when competing as larvae in the laboratory, in such a way that they reach the adult stage at different times (Budnik and Brncic 1983). Interestingly, the two species do not overlap in their population peaks when sampled in the field (figure 13.1).

In summary, there is enough evidence to suggest that the success of the *D. subobscura* colonization can be attributed mostly to the clear seasonal differences between this and the other species, particularly the synanthropic ones. Its population flush occurs in months in which other species that are present in large number are either experiencing a quite different ecology, such as the scaptomyzids, or as in the case of a few indigenous *Drosophila*, like *D. pavani*, which, although it is a polyphagous species, is, or is supposed to be, well adapted to exclusive local conditions. Therefore, the coexistence of *D. subobscura* with other species in Chile seems to be based both on the population reduction of its potential competitors during the flush, such as *D. simulans*, and on the population expansion during the time in which no valid competitors exist.

Long-Term Changes in the *Drosophila* Community in Central Chile

Collections of drosophilids in La Florida-Santiago, have been performed regularly since 1953. The records of those collections have revealed significant quantitative changes in the *Drosophila* fauna during that forty-year

TABLE 13.2.

Relative Frequencies (in %) of Flies Collected in Spring Since
1953 to 1992 in "La Florida" (Outskirts of Santiago)

Date	Flies Collected	D. pavani	D. immigrans	D. simulans + D. melanogaster	D. subobscura	S. denticauda	Other* Species
Nov. 1953	520	41.15	37.69	19.42	—	—	1.74
Nov. 1954	413	60.05	28.09	9.44	—	—	2.42
Nov. 1955	1397	32.29	59.63	4.72	—	1.22	2.14
Oct. 1957	497	46.08	34.41	12.68	—	2.01	4.82
Nov. 1960	522	41.38	40.04	8.43	—	4.21	5.94
Oct. 1961	708	55.79	16.24	5.08	—	5.79	17.10
Nov. 1967	297	20.53	29.97	41.75	—	0.67	7.08
Oct. 1971	992	13.81	38.51	16.13	—	16.83	14.72
Oct. 1974	238	42.44	12.18	15.97	—	25.21	4.20
Dec. 1975	710	27.04	10.42	59.58	—	0.14	2.82
Dec. 1979	162	12.34	25.92	17.90	29.63	3.09	11.12
Oct. 1980	526	0.95	1.71	0.76	16.35	43.73	36.50
Oct. 1982	361	—	—	4.43	87.53	4.71	3.33
Nov. 1984	325	4.00	20.30	14.77	58.15	—	2.78
Nov. 1985	720	3.61	1.80	7.91	85.00	0.42	1.26
Nov. 1986	599	0.83	3.34	3.17	89.65	1.17	1.84
Nov. 1987	1099	0.09	17.01	6.19	71.34	0.64	4.73
Nov. 1988	467	2.57	41.33	20.34	26.55	0.64	8.57
Oct. 1989	477	3.98	42.35	14.89	24.11	9.01	5.66
Nov. 1990	660	3.33	20.76	15.45	56.82	1.21	2.43
Nov. 1991	499	4.00	21.24	12.02	53.82	3.21	5.71
Nov. 1992	488	7.23	7.83	8.44	73.29	2.41	0.80

Drosophila araucana, D. busckii, D. buzzatii, D. funebris, D. hydei, D. mercatorum, D. nigricruria, D. repleta, D. virilis, Scaptomyza intermedia, S. melancholica, S. multispinosa.

period. Those changes can be observed in table 13.2, which summarizes the results of spring collections (October, November, and early December) since 1953 up to 1992. One of the interesting facts observed in table 13.2 is that, together with the striking success of *D. subobscura* since the time of its first collection in that place (1979), the endemic species *D. pavani* has dramatically changed in relative frequency. This species became a rare one after being the most abundant and dominant in many of the samples before 1978. Is this change a consequence of the introduction of *D. subobscura* or are both phenomena a by-product of other factors? In this context, remember that successful introductions of new species occur mostly into highly disturbed and degraded habitats and particularly in anthropogenic environments (Simberloff 1981; Herbold and Moyle 1986). I believe that the events related to the colonization by *D. subobscura* in Chile could be compatible with this statement. But introduced species could interact negatively with the species of the local community through resource competition.

Chromosomal Polymorphism in *D. pavani* and *D. immigrans*

Tables 13.1 and 13.2 indicate that six of the sixteen species existing in La Florida (Santiago-Chile) constitute more than 95% of the flies recorded in the samples. Two of the species, the cosmopolitan *D. immigrans* and the

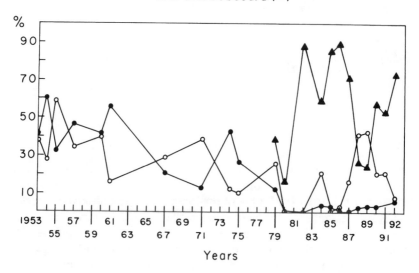

Relative frequency (in %) of *D. pavani* (●) *D. immigrans* (○)
and *D. subobscura* (▲)

Years

FIGURE 13.2.
Relative frequency (in %) of *D. pavani* (black circles), *D. immigrans* (open circles), and *D. subobscura* (black triangles) in spring season from November 1955 to November 1992 in La Florida.

Frequencies of inversions in *D. pavani.*

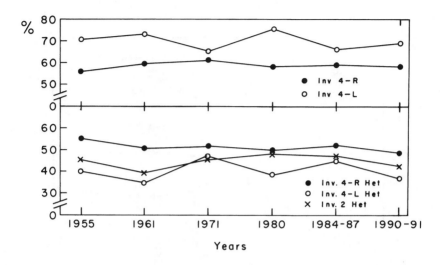

FIGURE 13.3.

Frequencies of the different inversions at homozygous state, in the right arm (4-R) and left arm (4-L) of the chromosome 4 of *D. pavani* in the locality of La Florida from November 1955 to November 1991 (upper part of the figure) and frequencies of heterozygous inversions in the chromosomes 4-R, 4-L, and 2 (lower part of the figure).

endemic *D. pavani*, are the ones that have changed more drastically in relative frequency since the colonization of *D. subobscura* (figure 13.2). Both species are chromosomally polymorphic owing to the existence of paracentric inversions. These polymorphisms have been studied in many of the Chilean populations, especially in the locality of La Florida, starting in 1953. Thus, it is of interest to report what has happened to the inversion frequencies in *D. pavani* and *D. immigrans* with the introduction of *D. subobscura*.

As shown in figures 13.3 and 13.4, no long-term changes have been detected in the chromosomal polymorphism of either species during the colonization of the central part of Chile by *D. subobscura*.

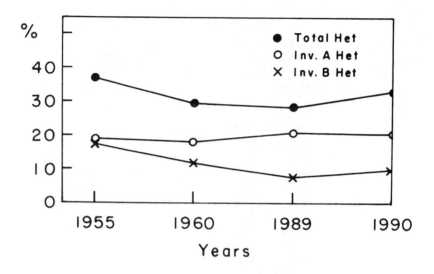

FIGURE 13.4.

Frequencies of the total heterozygotes and heterozygotes inversion A and B in chromosome 2 of *D. immigrans* from 1955 to 1990.

Concluding Remarks

Ayala, Serra, and Prevosti (1989) have defined the colonization process of *D. subobscura* in the Americas as "a grand experiment in evolution." I would point out that the experiment has just begun. Preliminary results are very exciting, but they are not enough for a full understanding of the colonization process and its consequences in the new territories. Nevertheless, these results do contribute to an understanding of similar colonization processes in other parts of the world, such as the European *D. ambigua* in North America (Beckenbach and Prevosti 1986), the North American *D. pseudoobscura* in Colombia and New Zealand (Parsons 1981), and the African *D. malerkotliana* in Brazil (Val and Sene 1980). The common characteristics of all these species are that they belong to the group of widespread species that are generalist regarding resource utilization and tolerant of different climates. Species of this kind are always good candidates to invade new territories through human passive transportation at considerable distance from their center of origin. In summary, they exhibit a capability to become more or less domesticated.

According to Dobhzhansky (1970), chromosomal polymorphism assumes different characteristics in different species of *Drosophila*.

In some species that have become widespread geographically, owing to their association with man the chromosomal polymorphism is little differentiated geographically, so that populations from remote and climatically unlike countries carry the same chromosomal variant.

The causal connections between the chromosomal polymorphism and the ability of a species to adapt to human-made environment is a matter of speculation. In Chile, the observed differences in the chromosomal polymorphism of the endemic species *D. pavani* with regard to the cosmopolitan *D. immigrans* and the widespread palearctic species *D. subobscura* represent a good example of Dobzhansky's seminal ideas on this subject.

REFERENCES

Ayala, F. J., L. Serra, and A. Prevosti. 1989. A grand experiment in evolution: The *Drosophila subobscura* colonization of the Americas. *Genome* 31:246–55.

Beckenbach, A. T. and A. Prevosti. 1986. Colonization of North America by the European species *Drosophila subobscura* and *D. ambigua*. *Am. Midl. Natur.* 115:10–18.

Begon, M. 1976. Temporal variations in reproductive conditions of *Drosophila obscura* Fallen and *D. subobscura* Collin. *Oecologia (Berlin)* 23:31–47.

Benado, M. and D. Brncic (1994). An eight-year phenological study of a local drosophilid community in Central Chile. *Z. f. zool. Systematik u. Evolutionforschung* 32:51–63.

Brncic, D. 1987. A review of the genus *Drosophila* Fallen (Diptera: Drosophilidae) in Chile with the description of *Drosophila atacamensis* sp. nov. *Rev. Chilena Ent.* 15:37–60.

Brncic, D. and M. Budnik. 1980. Colonization of *Drosophila subobscura* Collin in Chile. *Drosophila Info. Serv.* 55:20.

———. 1984. Experiments on sexual isolation between Chilean and European strains of *Drosophila subobscura*. *Experientia* 40:1014–16.

———. 1987a. Chromosomal polymorphism in *Drosophila subobscura* at different elevations in Central Chile. *Genetica* (The Hague) 75:161–66.

———. 1987b. Some interactions of the colonizing species *Drosophila subobscura* with local *Drosophila* fauna in Chile. *Genet. Iber.* 39:249–67.

Brncic, D., M. Budnik, and R. Guiñez. 1985. An analysis of a Drosophilidae community in Central Chile during a three year period. *Z. f. zool. Systematik u. Evolutionsforschung* 23:90–100.

Brncic, D., A. Prevosti, M. Budnik, M. Monclús, and J. Ocaña. 1981. Colonization of *Drosophila subobscura* in Chile I: First population and cytogenetic studies. *Genetica* (The Hague) 56:3–9.

Budnik, M. and D. Brncic. 1983. Preadult competition between colonizing populations of *Drosophila subobscura* and established populations of *D. simulans* in Chile. *Oecologia* (Berlin) 58:137–40.

Budnik, M., L. Cifuentes, and D. Brncic. 1991. Quantitative analysis of genetic differentiation among European and Chilean strains of *Drosophila subobscura*. *Heredity* 67:29–33.

Burla, H. and G. Bächli. 1991. A search for pattern in faunistical records of drosophilids species in Switzerland. *Z. f. zool. Systematik u. Evolutionsforschung* 29:176–200.

Burla, H. and W. Götz. 1965. Veränderlichkeit des chromosomalen Polymorphismus bei *Drosophila subobscura*. *Genetica* (The Hague) 36:83–104.

David, J. R. and C. Bocquet. 1975. Similarities and differences in the latitudinal adaptation of two *Drosophila* sibling species. *Nature* 257:588–90.

David, J. R. and L. Tsacas. 1980. Cosmopolitan, sub-cosmopolitan and widespread species: Different strategies within the *Drosophila* family (Diptera). *C. R. Soc. Biogeogr.* (Paris) 57:11–26.

Dobzhansky, Th. 1970. *Genetics of the Evolutionary Process*. New York: Columbia University Press.

Dobzhansky, Th. and C. Pavan. 1950. Local seasonal variations in relative frequencies of species of *Drosophila* in Brazil. *J. Animal Ecol.* 19:1–14.

Durrson, W. and J. Travis. 1991. The role of abiotic factors in community organization. *Am Natur.* 138:1067–91.

Fuenzalida, H. 1950. "Biogeografia." In Corp. Fomento de la Producción, ed., *Geografía Económica de Chile*, pp. 371–420. Santiago: Imp. Universitaria.

Herbold, B. and P. B. Moyle. 1986. Introduced species and vacant niches. *Am Natur.* 128:751–60.

Krimbas, C. 1992. The inversion polymorphism of *Drosophila subobscura*. In *Drosophila Inversion Polymorphism*, pp. 127–220. Boca Raton, Fla.: CRC Press.

Latorre, A., A. Moya, and F. J. Ayala. 1986. Evolution of mitochondrial DNA in *Drosophila subobscura*. *Proc. Natl. Acad. Sci.* 83:8649–53.

López, M. 1985. *Drosophila subobscura* has been found in the Atlantic coast of Argentina. *Drosophila Info. Serv.* 61:113.

Lumme, J., S. Lakovaara, D. Muona, and O. Jarvinen. 1979. Structure of a boreal community of drosophilids (Diptera). *Aguilo Ser. Zool.* 20:65–73.

McFarquhar, A. M. and F. W. Robertson. 1963. The lack of evidence for coadaptation in crosses between geographical races of *Drosophila subobscura*. *Genet. Res.* 4:104–31.

Mestres, F., G. Peguroles, A. Prevosti, and L. Serra. 1990. Colonization of America by *Drosophila subobscura:* Lethal genes and the problem of the O_5 inversion. *Evolution* 44:1823–26.

Misra, R. K. and E. C. A. Reeve. 1964. Clines in body dimensions in populations of *Drosophila subobscura*. *Genet. Res.* 5:240–56.

Ochando, M. D., L. Rey, and C. Alarcon. 1991. Mating speed and duration of copulation in Chilean *Drosophila subobscura* populations recently colonizers. *Evolucion Biológica* 4:201–9.

Ogaki, M. and E. Nakashima-Tanaka. 1966. Inheritance of radioresistance in *Drosophila*. *Mutation Res.* 3:438–43.

Parsons, P. A. 1981. Evolutionary ecology of Australian *Drosophila*; A species analysis. *Evol. Biol.* 14:297–350.

Parsons, P. A. and S. M. Stanley. 1981. Domesticated and widespread species. In M. Ashburner, H. L. Carson, and J. N. Thompson, eds., *The Genetics and Biology of Drosophila.* Vol. 3a, pp. 349–93. London: Academic Press.

Pfriem, P. 1983. Latitudinal variation in wing size in *Drosophila subobscura* and its dependence on polygenes of chromosome O. *Genetica* (The Hague) 65:221–32.

Prevosti, A. 1955. Geographical variability in quantitative traits in populations of *Drosophila*. *Cold Spring Harbor Symp. Quant. Biol.* 20:294–99.

Prevosti, A., M. P. Garcia, L. Serra, M. Aguadé, G. Ribó, and E. Sagarra. 1983. Association between allelic isozyme alleles and chromosomal arrangements in European populations and Chilean colonizers of *Drosophila subobscura*. In M. C. Ratazzi, J. G. Scandalios, and G. S. Whitt, eds., *Isozymes.* Vol. 10. *Genetics and Evolution*, pp. 171–91. New York: Liss.

Prevosti, A., L. Serra, M. Aguadé, G. Ribó, F. Mestres, J. Balaña, and M. Monclús. 1989. Colonization and establishment of the palearctic species *Drosophila subobscura* in North and South America. In A. Fontdevila, ed., *Evolutionary Biology of Transient Unstable Populations*, pp. 114–29. Berlin: Springer-Verlag.

Prevosti, A., L. Serra, G. Ribó, M. Aguadé, E. Sagarra, M. Monclús and M. P. Garcia. 1985. The colonization of *Drosophila subobscura* in Chile. II: Clines in the chromosomal arrangements. *Evolution* 39:838–44.

Rozas, J., M. Hernández, V. M. Cabrera, and A. Prevosti. 1990. Colonization of America by *Drosophila subobscura*: Effect of the founder event on the mitochondrial DNA polymorphism. *Mol. Biol. Evol.* 7:103–9.

Shorrocks, B. 1975. The distribution and abundance of woodland species in British *Drosophila* (Diptera: Drosophilidae). *J. Animal Ecol.* 44:851–64.

Simberloff, D. 1981. Community effects of introduced species. In T. H. Nitecki, ed., *Biotic Crisis in Ecological and Evolutionary Time*, pp. 53–81, London: Academic Press.

Val, F. C. and F. M. Sene. 1980. A newly introduced species in Brazil (Diptera, Drosophilidae) *Papeis Avulsos Zool. S. Paulo* (Brazil) 33:293–98.

Watanabe, H. 1979. *Drosophila* survey of Hokkaido. XXXVI: Seasonal changes in the reproductive condition of wild and domestic species of *Drosophila*. *J. Fac. Sci. Hokkaido Univ. Ser. 4 (Zool.)* 21:365–72.

PART FOUR

Genetic Load and Life Cycles

14

Genetic Load and Population Size in Northern Populations of *Drosophila subobscura*

Anssi Saura

Dobzhansky (e.g., 1964) wrote extensively on the nature of the genetic load in *Drosophila* populations. Adaptation utilizes a huge amount of genetic variation as raw material. Some gene combinations are well adapted and confer high fitness to their bearers, whereas in other combinations the same genes may prove to be detrimental, even lethal. He stressed that genetic death is the price that populations pay for adapting. This kind of death does not need to produce a cadaver. In reality, it often means that an individual does not reproduce or has low fitness. Geographic patterns of variation were Dobzhansky's favorite subject even in his Leningrad days (e.g., Dobzhansky and Sivertzev-Dobzhansky 1927). In fact, he never did do a study through the entire distribution area of a species. The closest he came to this was with *D. willistoni*. The following is an account of the Scandinavian studies on genetic load in the marginal populations of *Drosophila subobscura*.

Drosophila subobscura is the European counterpart of both *D. pseudoobscura* and *D. willistoni*. It has some seventy inversions (Krimbas 1992). The bibliography of *D. subobscura* reads like a list of first- and second-

This study has been financially supported through the Erik Philip-Sörensen Foundation and the Swedish Environmental Protection Board.

I was with Dobzhansky in 1970–71 at the Rockefeller University in New York as a graduate student of Professor Seppo Lakovaara and had a later postdoctoral period at Davis, California, in the winter and spring of 1975.

generation students of Dobzhansky. In a way it also illustrates the difference in the way science is done in Europe and in the United States. European science has always some provincial touch, here shown, e.g., through
the names of the five acrocentric chromosomes of *D. subobscura:* AEIOU
(Mainx, Koske, and Smital 1953). That is the motto of Emperor Frederick
III, short for *Austria Est Imperare Omnis Universo* (Austria will reign over
the entire universe). In Europe the geographic location of a laboratory
often determines the research problem. It is then only natural that somebody working in northern Scandinavia will be concerned with the northern
margins of distribution of a species.

  D. subobscura is expanding its area of distribution. It is essentially a European species that is also found in the Near East and Northern Africa.
Danko Brncic tells (in his paper) the history of its most spectacular range
expansion: the colonization of the Americas, "A Grand experiment in evolution" (Ayala, Serra, and Prevosti 1989). Since 1970 *D. subobscura* has also
pushed its northern margin of distribution some 500 km to the north in
Scandinavia, that is, from about the latitude 61° N to the level of the town
of Skellefteå and beyond (to 65° N, Saura, Eriksson, and Kohonen-Corish
1990). It is found all the way from the Baltic Sea to Moscow (Saura,
Johansson, and Kohonen-Corish 1994) and extends from there to Kazakhstan (Gornostaev 1993) and Iran through northern Sahara all the way to
the Canary Islands and Madeira. The margin of distribution is, in other
words, not a static border, on the one side of which you see flies while on
the other side you do not. It is a broad area, where the flies may be relatively abundant in one year. Next year they are all gone but may reappear
again the year thereafter.

Marginal Versus Central Populations

Central populations of animals and plants represent the regions of highest
genetic diversity. The center is also the territory in which the species has
arisen and from which it has spread elsewhere. This is the theory of centers
of cultivated plants put forward by Vavilov (1926) in Leningrad at the time
when Dobzhansky was working with Philiptchenko, who also expressed
similar ideas (See the paper by Golubovsky and Kaidanov.) These concepts
have had immense practical importance in plant breeding. The underlying
theory was developed further by da Cunha et al. (e.g., da Cunha, Burla, and
Dobzhansky 1950; da Cunha et al. 1959) and Carson (1955, 1959); da
Cunha et al. were working on *Drosophila willistoni.* Townsend (1952) and
Cordeiro et al. (1958) studied the amounts of inversion polymorphism and

lethal load in northern and southern marginal populations. da Cunha et al. (1950, 1959) suggested that the amount of adaptive polymorphism carried in a population is proportional to the variety of habitats that the population occupies. At the margins of the distribution area the species has a narrow niche, because it has not had the time to evolve adaptive polymorphism. This polymorphism would permit it both to expand its niche locally and to spread further geographically.

Carson's (1955, 1959) ideas were based largely on his extensive studies on *Drosophila robusta*. Central and marginal populations are characterized through different kinds of genetic adjustment. Marginal populations are small, semi-isolated, and inbred. Genetic drift may operate and further depress the already low levels of heterozygosity. On the other hand, inversions are few and recombination is free. Any adaptive gene combinations may become fixed and species formation is promoted. Central populations are characterized with restricted recombination. Coadapted gene blocks provide for genetic balance with heterotic buffering. Since the populations are large and effectively outbreeding, incipient speciation is unlikely in central populations.

The arguments of da Cunha et al. (1950, 1959) and Carson (1955, 1959) were mainly based on observations on inversion polymorphism. Muller (1927) had demonstrated the use of a balanced lethal technique in his discovery of the mutagenic activity of the X rays. Balanced lethal stocks, with and without inbreeding, showed that natural populations have enormous amounts of hidden variation. Dobzhansky (1955) used this apparently unlimited polymorphism in formulating the "balance hypothesis" of the genetic structure of populations. He contrasted this view to the classical essay on "Our load of mutations" (Muller 1950).

Genetic Load

Muller (1950) estimated the total mutational load of humanity. He envisioned that humanity could not tolerate a doubling of the total dose of radiation. His pessimistic view was immensely successful, but it failed to take into account the fact that the human rate of reproduction is far below the potential rate. People just voluntarily refrain from having as many children as they can.

Haldane (1937) had expressed the load concept earlier: if the fitness of a homozygote for a recessive allele is impaired, then the population suffers an impairment, which is equal to the mutation rate to this allele. Put another way, genetic load can be defined as the proportional decrease in average fitness of a population in relation to the fitness of the optimal genotype.

Haldane (1957) pointed out that we lag far behind the optimal genotype: natural selection entails a demographic cost. Lewontin and Hubby (1966) demonstrated experimentally the amount of variation in natural populations. They used, of course, the noble species *D. pseudoobscura* (Dobzhansky, personal communication).

The demonstration of the immense amounts of variation was clearly incompatible with the load concept. It led to the development of the neutrality hypothesis and to the debate on the adaptive value—if any—of the vast amounts of variation present in nature. The debate resulted in a vast amount of data and active research in the early 1970s. As it became apparent that neither the selectionist (i.e., supporters of various forms of balancing selection as the factor maintaining this polymorphism) nor the neutralist camp could expect an easy victory, and since all intermediate positions were possible, people lost interest in the debate and started doing other things. The history of the debate is still unwritten.

An important milestone in that debate was Wallace's (1970) book *Genetic Load* (see also Wallace 1991). He argued for truncation selection and presented models in which a genetic load may actually benefit the population through culling excess individuals.

Structure of *D. subobscura* Populations

Krimbas (1992) has recently reviewed the inversion polymorphism in *D. subobscura*. All five acrocentric chromosomes harbor inversions, and the total number of inversions in his list is 66. Carson (1955) proposed the *index of free recombination* (IFR—the amount of euchromatin not bound by inversions that may take part in recombination) as a measure of inversion polymorphism. Highest values of IFR (that is, the lowest amount of chromosomal polymorphism) are found in northern marginal populations (96.5 in southern Finland) and the lowest (most inversions) in the Balkans, Turkey, and the Caucasus (73). Sperlich (1973) has used IFR as a way of differentiating central populations—the northern Mediterranean region through Turkey and southern Russia—from submarginal: North Africa through Israel to Iran in the south and again central Europe through northern Russia. The marginal populations have the least amount of inversion polymorphism and comprise the populations in the British Isles, the Low Countries, northern Germany, and Scandinavia. In summary, inversion polymorphism is well differentiated geographically and geographic barriers are well evident (Prevosti, Ocaña, and Alonso 1975).

The first comprehensive enzyme polymorphism studies were made on the southwest Finnish marginal populations (Lakovaara and Saura 1971). The absolute margin of distribution was then very exactly within the city limits of Helsinki. The western suburbs had *D. subobscura* while extensive sampling to the east of them did not yield a single fly. Certain allele frequencies fluctuated widely in the small samples and gave evidence for drift or differential selection. We collected a larger sample from Scandinavia and the French and Italian populations (Saura et al. 1973) and showed that enzyme allele frequencies were relatively uniform over this discontinuous area. This is, of course, the general result in enzyme polymorphism studies with *Drosophila*. Pinsker and Sperlich (1979, 1981) have studied the correlations between enzyme and inversion polymorphism in detail and showed that any uneven allele frequencies are largely due to inversions.

Data are now accumulating on mitochondrial DNA restriction fragment polymorphism in natural populations (Afonso et al. 1990; Latorre et al. 1992). Interestingly, populations from the Canary Islands had the greatest amount of polymorphism. All populations, including the Swedish marginal population of Gävle, had two morphs.

Accordingly, we have three sets of data on different kinds of polymorphism in natural populations of *D. subobscura*. Inversion polymorphism shows a clear-cut clinal pattern, with extensive polymorphism in the central populations and gradually decreasing levels of variation toward the margins. Enzyme polymorphism is largely undifferentiated. The major exceptions are alleles associated with inversions. Uniform allele frequencies over a vast and discontinuous area of distribution suggest high levels of gene exchange among populations, of an order of several adult migrants per generation. Mitochondrial DNA shows little differentiation within populations (Latorre et al. 1992). There is, however, extensive between-population differentiation. This contradicts strongly the idea of unrestricted gene flow but agrees with the results of inversion studies.

D. subobscura has two siblings, *D. guanche* and *D. madeirensis* on the Canary Islands and Madeira, respectively. It is sympatric with these two species, which are clearly derived from it. Evidently, the Canary Island populations of *D. subobscura* are old and possess unique polymorphisms (e.g., Latorre et al. 1992). Hampton Carson (personal communication) suggested that *D. subobscura* may well be an insular species that has succeeded in conquering much of the Old World and is now in the process of doing the same to the New. This is an interesting idea that seems to fit much of the evidence.

Amounts of Genetic Load in Natural Populations

Genetic load can be studied through several methods. Historically the oldest one is inbreeding that discloses viability differences and also deleterious alleles. The only set of data available from central populations concerns three localities in Greece. The study showed an average of 3.5, 4.4, and 4.6 mutations per female (Dobzhansky, Boesiger, and Sperlich 1980). We have made two sets of observations through inbreeding, one involving a population in southern Finland (M.-L. Lokki unpublished) and one involving the Hillevik population in Sweden (L. Rogo and B. Johansson unpublished). The amount of mutations per female was 1.9 in the Finnish and 2.3 in the Swedish population.

The standard method of measuring genetic load in *D. subobscura* is based on the balanced lethal stock *Va/Ba* (Sperlich et al. 1977, figure 14.1). Flies of this strain have two dominant marker genes, *Va (Varicose)* and *Ba (Bare)* in the O chromosome. This is homologous with the 3R of *D. melanogaster* or 2 of *D. pseudoobscura*. Both *Va* and *Ba* are lethal when homozygous. The chromosome containing the *Va* marker has three inversions, the natural

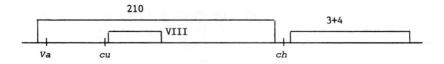

FIGURE 14.1.
The O chromosome containing the dominant marker *Va*, the recessive markers *cu* and *ch*, along with three inversions O_{3+4}, O_{210}, and O_{VIII}. *Va* is lethal in homozygous condition. In the balanced stock it is in combination with an O chromosome with the standard gene arrangement but containing the dominant marker *Ba*, also lethal when homozygous.

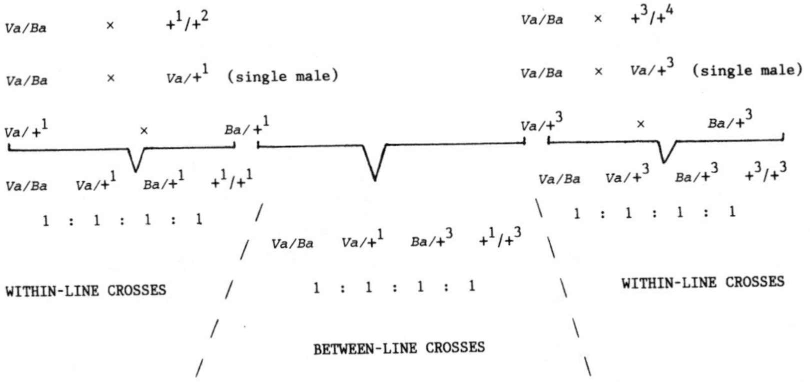

Va/Ba   × $+^1/+^2$ Va/Ba   × $+^3/+^4$

Va/Ba   × $Va/+^1$ (single male)   Va/Ba   × $Va/+^3$ (single male)

$Va/+^1$ × $Ba/+^1$ $Va/+^3$ × $Ba/+^3$

Va/Ba $Va/+^1$ $Ba/+^1$ $+^1/+^1$ Va/Ba $Va/+^3$ $Ba/+^3$ $+^3/+^3$

   1 : 1 : 1 : 1   1 : 1 : 1 : 1

Va/Ba $Va/+^1$ $Ba/+^3$ $+^1/+^3$

WITHIN-LINE CROSSES   1 : 1 : 1 : 1   WITHIN-LINE CROSSES

BETWEEN-LINE CROSSES

FIGURE 14.2.
Crossing procedure for testing the relative viability of individuals homozygous (within-line crosses) or randomly heterozygous (between-line crosses) for wild O chromosomes of *Drosophila subobscura*. Modified from Sperlich et al. 1977.

O_{3+4} and two radiation-induced ones, O_{210} and O_{VIII}. These inversions prevent recombination along the length of the O-chromosome in heterozygotes. When the inversion O_{3+4} is present in the population under study, the lethal frequencies have to be corrected to account for the recombination within this region. This is done, following Loukas, Krimbas, and Sourdes (1979), by $l_c = l_0/(1-0.339p)$, where l_c = corrected load, l_0 = observed load and p the frequency of the inversion O_{3+4} in the population. This correction does not need to be made for the Scandinavian populations, for the inversion O_{3+4} is not found there.

The crossing procedure (figure 14.2) involves the crossing of a wild-caught male or F_1 generation male of a wild-caught female with five *Va/Ba* females. One single *Va*/+ male F_1 offspring is crossed again to five *Va/Ba* females. In the F_2 generation two kinds of crosses are made. First, within-line crosses, in which the wild-type class is homozygous, and second, between-line crosses, where the wild-type class is heterozygous for two random O chromosomes, i.e., derived from two different wild flies. The females used for the F_2 crosses are always *Va*/+ type and the males *Ba*/+, because of the inversions in the *Va* chromosome. The expected frequencies of the different types of offspring in the F_3 generation are 1:1:1:1, both in within- and between-line crosses. Since the *Va/Ba* class does not contain any wild O-chromosomes, the viabilities of all other genotypes are compared with it.

Table 14.1 and figure 14.3 show the frequencies of lethals and semilethals in O chromosomes of *D. subobscura* from different natural populations. With the exception of the Gävle and Hillevik populations, marginal populations have a lethal load ranging from 0.1 to 0.2. Any realistic load figure must, of course, include the semilethals (viability from 10 percent to 50 percent), for any boundary between these two categories is arbitrary.

Evidently the marginal populations have a light load in comparison with the central ones. This is true also in Banzart (Bizerte) in North Africa and at Zernez in the Alps. Zernez is at an altitude of 1.5 km in the Inn valley in eastern Switzerland and represents an ecological rather than geographic margin.

TABLE 14.1.

Frequencies of Lethals and Semilethals in O-Chromosomes of D. subobscura *from Different Natural Populations*

Locality	Latitude	Lethals	Semilethals	Source
1. Umeå	63°50′	0.085	0	Saura et al. 1994
2. Hillevik	61°00′	0.250	0.16	Saura et al. 1994
3. Gävle	60°40′	0.309	0.118	Saura et al. 1990
4. Helsinki	60°10′	0.167	0.056	Saura et al. 1990
5. Tvärminne	59°50′	0.151	0.101	Saura et al. 1990
6. Sunne	59°45′	0.143	0.18	Pfriem and Sperlich 1982
7. Drøbak	59°40′	0.197	0.09	Sperlich et al. 1977
8. Ft. Augustus	57°10′	0.108	0.238	Pfriem and Sperlich 1982
9. Tübingen	48°31′	0.120	0.2	Pfriem and Sperlich 1982
10. Vienna	48°10′	0.549	0.1	Sperlich et al. 1977
11. Zernez	47°30′	0.127	0.2	Pfriem and Sperlich 1982
12. Barcelona	41°48′	0.385	0.112	Saura et al. 1990
13. Bordils	41°55′	0.290	0.076	Mestres et al. 1990
14. Formia	41°20′	0.266	0.2	Pfriem and Sperlich 1982
15. Ponza	40°55′	0.338	0.17	Pfriem and Sperlich 1982
16. Mt. Parnes	38°10′	0.481	0.148	Loukas et al. 1979
17. Alikainou	35°05′	0.356	0.079	Loukas et al. 1979
18. Cinisi	38°05′	0.174	0.22	Pfriem and Sperlich1982
19. Banzart	37°06′	0.134	0.135	Pfriem and Sperlich 1982

Note: The figures for populations 8 through 19 have been corrected to account for the presence of the O_{3+4} inversion.

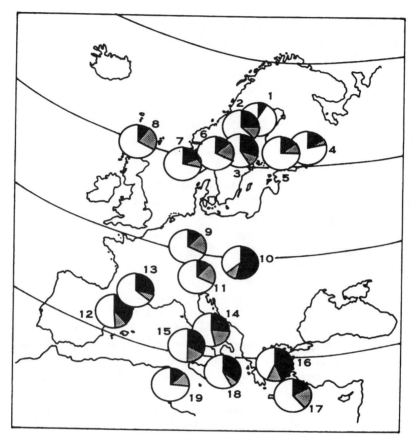

FIGURE 14.3.

Frequencies of lethals (black) and semilethals (shaded) in O-chromosomes of *D. subobscura* in different natural populations. The data are taken from table 14.1.

One may also use the *Va/Ba* technique to estimate the effects of wild O chromosomes from different geographic regions on viability. Pfriem and Sperlich (1982) showed that the mean viabilities of nonlethal homozygotes and random heterozygotes are lower in central than in marginal populations. They used the Sunne strain from Sweden, the Fort Augustus population from Scotland, and the Banzart population from North Africa as the marginal ones and compared them with the central populations. They also showed that the increase in viability through heterozygosity is higher in the northern marginal populations in comparison with others. Saura et al. (1990) made a similar comparison involving two Finnish, the Gävle population from Sweden, and the Barcelona population as the central one. The results are shown in table 14.2. Both in the within-line and between-line crosses the Barcelona and Gävle material are less viable than the two other marginal populations.

The proportion of the +/+ phenotype sinks in the northern material, when they are compared with the Mendelian expectation instead of the *Va/Ba* class; conversely the proportion of Barcelona +/+ category rises in this comparison.

In the between-line crosses, the lethals and semilethals are included. That is, the increased viability that one notices in the hetrozygous +/+ class is more pronounced than shown by the table, for the within-line material does not include lethals and semilethals.

All categories, heterozygous for different + chromosomes, are better than the corresponding homozygous +/+ categories. Interestingly, the Barcelona × Tvärminne (or vice versa) class is not significantly more viable than the Barcelona × Barcelona +/+ heterozygote class. When the Mendelian expectation is taken as the standard, the Barcelona × Tvärminne heterozygotes lose against the Barcelona × Barcelona class.

These results agree with those of Pfriem and Sperlich (1982). What is new is the observation that the combination of a northern and a central chromosome is not very viable; in fact, it is close to the between-line crosses from Barcelona. Pfriem and Sperlich (1982) showed a very marked increase in viability in crosses involving a central population (Cinisi in Italy) and a southern marginal population (Banzart from North Africa).

TABLE 14.2.

Viability of the +/+ Phenotype in the Within- and Between-Line Crosses as Compared with the Viability of the Va/Ba *Phenotype Class and That Expected if This Class Consisted of 25% of the Offspring*

	Tvärminne	Helsinki	Barcelona	Barcelona × Tvärminne	Gävle
Within lines, compared to:					
Va/Ba	1.005	1.015	0.955		0.951
25% of offspring	0.944	0.921	0.983		0.888
Bewteen lines, compared to:					
Va/Ba	1.107	1.149	1.028	1.039	1.073
25% of offspring	1.049	0.960	1.056	1.018	1.062
Total numbers	11,261	4443	8156	3780	8611

The Gävle and Hillevik populations stand out as having high loads and low increases in viability. Otherwise they are characteristically marginal populations. They have the inversion O_5 with a low frequency (Mestres et al. 1990, 1992).

Size of Marginal Populations

Dobzhansky and Wright (1941, 1943) developed the techniques to measure the effective size of a *Drosophila* population through the lethal allelism method as well as a measure of the dispersion rate through release-recapture. The latter was based on a visible mutant. In short, in a large population, the chance for allelism of lethals from a sample taken from local populations would be the same over a large area, while in small populations local differences would accumulate and lethals would be more allelic. Modern variants of the release-recapture method use micronized dust (Crumpacker 1974, Begon 1976).

Loukas, Krimbas, and Morgan (1980) and Begon, Krimbas, and Loukas (1980) used three independent methods to measure the population size in *D. subobscura* in Greece. The third method they used, in addition to lethal allelism and release-recapture, is based on the observation of change in allele frequencies that gives evidence for drift. They demonstrated widely fluctuating effective sizes with a harmonic mean around 100,000.

We have attempted to apply these methods to the interesting Gävle population. The release-recapture method has resulted in the recapture of 2 marked flies from a total of 90 released over several years. The lethals are not allelic (of a total of 43 lethals thus far). Allele frequencies drift in the fashion already described by Lakovaara and Saura (1971), but the total sample size of electrophoresed flies is so small (about 100 flies over several years) that comparisons are not very meaningful. A collective effort of one week in early September produces some 60–70 flies, which is less than in southern Finland and Sunne in Sweden but more than the 20–30 flies that one may catch in two months in Umeå. This does not translate into effective population sizes but indicates, nevertheless, that the difference in effective size between central and northern marginal populations is several orders of magnitude.

The Gävle and Hillevik Populations

Sampling was started in Gävle in the summer of 1986, in part because the area received a heavy but very local fallout (70,000–100,000 Bq/m²) in the late April of 1986. This was the first concentrated fallout of the Chernobyl event. The high load at Gävle soon became apparent (with frequencies of lethals 0.309 and semilethals 0.118 in the first study of Saura et al. 1990). Subsequent samplings have shown somewhat lower figures. The Hillevik area was chosen, for it received the highest fallout measured (200,000 Bq/m^2 and at some sites even more). The lethal load has remained there at a stable frequency of 0.25 in samples taken two years apart, while the frequency of semilethals has sunk from 0.16 to 0.1 (Saura, Johansson, and Kohonen-Corish 1994). Mushrooms, the most important larval food of *D. subobscura*, are notorious accumulators of radioactive isotopes of cesium. Gävle is an industrial town, and the collecting site is in the middle of the most industrially polluted area. Hillevik, on the other hand, is quiet and rural.

Can the heavy loads at Gävle and Hillevik be due to the Chernobyl event? Since incredible amounts of nonsense have been written about that, any *yes* must be duly reserved. Observing lethals in natural populations is not, of course, the most orthodox way of studying potential mutagenicity. The Chernobyl event was not a very sophisticated scientific experiment either. Ironically, Chernobyl is in the Ukraine, not very far from the home and school towns of Theodosius Dobzhansky.

I shall briefly try to set this study in the context of Dobzhansky's view on evolution. A central population has much of the genome tied through inversions into discrete units. Dobzhansky demonstrated that inversions are coadapted gene combinations that may have superior fitness in heterozygous condition. Inversions often contain lethals. These will, of course, also promote heterozygosity, for any homozygotes would be inviable or have inferior fitness.

Marginal populations have few inversions, but the lethals that they contain are, in comparison with central populations, more allelic. Since recombination is free in the margin, any fitness-related genetic variation is fixed with relative ease. Environmental uncertainty that obtains in the marginal area will counteract this fixation. A central population may accommodate a heavy load. Lethals may actually be beneficial, for they weed out excess individuals (Wallace 1970, 1991). A marginal population does not need culling unless it exploits a patchy resource. This may, however, exactly be what *D. subobscura* larvae do. They are, of course, subject to competition from other species utilizing the often rapidly perishing resource such as a mushroom.

Marginal populations are small and subject to low levels of gene flow. In the current models of speciation they are the main stage of evolutionary play.

REFERENCES

Afonso, J. M., A. Volz, M. Hernandez, H. Rutikay, A. M. Gonzalez, J. M. Larruga, V. Cabrera, and D. Sperlich. 1990. Mitochondrial DNA variation and genetic structure in Old World populations of *Drosophila subobscura*. *Mol. Biol. Evol.* 7:123–42.

Ayala, A., L. Serra, and A. Prevosti. 1989. A grand experiment in evolution: The *Drosophila subobscura* colonization of the Americas. *Genome* 31:246–55.

Begon, M. 1976. Dispersal, density and microdistribution in *Drosophila subobscura* Collin. *J. Animal Ecol.* 45:441–56.

Begon, M., C. B. Krimbas, and M. Loukas. 1980. The genetics of *D. subobscura* populations. XV: Effective size of a natural population estimated by three independent methods. *Heredity* 45:335–50.

Carson, H. L. 1955. The genetic characteristics of marginal populations of *Drosophila*. *Cold Spring Harbor Symp. Quant. Biol.* 20:276–87.

———. 1959. Genetic conditions which promote or retard the formation of species. *Cold Spring Harbor Symp. Quant. Biol.* 24:87–105.

Cordeiro, A. R., J. I. Townsend, J. A. Petersen, and E. C. Jaeger. 1958. Genetics of southern marginal populations of *Drosophila willistoni*. *10th Int. Congr. Genetics* 2:58–59.

Crumpacker, D. W. 1974. The use of micronized fluorescent dusts to mark adult *Drosophila pseudoobscura*. *Am. Midl. Natur.* 91:118–29.

da Cunha, A. B., H. Burla, and Th. Dobzhansky. 1950. Adaptive chromosomal polymorphism in *Drosophila willistoni*. *Evolution* 4:212–35.

da Cunha, A. B., Th. Dobzhansky, O. Pavlovsky, and B. Spassky. 1959. Genetics of natural populations. XXVIII: Supplementary data on the chromosomal polymorphism in *Drosophila willistoni* in its relation to its environment. *Evolution* 13:389–404.

Dobzhansky, Th. 1955. A review of some fundamental concepts and problems of population genetics. *Cold Spring Harbor Symp. Quant. Biol.* 20:1–15.

———. 1964. How do the genetic loads affect the fitness of their carriers in *Drosophila* populations? *Am. Natur.* 98:151–66.

Dobzhansky, Th., E. Boesiger, and D. Sperlich. 1980. *Beiträge zur Evolutionstheorie*. Jena: VEB Gustav Fischer.

Dobzhansky, Th. and N. P. Sivertzev-Dobzhansky. 1927. Die geographische Vairabilität von *Coccinella spetempunctata* L. *Biol. Zentralbl.* 47:556–69.

Dobzhansky, Th. and S. Wright. 1941. Genetics of natural populations. V: Relations between mutation rate and accumulation of lethals in populations of *Drosophila pseudoobscura*. *Genetics* 26:23–51.

———. 1943. Genetics of natural populations. X: Dispersion rates in *Drosophila pseudoobscura*. *Genetics* 28:304–40.

Gornostaev, N. G. 1993. New species and new records of Drosophilid flies Diptera Drosophilidae from Kazakhstan. *Entomol. Rev.* 71:79–82.

Haldane, J. B. S. 1937. The effect of variation in fitness. *Am. Natur.* 71:337–49.

————. 1957. The cost of natural selection. *J. Genet.* 55:511–24.

Krimbas, C. B. 1992. *Drosophila subobscura:* Biology, genetics and inversion polymorphism. In C. B. Krimbas and J. R. Powell, eds., *Inversion Polymorphism* in *Drosophila*, pp. 127–220. Boca Raton, Fla.: CRC Press.

Lakovaara, S. and A. Saura. 1971. Genic variation in marginal populations of *Drosophila subobscura. Hereditas* 69:77–82.

Latorre, A., C. Hernández, D. Martínez, J. A. Castro, M. Ramon, and A. Moya. 1992. Population structure and mitochondrial DNA gene flow in Old World populations of *Drosophila subobscura. Heredity* 68:15–24.

Lewontin, R. C. and J. L. Hubby. 1966. A molecular approach to the study of genic heterozygosity in natural populations. II: Amount of variation and degree of heterozygosity in natural populations of *Drosophila pseudoobscura. Genetics* 54:595–609.

Loukas, M., C. B. Krimbas, and K. Morgan. 1980. The genetics of *Drosophila subobscura* populations. XIV: Further data on linkage disequilibrium. *Genetics* 95:757–68.

Loukas, M., C. B. Krimbas, and J. Sourdis. 1979. The genetics of *Drosophila subobscura* populations. XII: A study of lethal allelism. *Genetica* 54:197–206.

Mainx, F., T. Koske, and E. Smital. 1953. Untersuchungen über die chromosomale Struktur europäischer Vertreter der *Drosophila obscura*-Gruppe. *Z. Vererbungslehre* 85:354–69.

Mestres, F., J. Balañá, C. Segarra, A. Prevosti, and L. Serra. 1992. Colonization of America by *Drosophila subobscura:* Analysis of the O_5 inversions from Europe and America and their implications for the colonizing process. *Evolution* 46:1564–68.

Mestres, F., G. Pegueroles, A. Prevosti, and L. Serra. 1990. Colonization of America by *D. subobscura:* Lethal genes and the problem of the O_5 inversion. *Evolution* 44:1564–68.

Muller, H. J. 1927. Artificial transmutation of the gene. *Science* 66:84–87.

————. 1950. Our load of mutations. *Am. J. Hum. Genet.* 2:111–76.

Pfriem, P. and D. Sperlich. 1982. Wild O chromosomes of *Drosophila subobscura* from different geographic regions have different effects on viability. *Genetica* 60:49–59.

Pinsker, W. and D. Sperlich. 1979. Allozyme variation in natural populations of *Drosophila subobscura* along a north-south gradient. *Genetica* 50:207–19.

————. 1981. Geographic pattern of allozyme and inversion polymorphism on chromosome O of *Drosophila subobscura* and its evolutionary origin. *Genetica* 57:51–64.

Prevosti, A., J. Ocaña, and G. Alonso. 1975. Distances between populations of *Drosophila subobscura*, based on chromosome arrangement frequencies. *Theor. Appl. Genet.* 45:231–41.

Saura, A., B. Johansson, E. Eriksson, and M. Kohonen-Corish. 1990. Genetic load in northern populations of *Drosophila subobscura. Hereditas* 112:283–87.

Saura, A., B. Johansson, and M. Kohonen-Corish (in press). 1994. Genetic load in marginal populations of *Drosophila subobscura*. *Hereditas*.

Saura, A., S. Lakovaara, J. Lokki, and P. Lankinen. 1973. Gene variation in central and marginal populations of *Drosophila subobscura*. *Hereditas* 75:33–46.

Sperlich, D. 1973. *Populationsgenetik*. Stuttgart: Gustav Fischer.

Sperlich, D., H. Feuerbach-Mravlag, P. Lange, A. Michaelidis, and A. Penzos-Daponte. 1977. Genetic load and viability distribution in central and marginal populations of *Drosophila subobscura*. *Genetics* 86:835–48.

Townsend, J. I. 1952. Genetics of marginal populations of *Drosophila willistoni*. *Evolution* 6:428–42.

Vavilov, N. I. 1926. Studies on the origin of cultivated plants. *Bull. Appl. Bot. Plant Breed., Leningrad* 16:1–248.

Wallace, B. 1970. *Genetic Load*. Englewood Cliffs, N.J.: Prentice-Hall.

———. 1991. *Fifty Years of Genetic Load/An Odyssey*. Ithaca, N.Y.: Cornell University Press.

15

Investigation of Genetic Variability in *Drosophila* Populations

Michael Golubovsky and Leonid Kaidanov

Theodosius Dobzhansky, like many outstanding representatives of Russian emigration, belongs to both cultures, the Russian and the Western. He and Timofeeff-Ressovsky transferred to the West the ideas and methodology of Russian evolutionary thought. Both scientists promoted the resurrection of genetics in Russia (USSR) after Lysenko's pogrom.

Dobzhansky's development as an evolutionist coincided with the appearance and flourishing of Russian evolutionary and population genetics (Dobzhansky 1980). Three main schools appeared in the 1920s. In Moscow N. K. Kol'tzov (1872–1940) organized the Institute of Experimental Biology. In 1921 he invited S. S. Chetverikov (1880–1959) and A. S. Serebrovsky (1882–1948) to join him. Nikolai Vavilov (1887–1943) and Yury A. Filipchenko (1882–1930) were leaders of the two Leningrad (St. Petersburg) schools. Dobzhansky as a young scientist was impressed by the intellectual level and personality of these eminent biologists and remembered Kol'tzov as "a man of multifarious interests and knowledge, of imposing presence and with eloquence of a spellbinding orator" (Dobzhansky 1980).

The authors are very grateful to M. M. Green, John Gibson, A. Korol, C. Krimbas, A. Saura, Nadine Plus, C. Biémont, and J. David for reading of the original versions of the manuscript, critical remarks, and advices and especially to Richard Newcomb for his comments and kind help in the preparation of the manuscript.

The genetic supervisor of Dobzhansky was Yury Filipchenko who in 1913 began teaching the first course in genetics in Russia at St. Petersburg University. Dobzhansky was an assistant professor in Filipchenko's department between 1924 and 1927. Filipchenko was a brilliant biologist combining both a profound knowledge of general biology with vast experimental skill. He wrote six textbooks and many reviews that, according to Dobzhansky (1980), had a great impact on evolutionary and genetical thought. He was the first person who clearly distinguished between microevolution and macroevolution and coined these concepts, which became famous after R. Goldschmidt published his books.

The main Russian contributions in *Drosophila* population research up to 1948 were often synchronous and parallel to Dobzhansky's studies (for references, see the reviews by Kaidanov 1989; Vorontsov and Golubovsky 1989). The problem of genetic load and its dynamics in natural population was central in Dobzhansky' studies from the 1930s until his death (Dobzhansky and Queal 1938; Anderson et al. 1975). Long-term studies on genetic variability in *Drosophila* populations were characteristic of Dobzhansky's approach. His last paper was dedicated to this question. It describes unpredictable and puzzling synchronous changes of some inversions in geographically remote populations. The authors concluded: "After three decades of study the problem remains unsolved . . . , and we have certainly come to appreciate the complexity of the forces at work in them" (Anderson et al. 1975).

In support of this prophetic conclusion we would like to present in general the main results of our own long-term studies on genetic variability both in natural and laboratory populations. These investigations were conducted independently in two laboratories and appeared to be connected with the action of the evolutionary forces due to mobile elements (ME).

Lethal Gene Pool in Natural
D. melanogaster Populations

One of the central points of Dobzhansky's contribution to evolutionary genetics was the study of the biological role of recessive lethals, as the most frequent class of drastic mutations. He tried to connect the occurrence and distribution of lethals with population structure, ecology, and the action of various forces of microevolution. Together with S. Wright, Dobzhansky studied the allelism of lethals in *D. pseudoobscura* extracted from local populations of the same or different regions. Dobzhansky and Wright found that lethals that have attained a high frequency within one population tend to be found in the neighboring ones. The high frequency of a particular lethal was first thought simply to be an accident of propagation and migration. Later in his book *Genetics of the Evolutionary Process*, published in 1970, Dobzhansky discussed the possibility of a greater fitness of heterozygotes for frequent lethals. In order to observe the regularities of lethal gene pool dynamics we decided to apply Dobzhansky's approach. Three populations from the Uman region (Ukraine) and two from the Caucasus region (Armenia) were chosen for long-term experiments.

About 40,000 diallelic crosses were performed for this purpose. The fate of lethal alleles of more than 100 genes of the second chromosome was investigated. We found two peculiar features regarding the dynamics of the lethal pool: the presence of a quasistationary state and the existence of a parallelism in the dynamics as well as the content of lethals in adjacent populations (Golubovsky 1980). The concentration of lethals that accumulated in heterozygotes was maintained at a relatively stable level in each population studied. Allelism tests showed that nearly half of the lethals sampled occurred singly, while the remaining lethals were found repeatedly being represented by two or more copies. Each year there was a considerable turnover of the gene pool on account of the transition of rare (single) lethals to the category of repeated ones and vice versa, and hence, the use of the term *quasistationary*.

Meanwhile, a systematic and simultaneous comparison of lethal sets revealed remarkable parallelism despite the constant allelic turnover. Sets of allelic lethals from different populations studied in the same season usually showed greater similarity than sets of lethals extracted in successive periods from the same population. Genetic drift, migration, or accidental factors cannot give rise to such parallelism. One has to search for similar mutagenic or selective forces acting in different populations. We have obtained indirect experimental data indicating that the parallelism may result from the action of extrinsic (biocenotic or ecosystem) forces, namely, host-parasite interactions (Golubovsky and Plus 1982; Alexandrov and Golubovsky 1983).

All living species in the ecosystem are constantly interacting with different viral agents at the cell, organismic, or population level. Viruses and viral DNA and RNA not only are powerful infectious and selective factors but also may be quite specific mutagens. According to long-term studies of S. M. Gershenson (1986), viral DNA and RNA induce unstable visible and lethal mutations in a definite group of loci characteristic for each type of nucleic acid tested. Especially important is that multilocus damages, which result in so-called multilethal chromosomes, were regularly observed. On the other hand multilethal chromosomes were isolated both in populations of the Ukraine and the Caucasus (Golubovsky 1980). Would it be possible that the multilethal chromosomes from nature are induced by viral foreign DNA? An experimental approach to answering this question involved a comparison of two types of lethal collections: lethals induced by viruses and exogenous DNA in the laboratory of Gershenson and lethals isolated from natural populations. The multilethal chromosomes from nature predominantly showed allelism with lethals induced by different viral DNA and RNA sources (Alexandrov and Golubovsky 1983). This indicates, together with data regarding the mutagenic action of infectious *Drosophila* picornaviruses (Golubovsky and Plus 1982), that viruses may be a powerful site-specific mutagenic agent responsible for the appearance of multisite lesions.

It follows also that owing to virus-induced mutagenesis in nature, similar multisite chromosomal damages can occur repeatedly in different populations as a result of a single mutational event. At the beginning of the 1980s additional evidence emerged supporting this conclusion. First, foreign viral agents may activate host MEs and induce unstable mutations. Second, site-specific multiple chromosomal lesions occur during a P-mediated hybrid dysgenesis (Berg, Engels, and Kreber 1980).

Dobzhansky was familiar with this long-term project from its beginning and encouraged it. Some twenty-five years ago he wrote (in Russian) to one of the authors, May 15, 1967, from New York: "I have just read your remarkable paper on autosomal lethals in No. 11 of *Genetika* 1966. I would like to congratulate you and wish further success in your studies in this extremely important and highly relevant field."

Long-Term Selection and Ordered ME-Mediated Variability

In order to analyze the genetic consequences of natural selection Dobzhansky combined the study of both natural and laboratory (cage) populations. The different aspects of his concept of balanced genetic load were tested in experiments with cage populations. During long-term experiments we analyzed the spectrum of genetic changes resulting from inbreeding on male sexual activity. The establishment of a low-activity (LA) strain by means of close inbreeding was started in 1965 from a sample of flies that originated from a natural population in Yessentuki

(Caucasus). Up to 1993 the LA stock had been inbred for 650 generations (Kaidanov 1980, 1990).

We have discovered the action of genetic forces that promote the survival of the inbred stock and exemplify Dobzhansky's concept of balanced genetic load. Three main phenomena have to be identified:

1. a highly increased level of mutability in inbred stocks, the accumulation of supervital mutations capable of suppressing lethals and semilethals, and the maintenance of a high potential for variability

2. the nonrandomly ordered transpositions of some copia-like mobile elements accompanied with increased variability

3. a considerable increase in genome instability of lines selected in minus direction

The manifestation of the first mechanism was demonstrated by a successful reverse selection of flies from the inbred LA stock. The possibility of successful selection in the plus and minus direction after many dozens or even hundreds of inbred generations demonstrated the high potential for hereditary variation in the strain studied. Using the standard CyL/Pm procedure we found in LA second chromosomes the whole spectrum of mutations from lethals to supervitals. The proportion of various classes of mutations was repeatedly estimated during the selection protocol (Kaidanov 1980, 1990).

In the high-activity (HA) strain selected simultaneously for a high male mating activity, the genetic load was minimal, whereas in the LA strain nearly 50% of the second chromosomes carried deleterious mutations. A considerable proportion of chromosomes in the highly active strain carried supervital mutations. This finding is consistent with the comparison of strains selected in opposite directions. By recombination analysis we have demonstrated the presence of suppressor mutations in the HA inbred stock. Owing to the presence of suppressor chromosomal segments, lethals become vitals and even supervitals when in homozygous condition. Our data showed that so-called "synthetic lethals" described first by Dobzhansky in 1946 and then demonstrated definitely by John Gibson are not curiosities but the usual components of genetic variability.

The behavior of different MEs in selected inbred stocks was studied by V. A. Gvozdev and his colleagues in the Institute of Molecular Genetics of the Russian Academy of Science. Distinct interstrain differences were found with respect to the number and localization of *mdg* copies. The original strains were quite stable over many years. Transposition frequencies of *mdg* elements according to rough estimates were close to 1:10 000.

The directed selection for increased adaptive properties resulted in the segregation of certain rare families with an increased adaptive capacity. It turns out that in these families the genomes are reconstructed: *mdg* elements appear at new sites and their number generally increases. Copies usually vanish from old sites, but in some cases a proportion of them may remain at their previous locations.

It has been demonstrated for the first time that this type of genome reconstruction involves a large number of MEs operating synchronously. The copies of MEs in new sites often appear together. The genome presumably contains "hot spots" that are targets for highly effective transposition of MEs during the selection for increased fitness. In the LA strain the number of *mdg* 1 was minimal. But in the substrains with increased viability, from one to seven new sites of *mdg* 1 appeared. At least two MEs, *mdg* 3 and *copia*, have an affinity for similar chromosomal regions. This leads to the conclusion that their transposition is coordinated. Such trends have not been observed for other *mdg* families. Copies of *mdg*4 or "*gypsy*" never change their localization in response to selection.

Lethal mutations occurred in LA stocks at a high frequency. Some LA chromosomes with an apparent lethal mutation were maintained as *CyL/lethal* heterozygotes in a balanced condition for nearly three years. Then they were screened for their disposition of MEs and for the presence of rearrangements. The result was striking. A considerable number (thirteen of thirty-three or 39.4 percent) of lethal-containing chromosomes carried chromosomal rearrangements, predominantly paracentric and pericentric inversions. Most of the rearranged chromosomes (nine of thirteen) carried a combination of several inversions, and among them 2 also had transpositions and 1 a translocation. Apparently in crosses of *CyL/Pm* strain to LA, the induction of complex multisite rearrangements had taken place.

It appears that one of the LA substocks contained about 30–35 copies of the *Hobo* element on all of its major polytene chromosome arms. In addition to rearrangements the lethal chromosomes maintained in balancers (*CyL/lethal*) showed an enormous level of mass transposition. In lethal chromosomes of LA strain, less than 30 percent of the original sites of *copia*-like MEs retain original position (Kaidanov et al. 1991). Such reshuffling of the genome is presumably the result of the activation of *Hobo* elements.

Gvozdev and Kaidanov in 1986 put forward the hypothesis concerning the existence of a system of adaptive transpositions of MEs, which is based on the regulatory effects of MEs on the array of structural genes (Pasyukova et al. 1986). The long terminal repeats of *copia*-like MEs contain both termination sites and enhancers. This gives them the capacity to regulate gene action.

The discovery of ordered or concerted transpositions of MEs observed in this and other Russian studies (Gerasimova et al. 1984) made it necessary to reevaluate earlier conventional views about the nondirectional nature of hereditary variation. We probably have to recognize that selection can control the formation of ordered migration of mobile elements in the genome. Hot spots of the genome are targets that quickly become occupied by MEs when selection is directed toward intensification of vital functions (Kaidanov 1989: Kaidanov et al. 1991).

Two Parts of a Genome and Two Parts of Genetic Variability

We can now describe a new approach to the understanding of sources and factors inducing genetic variability. Classical mutations evidently make up only part of the genetic variability in natural populations. It was suggested to discriminate in the eukaryotic genome between the two subsystems: the *obligate* and the *facultative* (Golubovsky 1985). This subdivision is a natural one.

The obligate part includes the set of genetic loci (and their clusters) located in chromosomes and in DNA of cellular organelles. In classical genetics each gene has a definite position on a genetic map. It is the skeleton of the genome, its structural or constitutive memory. But the genome also includes different kinds of facultative elements, varying in number and topography from cell to cell and from individual to individual. The facultative part of the genome includes the hierarchy of elements from highly repeated and satellite (st) DNA in plasmids, B-chromosomes, and certain (relatively stable) cytobionts present both in the nucleus and the cytoplasm.

This natural structural subdivision allows a discrimination between changes within obligate elements (changes of their structure, number, and position) and changes within the facultative elements. Only the first class of changes corresponds to classical mutations. Various changes of different facultative elements may be called *variations*. Variations are the most frequent class of hereditary changes, because the facultative elements are susceptible to a wide spectrum of environmental changes.

Mutations or direct damage of genetic loci occur mainly following the strong action of environmental factors (e.g., radiation, chemical mutagenesis, defects in the repair system). The sets of facultative elements respond initially to the weak action of environmental factors (in a broad sense: physical, biotic, and genetic). The interaction between the components of the system *environment-facultative-obligate elements* is the major source of most genetical changes occurring de novo in nature. This idea is shown in Figure 15.1 where arrows indicate the nature of the links while their width corresponds to the intensity of their force. The subsystem of facultative elements is the first one that reacts to stressful environmental changes (McClintock 1978).

Many variations (changes in the number of highly repeated DNA, changes in number and topography of amplified sequences and MEs, and the presence or absence of cytobionts) may have no visible or physiological effects detectable by usual genetic methods. At the same time such variations as transposition of MEs may induce insertional mutations and rearrangements of chromosomes with definite physiological consequences such as sterility.

Changes in number or cell topography of cytobionts (the usual kind of facultative elements in the eukaryotic cells), for instance, the number of RNA-containing *sigma* viruses in the cytoplasm of different *Drosophila* species, may have important physiological consequences such as lethality in the presence of carbon dioxide. The relationships between the obligate and the facultative elements of the genome have to be studied in the terms of molecular population genetics. Changes in the frequency of specific facultative elements may occur in stem cells during ontogenesis, for example,

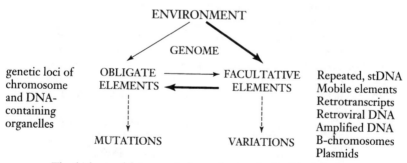

The thickness of the arrows indicates the magnitude of the influence.

FIGURE 15.1.
Two types of genetic elements in the eukarotic genomes and two types of genetic changes: mutations and variations.

a blockage of *sigma* virus replication by heat stress. On the phenotypical level these changes may look like the inheritance of developmentally acquired (induced) traits (Golubovsky 1985; Landman 1991).

There is a constant flux between obligate and facultative elements. For instance insertion mutations can be presented as transition of facultative MEs into the obligate parts of a genome. On the other hand amplification phenomena, which may result in additional microchromosomes, may be taken as the transition of obligate elements to facultative ones.

Roman B. Khesin (1922–1985) put forth the view that the eukaryotic genome must be viewed as a population of self-reproducing DNA and RNA molecules. Moreover, having in mind the regular horizontal transfer of transposons and plasmids in microorganisms, the increasing number of similar facts in *Drosophila* species (Kidwell 1992), and the wide distribution of the same retroviral sequences among remote animal species, we may agree with Khesin's paradoxical conclusion made in his comprehensive volume (Khesin 1984) that it is possible to conceive of a gene pool encompassing in its potential all living organisms.

In one of his last letters to one of the authors Dobzhansky remarked how incorrectly the molecular geneticists saw population genetics in the 1970s. They thought of it as an exhausted field, something like a squeezed lemon. Now, almost twenty years later, we see the beginning of a promising synthesis of population and molecular genetics.

REFERENCES

Alexandrov, Yu. N., M. D. Golubovsky. 1983. The multisite mutations induced by viruses and foreign DNA can spread in natural populations of *Drosophila*. *Drosophila Info. Serv.* 59:10–12.

Anderson W., Th. Dobzhansky, O. Pavlovsky, I. Powell, and D. Yardley. 1975. Genetics of natural populations. Three decades of genetic change in *Drosophila pseudoobscura*. *Evolution* 29:24–36.

Berg, R. L., W. R. Engels, and R. A. Kreber. 1980. Site-specific X-chromosome rearrangements from hybrid dysgenesis in *Drosophila melanogaster*. *Science* 210:427–29.

Dobzhansky, Th. 1980. The birth of the genetic theory of evolution in the Soviet Union in the 1920s. In E. Mayr and W. B. Provine, eds., *The Evolutionary Synthesis*, pp. 229–42. Cambridge: Harvard University Press.

Dobzhansky, Th. and M. L. Queal. 1938. Genetics of natural populations. II: Genetic variation in population of *Drosophila pseudoobscura* inhabiting isolated mountain regions. *Genetics* 23:463–84.

Gerasimova, T. I., L. V. Matyunina, Yu. V. Iluin, and G. P. Georgiev. 1984. Simultaneous transpositions of different mobile elements: Relation to multiple mutagenesis in *Drosophila melanogaster. Mol. Gen. Genet.* 194:517–22.

Gershenson, S. M. 1986. Viruses as environmental mutagenic factors. *Mutation Res.* 167:203–13.

Golubovsky, M. D. 1980. Mutational process and microevolution. *Genetica* 52/53:139–49.

———. 1985. Organisation of the genotype and kinds of hereditary variability in eukaryotes. *Adv. Modern Biol.* (Moscow) 100 3(6):323–39.

Golubovsky, M. D. and N. Plus. 1982. Mutability studies in two *Drosophila melanogaster* isogenic stocks endemic for *C. picornavirus* and virus free. *Mutation Res.* 103(1):29–32.

Kaidanov, L. Z. 1980. The analysis of genetic consequencies of selection and inbreeding in *Drosophila melanogaster. Genetics* 52/53:165–81.

———. 1989. Animal population genetics. *Sov. Sci. Rev. F. Physiol. Gen. Biol.* 3:201–56. London: Harwood.

———. 1990. The rules of genetic alteration in *Drosophila melanogaster* inbred lines determined by selection. *Archieves des Science Biol.* (Belgrade) 42:131–48.

Kaidanov, L. Z., V. N. Bolshakov, P. N. Tsygintsev, and V. A. Gvozdev. 1991. The sources of genetic variability in highly inbred long-term selected strains of *Drosophila melanogaster. Genetica* 85:73–78.

Khesin, R. B. 1984. *Genome Inconstancy*, p. 472. Moscow: Nauka (in Russian).

Kidwell, M. G. 1992. Horizontal transfer of P elements and other short inverted repeat transposons. *Genetica* 86:275–86.

Landman, O. 1991. Inheritance of acquired characteristics. *Ann. Rev. Genet.* vol. 25:1–20.

McClintock, B. 1978. Mechanisms that rapidly reorganize the genome. *Stadler Symp.* 10:25–47.

Pasyukova, E. G., E. S. Belyaeva, G. L. Kogan, L. Z. Kaidanov, and V. A. Gvozdev. 1986. Concerted transpositions of mobile genetic elements coupled with the fitness changes in *Drosophila melanogaster. Mol. Biol. Evol.* 3:299–312.

Vorontsov, N. N. and M. D. Golubovsky 1989. Population and evolutionary genetics in the USSR in Vavilov's time. In V. K. Shymnyi, ed., *Vavilov's Heritage in Modern Biology*, pp. 270–98. Moscow: Nauka.

16

Genetics and Ecology of Natural Populations

Antonio Fontdevila

The time for superficial ecological studies is over.
—Endler 1986

Dobzhansky's Turning Point in Evolutionary Biology

There are crucial times in one's life. My twentieth birthday was one of them, not only because I was coming of age among my student peers, but mainly because they showed the greatest sensitivity in appreciating my incipient drive toward evolution with the most thoughtful present, a Spanish edition of *Genetics and the Origin of Species*. The book appeared to me as such a strong argument for evolution that my intellectual life has been imprinted ever since. What impressed me most was Dobzhansky's ability in presenting the role of genetics in clearing up the somewhat confusing connection between the world of population variability and the origin of species. Time went by, and I have incorporated new ideas on the origin of species during these thirty years that followed my crucial birthday, and yet, my appreciation toward Dobzhansky's contribution to evolution has never failed. Later, as a postdoctoral student, I had the privilege of interacting for two years with Doby and to appreciate his excellence as both a scientist and a humanist.

Viewed in retrospect, some of the selective mechanisms posited by Dobzhansky to explain the maintenance of genetic polymorphisms, namely, het-

The research reported here in which I participated has been financially supported by the following agencies: CICYT and DGICYT (Ministry of Education, Spain) grants 2920/76; 0910/81; 2825/83; PB85/0071 and PB8903/25, and scholarships from Generalitat de Catalunya.

erosis, may seem too naive for young generations of evolutionists. I wish them to remember only that we are still discussing, even at the molecular level, concepts put forth by Dobzhansky, such as balancing selection, heterosis or hybrid vigor, and reproductive isolation, which are even now far from being fully understood. The biological concept of species, with all its unavoidable pitfalls and its lack of generality, is one of the most fruitful attempts ever made to understand the passage from the population to the species level (Dobzhansky 1935). Actually, this is a genetic concept that makes use of the concept of the Mendelian population as the arena where genes interact and combine and, in this sense, is based on population thinking, long rooted in the Darwinian paradigm. Dobzhansky was a Darwinian who incorporated genetical thinking into the theory of natural selection.

Others had similar ideas, but it is doubtful that any of them had as much influence as Dobzhansky on generations of evolutionists. He was an heir of the naturalist Russian school that developed a great interest in evolutionary genetics in the 1920s, a time when in England few were interested in this discipline (Dobzhansky 1980). Among Russian population geneticists, Chetverikov was the most influential on Dobzhansky. He showed that natural populations host a guild of genetic variability produced by mutation and that evolution is brought about by natural selection acting on this variability. This theory was formalized mathematically by Fisher, Wright, and Haldane early in the 1930s. Dobzhansky benefited from these ideas, and his lust for nature drove him to devise a vast plan to produce evidence of the operation of natural selection in the wild. He was tremendously successful and his data on chromosome and genic polymorphisms in natural populations were convincing to many evolutionist of his time. Yet, his observations opened up a host of new questions and unveiled the inherent difficulties in accurately measuring selection in nature.

Measuring Selection in the Wild

Pioneering studies such as that of Dobzhansky and Levene (1948) make use of the deviations of adult genotypes from the Hardy-Weinberg frequencies to infer selection. Aside from the fact that this test has low potency, it is known that selection is not the only cause of deviations from Hardy-Weinberg (Workman 1969) and that certain kinds of selection cannot be detected with the test (Lewontin and Cockerham 1959). Most important of all, the expected frequencies were calculated from adults that have passed through an adult selection phase. In this case, deviations from

Hardy-Weinberg cannot be interpreted in terms of only one component of selection (namely, larval viability), and they often indicate the joint operation of selection on different life history traits (Wallace 1958; Lewontin and Cockerham 1959; Novitski and Dempster 1958). Moreover, Prout (1965, 1969) has formally shown that if selection is acting early (e.g., larval viability) and late (e.g., mating success) in the life cycle, selection coefficients for each fitness component cannot be estimated from changes in zygotic frequencies in only one generation. The most we can say from the Dobzhansky and Levene (1948) and other similar studies (Lewontin and White 1960; Richmond and Powell 1970) is that some kind of selection was operating on inversion polymorphism in the wild population of *D. pseudoobscura* under study. The heterozygote excess found by them cannot be taken as proof of heterosis since this apparent excess is produced whenever heterozygote fitness is greater than the geometric mean of homozygote fitness (Lewontin 1974).

The previous paragraph is just a brief discussion of some problems encountered when one is attempting to detect selection in nature. The sampling of all life cycle stages seems the most appropriate method of estimating fitness component selection, but it is not always possible experimentally. Bundgaard and Christiansen (1972) developed a method of estimating fitness components that has been applied in several experimental populations of *Drosophila*. However, in natural populations, sampling in all life stages has become elusive, and only a limited number of fitness components (e.g., gametic selection, mating success, zygotic selection) have been studied by means of sample comparisons among adult males, sterile females, fertile females, and their offspring (Christiansen and Frydenberg 1973). This method has been applied in natural populations of several animals and plants (see Endler 1986 for review and references therein), but not in *Drosophila*, probably because of sampling difficulties in nature. Population ecology of most of the *Drosophila* species is poorly known and difficult to study, cactophilic *Drosophila* species being an interesting exception that our research group has been trying to exploit during the last fifteen years toward the objective of measuring natural selection in the wild.

Why *Drosophila buzzatii*?

In the 1960s some voices were raised to integrate ecological parameters into population genetics, but it was in the 1970s when this integration flourished in a discipline named population biology that a new approach was instituted (Christiansen and Fenchel 1977; Brussard 1978). This inclusive view of population biology implies a broader definition of fitness in which each genotype is not related to a unique fitness value. On the other hand, fitness is an attribute of each genotype-environment combination. This approach opened up the possibility of relating genetic parameters with (1) intrapopulation variables such as population growth and fluctuations, age distributions, and breeding structure; and (2) interpopulation (interspecific) interactions, among them competition, parasitism, and predation.

My interest in *Drosophila buzzatii* was not coincidental. In the mid 1970s I chose this species as material for my future research after a long period of reflexive thinking on some of the problems faced by *Drosophila* population geneticists when trying to understand the operation of selection in nature. That is, *Drosophila* is an excellent material with which to study population genetics, but most of the genetically studied *Drosophila* species are rather ecologically intractable, mostly because their trophic niches are difficult to define and poorly known. This is the case with *D. melanogaster* and, above all, with other species that have been used to understand the dynamics of their natural populations, such as *D. subobscura* and *D. pseudoobscura*. Unfortunately, many *Drosophila* species with a defined niche have a rather restricted geographical distribution that makes its study under natural conditions available only to right-on-the-spot scientists. The case of *D. buzzatii*, being subcosmopolitan (David and Tsacas 1980) and associated with cacti in nature (Fontdevila 1982), provides a compromise and a workable alternative.

D. buzzatii is a cactophilic species that feeds upon the rotting stems and fruits of several cacti. The species originated in subamazonian South America, probably in northwest Argentina (Fontdevila 1982), but in recent historical times it has colonized some Atlantic archipelagoes, Africa, the Mediterranean basin, and Australia (for a review, see Fontdevila 1991 and references therein). In Argentina the *D. buzzatii* host plants include several species of the genus *Opuntia* (e.g., *O. quimilo*, *O. sulphurea*, *O. vulgaris*) and also some columnar cacti (e.g., *Cereus validus*, *Trichocereus terschekii*, *Trichocereus pasacana*) (Hasson, Naverra, and Fontdevila 1992). However, in the colonized areas *D. buzzatii* feeds on a limited number of *Opuntia* species, mainly *O. ficus-indica* in the Mediterranean basin (Fontdevila et al. 1981) and *O. stricta* in Australia (Barker and Mulley 1976).

This species offers an excellent opportunity to measure natural selection in the wild. Our work has been performed mainly on two populations, one in an original area, Arroyo Escobar, Argentina, and the other in a colonized locality, Carboneras, Spain (for a description of these localities, see Hasson et al. 1991 and Ruiz et al. 1986, respectively). These studies have consisted mainly in: (1) detecting the genetic changes through all possible life cycle phases in order to measure fitness components such as larval and pupal viability, fecundity, mating success (virility), and longevity (figure 16.1); and (2) establishing correlations between fitness components; phenotypic characters, such as body size; and genotypes, namely, allozyme and chromosome polymorphisms. The fitness component approach has unveiled the operation of natural selection for the chromosome polymorphism in several fitness components. The most striking effect was for larval viability selection in Carboneras (Ruiz et al. 1986), where selection coefficients were higher than 10% in most cases. Heterokaryotypes were, in general, superior, but this cannot explain the maintenance of this polymorphism since fitness coefficients do not conform to the triangular inequality (Lewontin, Ginzburg, and Tuljapurkar 1978). The detected directional effects of viability selection suggest the presence of trade-offs among fitness components in order to explain the stability of the chromosome

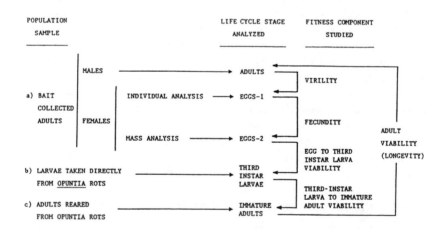

FIGURE 16.1.

Experimental scheme for the analysis of fitness components carried out in a natural population of *Drosophila buzzatii* (Carboneras, Southeast Spain). Selection is detected through the comparison of gene and genotypic frequencies between consecutive life cycle stages (connected by arrows). Eggs-1 sample: Individual females were allowed to lay eggs, and one larva from each individual progeny was analyzed (individual analysis). Eggs-2 sample: A group of females were allowed to lay eggs, and a sample of this progeny was analyzed (mass analysis).

(From Ruiz et al. 1986. Reprinted with permission from *Evolution.*)

polymorphism. Fecundity and longevity components are responsible for significant directional changes of some rearrangements that counteract the effects of viability. However, our measures of fecundity were performed with natural collected flies but with artificial media. This procedure was able to detect some significant differences in fecundity, but the possibility exists that fecundity differences are amplified in nature owing to natural diversity in oviposition behavior depending in part on heterogeneity, partitioning, and/or frequency of resources. Our estimates of fecundity were obtained from a sample of females of all ages.

Since our data indicate longevity selection, some rearrangements will be more represented among young females than others. This will produce an apparent superior fecundity of those rearrangements. Moreover, with our samples of several hundreds of individuals we do not expect to detect differences of less than 4% for the conventional level of significance of 0.05. The potency of our experimental methods is limited by the realization that our statistics measures only half of the true differences between karyotype frequencies (Ruiz et al. 1986). All in all, our data suggest that fecundity can counteract viability in part, but these considerations show the difficulty of detecting small fecundity differences in nature, not only because experimental procedures do not allow us to separate fecundity easily from other components such as longevity, but also because the necessary sample sizes are impracticable. Since fecundity may be one of the most important fitness components (Lewontin 1974:64), we are confronted with an important difficulty.

The comparative approach between two populations has produced new insights to the reaction norms of different karyotypes for life history traits (figure 16.2). Fontdevila (1989) has demonstrated that Arroyo Escobar may be taken as the original source of the founders of Carboneras, and thus, we can relate both populations by colonization. Life history traits can be classified following between-population comparisons. Second chromosome rearrangements may present the same reaction norm for some fitness components in both populations. This is the case of fecundity, which shows a remarkable constancy in its selective pattern, the same rearrangement (2jz3) being selected against in both populations. This disadvantage can be related to a greater longevity associated with this rearrangement, as stated above. Nonetheless, fecundity effects seem to be real and consistent, in spite of the difficulty of their detection. Considerations of the colonization process are in accordance with the observed data. Thus, fecundity can be one component of high selective value in a colonization process where r-selection is strong and one expects a decrease of those rearrangements that

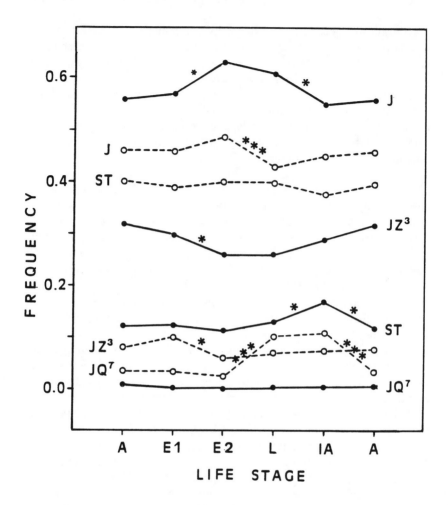

FIGURE 16.2.

Cyclic frequency changes of second chromosome arrangements in five consecutive life stages of one original (Arroyo Escobar, Argentina: continuous line) and one colonizing (Carboneras, Spain: dotted line) population of *Drosophila buzzatii*. A: collected adults; E1: eggs-1 sample; E2: eggs-2 sample; L: larvae from rots; IA: immature emerged adults. Statistical significance between stage frequencies: *: $p<0.05$; **: $p<0.01$; ***: $p<0.001$.

(From Fontdevila 1989. Reprinted with permission from Springer-Verlag.)

confer lower fecundity in colonized populations. This is exactly what is observed.

Other fitness components seem to behave differently in different populations. No significant changes in chromosome polymorphism associated with larval viability have been detected in Arroyo Escobar, but they are very important in Carboneras. On the other hand, pupal viability is highly selected in the original population, but no effect is found in the colonized population. Several explanations for this are possible. Pupal viability differences can be explained by the action of different incidence of parasitism depending on season and geographical area. For example, parasitism is rare in the British Islands (Begon and Shorrocks 1978) but frequent in the European continent (Janssen et al. 1988). It is also frequent in some periods of the year in the Tunisian *Drosophila* community associated with *Opuntia* (see Carton et al. 1986 and references therein) where there is a high degree of specificity between parasitoids and *Drosophila* species.

The importance of changes in parasitism due to colonization remains to be studied. The differential effect of a chromosome rearrangement in different environments can be accounted for by phenotypic plasticity, but colonization can also induce changes in the genetic content of inversions due to founder effects. The majority of larval viability effects in Carboneras can be accounted for by the 2jq7 rearrangement, an inversion that has significantly increased its frequency in colonized populations compared with its localized and low frequency (0.015) in Arroyo Escobar, the only locality in South America where it is present (Fontdevila et al. 1982). Yet, the fact that 2jq7 is not present in several colonized populations, especially in islands and marginal localities (Fontdevila et al. 1981), suggests that stochastic events associated with founder populations must have played an important role in its present and past establishment. The relative roles of preadaptation and chance of a second chromosome effect on larval viability can be assessed only by studying original and colonized populations in more detail.

Can We Improve the Detection of Selection in Nature?

The preceding paragraph illustrates some of the complexities of natural selection in maintaining genetic variability and also in contributing to changes in this variability under an environmental shift. The experimental approach of studying changes in gene frequencies among life cycle stages does not detect either certain kinds of selection (e.g., heterosis) or selection intensities under 10% with the usual sample sizes possible in realistic field experiments. However, these preliminary data have raised a series of interesting questions. Namely, what are the trade-offs necessary to explain the genetic equilibrium? If larval viability is the only clear-cut directional fitness component, how can we design experiments to find the antagonistic components? Can we improve the detection of fecundity and separate it from other components, namely, longevity or habitat selection? In spite of the lack of statistical association of mating success with the *D. buzzatii* chromosome polymorphism, can we use a new approach to unveil possible effects? We need a new approach in order to answer these questions. Another shortcoming of the preceding approach is that no mechanism of selection (e.g., physiological, behavioral, or ecological) is suggested. Since selection operates upon the phenotype, we need to know what the phenotypic trait is that responds to selection.

Body size is a phenotypic trait that is correlated with many life history traits, possibly because it is an overall representation of general metabolic efficiency (Roff 1986). We have strong evidence that body size is positively correlated with longevity, mating success, and fecundity (Santos et al. 1992) in the population of *D. buzzatii* of Carboneras and also in the population of Arroyo Escobar for longevity (Hasson et al. 1993). The genetic consequences of these correlations are under investigation, but significant additive genetic variance for body size is known to exist in nature (Prout and Barker 1989; Ruiz et al. 1991).

Our data indicate that body size is correlated with karyotype frequencies, small size being associated with genotypes containing the 2st rearrangement (Ruiz and Santos 1989; Hasson et al. 1992). Mating probability in nature is higher for large individuals (figure 16.3) and is age dependent, data suggesting that larger flies mate more often earlier in life (Santos et al. 1988, 1992). Sexual selection in relationship to body size has been demonstrated several times in the laboratory and in nature (see Ruiz and Santos 1989 for a discussion). In our studies in Carboneras we have directly measured the phenotype (body size) and the genotype (karyotype) of both parents by sampling mating pairs in nature. Phenotypic and genotypic mating distributions were compared with distributions obtained from a random sample of nonmating individuals. Size comparisons between mating and nonmating individuals gave highly significant estimates of directional sexual selection in favor of large size. Moreover, comparisons between karyotypic distributions across size classes are statistically different (figure 16.4), confirming the correlation between size and karyotype. On the other hand,

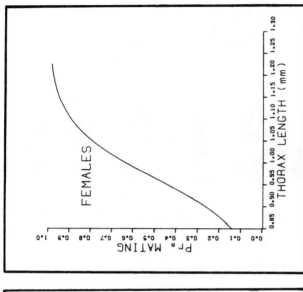

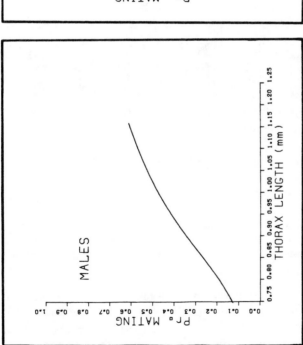

FIGURE 16.3.

Nonparametric cubic-spline estimates of sexual selection fitness functions of wild *Drosophila buzzatii* males and females in relation to thorax length.

(From Santos et al. 1992. Reprinted with permission from the *Journal of Evolutionary biology.*)

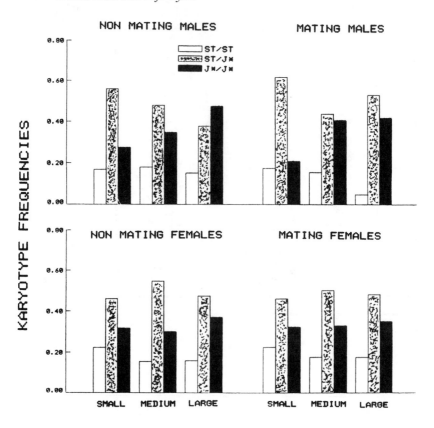

FIGURE 16.4.
Second chromosome karyotype frequencies in samples of mating and nonmating *D. buzzatii* adults, classified according to body size. The total G value for the comparison of karyotype and arrangement frequencies among body size classes is statistically significant. Notice that the 2st arrangement is associated with small size. From Ruis and Santos 1989 Reprinted with permission from Springer Verlag

comparisons of genotype distributions between mating and nonmating individuals by means of G statistics gave no statistically significant differences in data collected in three different years (Barbadilla et al., in press). This seems to indicate that mating success is insensitive to karyotypes. However, as indicated above, the potency of this test is very low for genotypic responses under 10% with sample sizes of several hundred individuals.

We can predict the exact change of each rearrangement by using a model developed by Kimura and Crow (1978), in which we obtain the expected genetic response resulting from phenotypic selection and the correlation between phenotype and genotype. Barbadilla et al. (in press) have obtained estimates of karyotype changes due to mating success that are not significantly different from the values found from observing mating pairs in nature. Therefore, the hypothesis that selection is acting cannot be rejected. Nonetheless, all confidence intervals of these estimates include the zero value, and the absence of selection cannot be rejected either. So, what can we infer from these experiments? The most sensible inference is that, again, with practical sample sizes we cannot detect gene frequency changes of less than 10%, about 1% in our case of sexual selection. The take-home lesson is that a lack of statistical significance means nothing when selection is very weak and there is need of more integrative models.

The Importance of Population Structure

The preceding paragraphs show that only high intensities of directional selection are able to be detected through observation of gene frequency changes across life cycle stages. Other approaches that try to observe selection across generations are even less sensitive because (1) they cannot distinguish many kinds of selection (e.g., antagonistic trade-offs), (2) they are plagued by statistical difficulties, and they need much more sampling effort, spanning several generations. Nevertheless, the method of correlating phenotypes that respond to selection with genotypes allows us to infer the operation of selection in nature, even when its intensity is insensitive to direct statistical tests. All in all, these approaches rely on certain assumptions that, like random mating and environmental homogeneity, depend on the breeding and the ecological structure of populations.

I want to illustrate the importance of population structure to explain the maintenance of genetic variability, summarizing our present knowledge on mating and breeding structure in our population of Carboneras. By directly sampling mating pairs in nature we have been able to prove that mating is random (Quezada-Diaz et al. 1992; Barbadilla et al., in press) with reference to all genetic markers utilized (i.e., chromosome rearrangements and several allozyme systems: Est-2, Adh-1, Pept-2 and Aldox). This finding contradicts the hypothesis put forth by Thomas and Barker (1990) that Australian populations of *D. buzzatii* are inbred. According to these authors, each rotting cladode can sustain several generations (two to three) of flies that tend to stay and mate among themselves in the same rot. Since the number of effective parental flies colonizing each rot is low (around ten) (Santos, Ruiz, and Fontdevila 1989; Thomas and Barker 1990), this founder event coupled with limited dispersal among rots would produce large amounts of inbreeding that could explain, in part, the population-wide heterozygote deficiency ($F_{IT} > 0$) found in Australian populations (Barker and Mulley 1976).

Our ecological and genetical experience in Carboneras is, however, very different. First, using pH as an indicator of cladode decomposition, we know that cladodes rot very fast and there is a synchronic correlation between rot stage and larval instar, namely, first instar larvae inhabit cladodes in early stage of rotting (pH=6.5), pupae are found in cladodes that have almost completed their rotting (pH=8.0–8.5), whereas second and third instar larvae are found in intermediate rot stages (Peris unpublished). This agrees with evidence that *D. buzzatii* females lay eggs only in the initial stages of cladode rotting (Starmer et al. 1986). Second, no overall heterozygote deficiency is observed either in recruitments from rots ($F_{IS} < 0$) or in the whole population ($F_{IT} < 0$); rather, an excess of heterozygotes is the rule for most allozymes (Quezada-Diaz 1993) and chromosome rearrangements tested (Santos, Ruiz, and Fontdevila 1989). This information coupled with direct evidence of random mating strongly suggests that in Carboneras there is only one round of rot colonization by a limited number of females, which produces an excess of heterozygotes in each rot (Robertson 1965) and is followed by active dispersal and random mating of the whole population.

Other colonizing populations, such as those in Australia, may have a more dispersed plant distribution than in Carboneras. This and the presence of other *Opuntia* species with longer rotting processes (Barker et al. 1983) could favor intrapopulation differentiation through pronounced subdivision mediated by drift effects. This would explain the pervasive deficiency of allozyme heterozygotes found in many Australian (Barker and Mulley 1976), New World, and Old World populations (Sanchez 1986), but other explanations can be advanced (see below).

The lack of population subdivision in Carboneras does not preclude that habitat heterogeneity may exist and play a role in genetic differentiation. Recent work on yeast abundance and diversity has shown that among-rot heterogeneity exists only for the latter but not for the former (Peris et al. unpublished). The apparent heterogeneity in among-rot abundance is due to differences in rotting stage, and when the analysis is corrected for temporal succession, most of yeast heterogeneity is apportioned to among-rot diversity differences. In other words, female flies will always encounter a similar yeast abundance on those young rots ready for oviposition. This succession effect on heterogeneity was suggested in part by Starmer (1982). The among-rot heterogeneity in yeast diversity may have important consequences for larval selection. Early work by Fogleman, Starmer, and Heed (1981, 1982) demonstrated that larvae of the cactophilic species *D. mojavensis* are capable of yeast discrimination in feeding, and Vacek (1982) showed in *D. buzzatii* a correlation between selective feeding and both developmental time and viability. The observed correlation between viability and karyotype described above might allow us to perform an analysis similar to that of mating success. Even if no selective feeding exists, among-rot differences in yeast diversity may favor different karyotypes in each rot and contribute to the maintenance of chromosome polymorphism.

When discussing measures of fecundity, I pointed out the difficulty of separating habitat selection for oviposition from number of eggs laid. Random oviposition is an assumption in our analysis. Under experimental conditions the evidence is ambiguous. Some results (Vacek 1982; Barker et al. 1986) seem to indicate that different *D. buzzatii* female genotypes do not have feeding preferences but do select yeasts for oviposition. Nonetheless, under natural conditions, abundance of yeasts in the early stage seems to be the cue for oviposition. More work is needed, however, regarding oviposition behavior in nature, but in view of the high mobility of *D. buzzatii* and the fact that rots are quite ephemeral, habitat selection for oviposition may be unimportant for polymorphism maintenance in Carboneras.

Future Prospects in Population Biology of Drosophila

All the reported information on the dynamics of natural populations of *D. buzzatii* illustrates the complexity of the operation of selection in the wild. The *Drosophila* community in Carboneras consists of at least four main species: *D. buzzatii*, *D. melanogaster*, *D. simulans*, and *D. hydei*, which feed and breed on two main resources, rotting fruits and cladodes of *Opuntia ficus-indica*. We know that *D. buzzatii* is almost the only species emerging from cladodes. Yet, considering that our studies have dealt with only one species (*D. buzzatii*) and one resource (cladodes), our understanding of the population dynamics may be very limited. Community ecology has traditionally considered conspecific populations as groups of biologically identical individuals, and population biology has disregarded, in many instances, the role of species interaction in shaping adaptation and variability inside populations (Lomnicki 1988). In particular, the fact that natural selection operates through differential reproduction among conspecific individuals casts serious doubts on the understanding of community evolution and structure when population dynamics is disregarded. In fact, an integration between levels, from population to community, is needed. In most instances, this endeavor may be impossible owing to the complexity of nature, but in some simplified communities an inclusive approach may be possible. *Drosophila* communities of cactophilic *Drosophila* offer an opportunity in this respect, because of the workable number of species and resources they include.

The community of Carboneras can be considered as a paradigm for these inclusive studies. Using all the reported evidence, one can advance several workable hypotheses to explain the maintenance of genetic polymorphisms. Let me illustrate two of them: the trade-off hypothesis and the founder-spatial variation hypothesis.

The trade-off hypothesis is rooted in the observation that there are negative genetic correlations among fitness components. These negative correlations make it possible that abundant genetic variance exists for life history traits even in the absence of fitness variance in equilibrium populations. Theoretically these antagonistic relationships not only explain the maintenance of polymorphisms in nature but also unveil present or past ecophysiological constraints. Charlesworth (1990) has suggested that these constraints may bridge the gap between optimization and quantitative genetic models. Unfortunately, the evidence for trade-off mechanisms in nature is scarce and contradictory. For example, negative correlations between rate of development and longevity have been widely documented at the level of class and family (Charnov and Berrigan 1990). However, experiments with *D. melanogaster* have produced contradictory results (Sevenster 1993) that have been ascribed to experimental conditions that are too far removed from the natural environments optimal for unveiling trade-off patterns. Life history traits, such as longevity and developmental rate, are correlated with metabolic components and, consequently, with body size. These correlations confer on body size an important role in fitness. The fitness value of size is, however, dependent on ecological factors, such as resource availability. The theory predicts that genetically large individuals develop slower and live longer than small ones. When resources are abundant, fast individuals (small and short lived) are superior in the juvenile phase (larval) if there is competition. On the other hand, they are inferior when resources are scarce and no competition exists, because then large, long-lived individuals have an increased probability of finding breeding substrates.

This trade-off mechanism may explain the stability of size and the maintenance of genetic polymorphisms related to size (e.g., chromosome polymorphism). Moreover, the same mechanism may explain the coexistence of species having fast and slow development, when the availability of resources varies with time (Sevenster 1993). The interspecific approach is most powerful in detecting life history constraints because interspecific genetic diversity is greater than intraspecific variability and represents a set of alternative strategies that have been optimized in the same environment. It is not easy to understand these trade-offs without approaching a community study. Antagonistic correlations are dependent not only upon patchiness and availability of resources but also upon the presence of intraspecific and interspecific competition, which depends, in turn, on interactions with parasites and predators.

The <u>founder-spatial variation hypothesis</u> is based on the observation that individuals mate at random in each generation and that each cladode (subniche) is colonized at random by a small number of gravid females. The deposited zygotes (larvae) are differentially selected in each subniche because of among-cladode heterogeneity (e.g., variance in yeast diversity). The genetic diversity of emerged adults is the combined result of founder effects and spatial heterogeneity. This hypothesis is similar to some classical models of spatial variation (Levene 1953; Dempster 1955) in that it implies that premating adults from all subniches mix completely each generation, but it differs in that the number of founders is small and variable and that the number of emerged adults varies among subniches (hard selection). Santos (in press) has investigated theoretically the recruitment diversity under this model and found ample conditions for the maintenance of polymorphism, similar to those found by Levene (1953). Most interesting is the fact that the random variance of founders increases the mean fixation index among subniches (F_{ST}), thereby increasing the individual fixation index for the whole population (F_{IT}). This may explain the deficiency of heterozygotes observed in natural situations without advocating any kind of inbreeding.

The population dynamics in Carboneras can be investigated within the framework of the founder-spatial variation hypothesis. Until now we have studied part of this dynamics only in reference to cladodes, but we know almost nothing about the adaptive mechanisms when both fruits and cladodes occur simultaneously in the field. For *D. buzzatii*, a colonizer species, fruits of *Opuntia ficus-indica* may represent a new resource since it is doubtful that this species utilizes fruits of other *Opuntia* species in original populations. The study of fruit dynamics opens up the possibility to study not only the incidence of interspecific competition in the coexistence of all *Drosophila* species in Carboneras but also the relationship between intraspecific and interspecific competition in the maintenance of genetic variability in its populations. Intraspecific competition allows us to explain this maintenance of genetic variability because maximal resource utilization is useful in enlarging the trophic niche. This is referred to as the niche-variation hypothesis (Soule and Stewart 1970). Nevertheless, when resources are limiting, individual variation is reduced because natural selection favors an all-purpose phenotype, able to utilize a large range of resources. This explanation of variability at the exclusive population level may be rather simplistic because it does not take into account other factors of interaction with other species, in particular, interspecific competition. Classically, interspecific competition is viewed as a promoter of exclusion (Hardin 1960), but in practice coexistence is observed by means of a restriction in the range of species resources. Consequently, interspecific competition may reduce intrapopulation variability without resource limitations.

All these considerations are based on the general agreement that competition exists, but its detection in nature has proved to be elusive. Interspecific competition is inferred in the majority of study cases (Schoener 1983; Connell 1983), but its intensity is dependent on the trophic level (Hairston, Smith, and Slobodkin 1960) Thus, in herbivores, predators and parasites are the limiting factors of populations, diminishing their population sizes below those limits in which resource competition is important (Huston 1979). Strategies of many *Drosophila* species have been qualified as herbivore, yet it has been shown that frugivorous species compete as larvae (Slobodkin, Smith, and Hairston 1967). Work in *Drosophila* communities is controversial. There are some evidences of niche separation among species breeding in flowers (Pipkin, Rodriguez, and León 1966) and in cactus (Ruiz and Heed 1988). Most studies with frugivorous species report some kind of specialization by season, substrate, or daily activity, but species niche overlap is, generally, large (Dobzhansky and Pavan 1950; Pavan, Dobzhansky, and Burla 1950; Pipkin 1953, 1965; Birch and Battaglia 1957; Martinez Pico, Maldonado, and Levins 1965). Recently, Sevenster (1993) has performed a thorough analysis in a neotropical community of frugivorous *Drosophila* and found that, although competition can occur, there is no evidence that competition influences community structure. When compared with randomly composed communities, this community does not show greater specialization, overlap, or reduction in use of a resource type, as competition theory would predict.

All in all, serious doubts exist on competition being the major force in shaping community structure. Not only other biotic interactions, such as predation (Paine 1966) or parasitism (Carton et al. 1986), may be more important, but also abiotic perturbations (Connell 1978) and certain forms of spatial and temporal heterogeneity are worth our attention (Shorrocks et al. 1984; Chesson 1985). In particular, an aggregation distribution of a superior species in a patchy, ephemeral resource allows the occupation of empty or low-density patches by other inferior species. This aggregation model may explain interspecific coexistence in many invertebrates (Shorrocks, Atkinson, and Charlesworth 1979) and may contribute to the maintenance of intraspecific variability in a heterogeneous environment without resource partition.

Measuring natural selection in the wild is not an easy task, but it becomes more difficult when we try to detect selection at the gene level. We have progressed in the understanding of the tempo and mode of selective mechanisms, but there is a long way to go. Dobzhansky was aware of these difficulties and pushed to a maximum the tools of his time to unveil selection mechanisms. We owe him many seminal insights in the understanding of natural variability, especially his concepts of balancing selection and coadaptation, which go much further than the simple heterosis concept. We are beginning to understand how trade-offs are instrumental in creating the balanced concept and how we can utilize correlations between genotype and phenotype to detect selection with workable sample sizes. The need for increasing our potency to test selection is a must in future studies.

Perhaps the least developed field in the understanding of natural selection relates to studies of population and community structure. I believe that recent advances in this field and the increasing need to integrate population and community levels will result in a new comprehension of the interplay between selection and structure to shape variability in nature. Dobzhansky was also interested in the ecology of populations and communities of *Drosophila*. His cooperation with field geneticists and ecologists produced some pioneering works that attempted to define *Drosophila* community structure (Dobzhansky and Pavan 1950; Pavan, Dobzhansky, and Burla 1950; da Cunha, Dobzhansky, and Sokoloff 1951; da Cunha, Shehata, and Oliveira 1957). His desire to measure population parameters such as dispersal, size, and niche subdivision was evident in his collaborative work with S. Wright (Dobzhansky and Wright 1941, 1942, 1947; Wright, Dobzhansky, and Hovanitz 1942) and H. Levene (Dobzhansky and Levene 1948). Much remains to be explored in these areas, but recent progress in ecological genetic techniques that make use of molecular markers is very promising. The capability of genetically marking individuals in a population makes it possible to study genealogies in relationship to individual resources, clearing up the long-held confusion between allozygotes and autozygotes when one is trying to measure inbreeding in nature. Endler (1986) has advocated more field work to answer ecological questions. However, this work must take advantage of the new ideas and methods now available to answer the whys and hows of natural selection. Dobzhansky would have done it this way.

REFERENCES

Barbadilla, A., A. Ruiz, M. Santos, and A. Fontdevila, (in press). Mating pattern and fitness component analysis associated to inversion polymorphism in a natural population of *Drosophila buzzatii. Evolution.*

Barker J. S. F. and J. C. Mulley. 1976. Isozyme variation in natural populations of *Drosophila buzzatii. Evolution* 30:213–33.

Barker, J. S. F., G. L. Toll, M. Miranda, and H. J. Phaff. 1983. Heterogeneity of the yeast flora in the breeding sites of cactophilic *Drosophila. Can. J. Microbiol.* 29:6–14.

Barker J. S. F., D. C. Vacek, P. D. East, and W. T. Starmer. 1986. Allozyme genotypes of *Drosophila buzzatii:* Feeding and oviposition preferences for microbial species, and habitat selection. *Aust. J. Biol. Sci.* 39:47–58.

Begon M. and B. Shorrocks. 1978. The feeding- and breeding-sites of *Drosophila obscura* Fallen and *D. subobscura* Collin. *J. Nat. Hist.* 12:137–51.

Birch, L. C. and B. Battaglia. 1957. The abundance of *Drosophila willistoni* in relation to food in natural populations. *Ecology* 38:165–66.

Brussard, P. ed. 1978. *Ecological Genetics: The Interface.* New York: Springer-Verlag.

Bundgaard, J. and F. B. Christiansen. 1972. Dynamics of polymorphisms, I: Selection components in an experimental population of *Drosophila melanogaster. Genetics* 98:849–69.

Carton, Y., M. Bouletreau, J. J. M. van Alphen, and J. C. van Lenteren. 1986. The *Drosophila* parasitic wasps. In M. Ashburner, H. L. Carson, and J. N. Thompson, eds., *The Genetics and Biology of Drosophila 3e*, pp. 347–94. London: Academic Press.

Charlesworth, B. 1990. Optimization models, quantitative genetics, and mutation. *Evolution* 44:520–38.

Charnov, E. L. and D. Berrigan. 1990. Dimensionless numbers and life history evolution: Age of maturity versus adult lifespan. *Evol. Ecol.* 4:273–75.

Chesson, P. L. 1985. Coexistence of competitors in spatially and temporally varying environments: A look at the combined effects of different sorts of variability. *Theor. Popul. Biol.* 28:263–87.

Christiansen, F. B. and O. Frydenberg. 1973. Selection component analysis of natural polymorphisms using population samples including mother-offspring combinations. *Theor. Popul. Biol.* 4:425–45.

Christiansen, F. B. and T. M. Fenchel, eds. 1977. *Measuring Selection in Natural Populations.* New York: Springer-Verlag.

Connell, J. H. 1978. Diversity in tropical rain forest and coral reefs. *Science* 199:1302–10.

———. 1983. On the prevalence and relative importance of interspecific competition: Evidence from field experiments. *Am. Natur.* 122:661–96.

da Cunha, A. B., Th. Dobzhansky, and A. Sokoloff. 1951. On food preferences of sympatric species of *Drosophila. Evolution* 5:97–101.

da Cunha, A. B., A. M. E. T. Shehata, and W. de Oliveira. 1957. A study of the diets and the nutritional preferences of tropical species of *Drosophila. Ecology* 38:98–106.

David, J. R. and J. R. Tsacas. 1980. Cosmopolitan, subcosmopolitan and widespread species: Different strategies within the drosophilid family (Diptera). *C. R. Soc. Biogeogr.* 57:11–26.

Dempster, E. 1955. Maintenance of genetic heterogeneity. *Cold Spring Harbor Symp. Quant. Biol.* 20:25–32.

Dobzhansky, Th. 1935. A critique of the species concept in biology. *Phil. Sci.* 2:344–55.

———. 1941. *Genetics and the Origin of Species.* 2d. ed. New York: Columbia University Press.

———. 1980. The birth of the genetic theory of evolution in the Soviet Union in the 1920s. In E. Mayr and W. B. Provine, eds., *The Evolutionary Synthesis*, pp. 229–42. Cambridge: Harvard University Press.

Dobzhansky, Th. and H. Levene. 1948. Genetics of Natural Populations. XVII:Proof of operation of natural selection in wild populations of *Drosophila pseudoobscura. Genetics* 33:537–47.

Dobzhansky, Th. and C. Pavan. 1950. Local and seasonal variations in relative frequencies of species of *Drosophila* in Brazil. *J. Animal Ecol.* 19:1–14.

Dobzhansky, Th. and S. Wright. 1941. Genetics of natural populations. V:Relations between mutation rate and accumulation of lethals in populations of *Drosophila pseudoobscura. Genetics* 26:23–51.

———. 1942. Genetics of natural populations. VII: The allelism of lethals in the third chromosome of *Drosophila pseudoobscura. Genetics* 27:363–94.

———. 1947. Genetics of natural populations. XV: Rate of difussion of a mutant gene through a population of *Drosophila pseudoobscura. Genetics* 32:303–24.

Endler, J. A. 1986. *Natural Selection in the Wild.* Princeton: Princeton University Press.

Fogleman, J. C., W. T. Starmer, and W. B. Heed. 1981. Larval selectivity for yeast species by *Drosophila mojavensis* in natural substrates. *Proc. Natl. Acad. Sci.* 78:4435–39.

————. 1982. Comparisons of yeast florae from natural substrates and larval guts of southwestern *Drosophila. Oecologia* (Berlin) 52:187–91.

Fontdevila, A. 1982. Recent developments on the evolutionary history of the *Drosophila mulleri* complex in South America. In J. S. F. Barker and W. T. Starmer, eds., *Ecological Genetics and Evolution.* Sydney: Academic Press.

————. 1989. Founder effects in colonizing populations: The case of *Drosophila buzzatii.* In A. Fontdevila, ed., *Evolutionary Biology of Transient Unstable Populations.* Berlin: Springer-Verlag.

————. 1991. Colonizing species of *Drosophila.* In G. M. Hewitt, A. W. B. Johnston, and J. P. W. Young, eds., *Molecular Techniques in Taxonomy,* pp. 249–69. Berlin: Springer-Verlag.

Fontdevila, A., A. Ruiz, G. Alonso, and J. Ocaña. 1981. The evolutionary history of *Drosophila buzzatii.* I: Natural chromosomal polymorphism in colonized populations in the old world. *Evolution* 35:148–57.

Fontdevila, A., A. Ruiz, J. Ocaña, and G. Alonso. 1982. The evolutionary history of *Drosophila buzzatii.* II: How much has chromosomal polymorphism changed in colonization? *Evolution* 36:843–51.

Hairston, N. G., F. E. Smith, and L. B. Slobodkin. 1960. Community structure, population control and competition. *Am. Natur.* 94:421–25.

Hardin, G. 1960. The competitive exclusion principle. *Science* 131:1292–97.

Hasson, E., J. C. Vilardi, H. Naveira, J. J. Fanara, C. Rodriguez, O. Reig, and A. Fontdevila. 1991. The evolutionary history of *Drosophila buzzatii.* XVI: Fitness component analysis in an original natural population from Argentina. *J. Evol. Biol.* 4:209–25.

Hasson, E., J. J. Fanara, C. Rodriguez, J. C. Vilardi, O. A. Reig, and A. Fontdevila. 1992. The evolutionary history of *Drosophila buzzatii.* XXIV: Second chromosome inversions have different average effects on thorax length. *Heredity* 68:557–63.

Hasson, E., J. J. Fanara, C. Rodriguez, J. C. Vilardi, O. A. Reig, and A. Fontdevila. 1993. The evolutionary history of *Drosophila buzzatii.* XXVII: Thorax length is positively correlated with longevity in a natural population from Argentina. *Genetica* 92:61–65.

Hasson, E., H. Naveira, and A. Fontdevila. 1992. The breeding sites of Argentinian cactophilic species of the *Drosophila-mulleri* complex (Subgenus *Drosophila,* repleta group. *Rev. Chilena de Hist. Natural* 65:319–26.

Huston, M. 1979. A general hypothesis of species diversity. *Am. Natur.* 113:81–101.

Janssen, A., G. Driessen, M. de Haan, and N. Roodbol. 1988. The impact of parasitoids on natural populations of temperate woodland *Drosophila. Neth. J. Zool.* 38:61–73.

Kimura, M. and J. Crow. 1978. Effect of overall phenotypic selection on genetic change at individual loci. *Proc. Natl. Acad. Sci.* 75:6168–71.

Levene, H. 1953. Genetic equilibrium when more than one ecological niche is available. *Am. Natur.* 87:311.

Lewontin, R. C. 1974. *The Genetic Basis of Evolutionary Change.* New York: Columbia University Press.

Lewontin, R. C. and C. C. Cockerham. 1959. The goodness-of-fit test detecting selection in random mating populations. *Evolution* 13:561–64.

Lewontin, R. C. and M. J. D. White. 1960. Interaction between inversion polymorphisms of two chromosome pairs in the grasshopper *Moraba scurra. Evolution* 14:116–29.

Lewontin, R. C., L. R. Ginzburg, and S. D. Tuljapurkar. 1978. Heterosis as an explanation for large amounts of genic polymorphism. *Genetics* 88:149–70.

Lomnicki, A. 1988. *Population Ecology of Individuals.* Princeton: Princeton University Press.

Martinez Pico, M., C. Maldonado, and R. Levins. 1965. Ecology and genetics of Puerto Rican *Drosophila:* I: Food preferences of sympatric species. *Caribb. J. Sci.* 5:29–36.

Novitski, E. W. and E. R. Dempster. 1958. An analysis of data from laboratory populations. *Genetics* 43:470–79.

Paine, R. T. 1966. Food web complexity and species diversity. *Am. Natur.* 100:65–75.

Pavan, C., Th. Dobzhansky, and H. Burla. 1950. Diurnal behavior of some neotropical species of *Drosophila. Ecology* 31:36–43.

Pipkin, S. B. 1953. Fluctuations in *Drosophila* populations in a tropical area. *Am. Natur.* 87:317–22.

———. 1965. The influence of adult and larval food habits on population size of neotropical ground-feeding *Drosophila. Am. Midl. Natur.* 74:1–27.

Pipkin, S. B., R. L. Rodriguez, and J. León. 1966. Plant host specificity among flower-feeding neotropical *Drosophila* (Diptera: Drosophilidae) *Am. Natur.* 100:135–56.

Prout, T. 1965. The estimation of fitness from genotypic frequencies. *Evolution* 19:546–51.

———. 1969. The estimation of fitness from population data. *Genetics* 63:949–67.

Prout, T. and J. S. F. Barker. 1989. Ecological aspects of the heritability of body size in *Drosophila buzzatii. Genetics* 123:803–13.

Quezada-Diaz, J. E. 1993. Estructura poblacional y patrón de apareamientos de la especie cactófila *Drosophila buzzatii.* Ph.D. thesis. Universitat Autonoma de Barcelona, Spain.

Quezada-Diaz, J. E., M. Santos, A. Ruiz, and A. Fontdevila. 1992. The evolutionary history of *Drosophila buzzatii.* XXV: Random mating in nature. *Heredity* 68:373–79.

Richmond, R. C., and J. R. Powell. 1970. Evidence of heterosis associated with an enzyme locus in a natural population of *Drosophila. Proc. Natl. Acad. Sci.* 67:1264–67.

Robertson, A. 1965. The interpretation of genotypic ratios in domestic animal populations. *Animal Prod.* 7:319–24.

Roff, D. A. 1986. Predicting body size with life history models. *BioScience* 36:316–23.

Ruiz, A. and W. B. Heed. 1988. Host-plant specificity in the cactophilic *Drosophila mulleri* species complex. *J. Animal Ecol.* 57:237–49.

Ruiz, A. and M. Santos. 1989. Mating probability, body size and inversion polymorphism in a colonizing population of *Drosophila buzzatii.* In A. Fontdevila, ed., *Evolutionary Biology of Transient Unstable Populations,* pp. 96–113. Berlin: Springer-Verlag.

Ruiz, A., A. Fontdevila, M. Santos, M. Seoane, and E. Torroja. 1986. The evolutionary history of *Drosophila buzzatii.* VIII: Evidence for endocyclic selection acting on the inversion polymorphism in a natural population. *Evolution* 40:740–55.

Ruiz, A., M. Santos, A. Barbadilla, J. E. Quezada-Díaz, E. Hasson, and A. Fontdevila. 1991. Genetic variance for body size in a natural population of *Drosophila buzzatii. Genetics* 128:739–50.

Sanchez, A. 1986. Relaciones filogenéticas en los clusters buzzatii y martensis (grupo repleta) de *Drosophila. Ph.D.* dissertation. Universitat Autònoma de Barcelona, Spain.

Santos, M. (in press). Heterozygote deficiencies under Levene's subdivision structure. *Evolution.*

Santos, M., A. Ruiz, A. Barbadilla, J. E. Quezada-Diaz, E. Hasson, and A. Fontdevila. 1988. The evolutionary history of *Drosophila buzzatii.* XIV: Larger flies mate more often in nature. *Heredity* 61:255–62.

Santos, M., A. Ruiz, and A. Fontdevila. 1989. The evolutionary history of *Drosophila buzzatii.* XIII: Random differentiation as a partial explanation of chromosomal variation in a structured natural population. *Am. Natur.* 133:183–97.

Santos, M., A. Ruiz, J. E. Quezada-Diaz, A. Barbadilla, and A. Fontdevila. 1992. The evolutionary history of *Drosophila buzzatii.* XX: Positive phenotypic covariance between field adult fitness components and body size. *J. Evol. Biol.* 5:403–22.

Schoener, T. W. 1983. Field experiments on interspecific competition. *Am. Natur.* 122:240–85.

Sevenster, J. G. 1993. The community ecology of frugivorous *Drosophila* in a neotropical forest. Ph.D. dissertation. University of Leiden.

Shorrocks, B., W. Atkinson, and P. Charlesworth. 1979. Competition on a divided and ephemeral resource. *J. Animal Ecol.* 48:899–908.

Shorrocks, B., J. Rosewell, K. Edwards, and W. Atkinson. 1984. Interspecific competition is not a major organizing force in many insect communities. *Nature* 310:310–12.

Slobodkin L. B., F. E. Smith, and N. G. Hairston. 1967. Regulation in terrestrial ecosystems, and the implied balance of nature. Am. Natur. 101:109–24.

Soule, M. and B. R. Stewart. 1970. The "niche-variation" hypothesis: A test and alternatives. *Am. Natur.* 104:85–97.

Starmer, W. T. 1982. Analysis of the community structure of yeasts associated with the decaying stems of cactus. I: *Stenocereus gummosus*. *Microb. Ecol.* 8:71–81.

Starmer, W. T., J. S. F. Barker, H. J. Phaff, and J. C. Fogleman. 1986. Adaptations of *Drosophila* and yeasts: Their interactions with the volatile 2-propanol in the cactus-microorganism-*Drosophila* system. *Aust. J. Biol. Sci.* 39:69–77.

Thomas, R. and J. S. F. Barker. 1990. Breeding structure of natural populations of *Drosophila buzzatii:* Effects of the distribution of larval substrates. *Heredity* 64:355–65.

Vacek, D. C. 1982. Interactions between microorganisms and cactophilic *Drosophila* in Australia. In J. S. F. Barker and W. T. Starmer, eds., *Ecological Genetics and Evolution.* Sydney: Academic Press.

Wallace, B. 1958. The comparison of observed and calculated zygotic distributions. *Evolution* 12:113–14.

Workman, P. L. 1969. The analysis of simple genetic polymorphisms. *Hum. Biol.* 40:260–79.

Wright, S., Th. Dobzhansky, and W. Hovanitz. 1942. Genetics of Natural Populations. VII: The allelism of lethals in the third chromosome of *Drosophila pseudoobscura*. *Genetics* 27:363–94.

17

Adaptation and Density-Dependent Natural Selection

Laurence D. Mueller

The field of evolutionary ecology attempts to bridge the boundaries of two disciplines in order to develop a coherent picture of the way in which the natural environment affects the evolutionary process. Traditionally the ecologist has viewed populations having properties that were fixed in time and thus evolutionary change was implicitly ignored. Meanwhile, the population geneticist dealt with simplified and static environments that were also viewed as constant under the theoretical and experimental methods employed by that discipline. Ultimately, a serious attempt would be made to unite these disciplines. However, we see that those involved in this union have often brought with them the viewpoints characteristic of their early training: either ecology or population genetics.

Here I review the development of one important topic in evolutionary ecology, density-dependent natural selection. While its most important theoretical development was accomplished by ecologists, notably Robert MacArthur, Dobzhansky was instrumental, in my view, in developing a research paradigm that could be used to experimentally test the predictions of density-dependent selection. I first summarize the contributions of MacArthur and Dobzhansky to this field and then review my own work in this area and how it has been molded by each of these scientists.

I thank F. J. Ayala and T. Prout for useful comments and NIH grant AG 09970 for financial support.

R. H. MacArthur

MacArthur was trained in both mathematics and ecology, under G. E. Hutchinson, and was keenly interested in developing an evolutionary perspective to ecology. His first major effort in this area was his development of theories analogous to Fisher's fundamental theorem of natural selection in an ecological context. To accomplish this MacArthur dealt with an aspect of ecology that was both related to individual fitness and had a well-developed theory. Density-dependent population growth was the obvious connection. MacArthur's exposition of this theory (MacArthur 1962) does not actually accomplish this union. His theory starts by asserting that fitness will be proportional to the carrying capacity of a genotype rather than derive this from more basic assumptions. Later, more careful developments of this theory would explicitly show that fitness was assumed to be equivalent to density-dependent growth rates (Roughgarden 1971). From this point one could then achieve predictions that natural selection would lead to the increase of either r or K depending on the environment (Roughgarden 1971). Roughgarden (1976) has also shown that natural selection will locally maximize population size in a manner analogous to the maximization of mean fitness, although the generality of this result is subject to qualifications (Prout 1980).

The theory of density-dependent natural selection received its most important development in MacArthur and Wilson's book *The Theory of Island Biogeography* (1967). In this book the earlier themes of MacArthur are expanded and the evolutionary outcome of natural selection in populations kept at very high or very low densities is discussed. The tendency to focus on the predictions at extreme densities is characteristic of much of MacArthur's work, and these extremes were called r-selection (selection at low density) and K-selection (selection at high density).

The discussion of these theories by MacArthur and Wilson is largely qualitative, and this sets the tone for much of the future research in this area. So, for instance, in environments that result in K-selection, they state, "Evolution here favors efficiency of conversion of food into offspring—there must be no waste." (MacArthur and Wilson 1967:149). They acknowledge that Dobzhansky had made some similar qualitative predictions some seventeen years earlier (Dobzhansky 1950). The reason they discuss this theory is that they are interested in exploring the type of evolution that will ensue following a colonization event. In the broader scheme they are most interested in understanding patterns of life history differences among many species of plants and animals. Indeed the examples they discuss as a potential illustration of their theory are differences in clutch size among species of birds in the tropics and temperate climates.

Subsequent applications of these ideas would assert that almost all important differences in life histories between organisms as different as whales and bacteria (Pianka 1970, 1972) could be understood with the theory of *r*- and *K*-selection. Many of the empirical tests of these theories would also involve the comparison of closely related species that were believed to have experienced different density regimes since they speciated. Many of these experiments and their conceptual flaws are reviewed by Stearns (1976, 1977).

The large number of studies aimed at testing *r*- and *K*-selection were initiated by scientists largely with backgrounds in ecology and involved interspecific comparisons. This research closely followed the backgrounds and methods first explored by MacArthur and Wilson. What was lacking in these studies was the ability to demonstrate genetic differentiation that was unambiguously due to the populations density environment.

Th. Dobzhansky

Of great interest to Dobzhansky were the causes of changes in certain inversion polymorphisms in *Drosophila pseudoobscura*. Data from thirty-one years of sampling in Piñon Flats, for instance are summarized in Dobzhansky (1970; see also Anderson 1989). These data show pronounced cycles that repeated on a yearly basis. Thus, some aspect of a seasonal environment would seem to be involved. These studies were also clearly examining genetic variation within populations, and thus environmentally induced changes in the relative fitness of genotypes and natural selection were possible candidates for explaining these changes.

Dobzhansky's own work on these populations suggested that chromosomally polymorphic genotypes were more fit than monomorphic genotypes (Beardmore, Dobzhansky, and Pavlosky 1960; Dobzhansky and Pavlovsky 1961; Dobzhansky 1964). But these results would not explain the fluctuation in chromosome frequencies. Dobzhansky (1970), in a review of this work late in his life, suggests that changes in population density that occur with some regularity in the Piñon Flats population may be important in controlling this cycle. He notes that experimental results of Birch (1955) suggest that fitness varies with density and that the regular changes in density in Piñon Flats are consistent with the fitness differences described by Birch.

In 1950 Dobzhansky speculated about the evolutionary events that could have given rise to the great diversity of species seen in the tropics

(Dobzhansky 1950). The discussion in this paper is decidedly ecological. He first notes the "Gause principle"—that two species with similar ways of life will not coexist indefinitely. From this he develops the idea that in the tropics the different types of habitats available for organisms is very diverse, in part owing to the relatively benign climate in the tropics. This observation coupled with Gause's principle led Dobzhansky to suggest that for species to be successful in the tropics requires specialization and this then results in a wide diversity of species. This specialization is a product of the intense competition between species for the available habitats, which are many, in the tropics.

As noted by MacArthur and Wilson, Dobzhansky's view of the important evolutionary factors in the tropics are similar to their own, which is developed from the theory of r- and K-selection, although it is clear that Dobzhansky did not explicitly consider population dynamics in reaching his conclusions. Nevertheless, the ecological tone of Dobzhansky's arguments are clear.

Demographic ideas are addressed directly in Dobzhansky, Lewontin, and Pavlovsky (1964). In this paper age-specific survival and fertility measurements at low density are made on two different karyotypes of $D.$ $pseudoobscura$. This study is, to my knowledge, the first such measurements of r_{max} on different genotypes within the same species. The subtleties of using demographic theory first outlined by Lotka (1925) to compute fitness are not addressed by Dobzhansky. However, this is an additional example of the application of theory, previously in the domain of ecology, to population genetics.

Density-Dependent Natural Selection

Chromosome Extraction Lines and Population Growth Rates

My early research in this field focused on the ecological characteristics of populations of $D.$ $melanogaster$ made homozygous for whole second chromosomes (Mueller and Ayala 1981 a, b, c). These genotypes had been extensively studied by population geneticists to ascertain the fitness consequences of homozygosity at many loci. Much of this work has been reviewed by Lewontin (1974), and this technique had been used numerous times by Dobzhansky (Dobzhansky, Krimbas, and Krimbas 1960; Dobzhansky and Spassky 1954, 1963, 1968).

While my studies permitted an assessment of the utility of population growth rates as a measure of fitness (Mueller and Ayala 1981a) they did not yield insights into the process of density-dependent natural selection. The reason is that genotypes homozygous for whole second chromosomes are obviously not the products of natural selection and thus may have genetic correlations and other properties that would not be found in "naturally" occurring genotypes. For this reason I initiated a controlled experiment in which replicate populations of *Drosophila* were permitted to evolve at different population densities. In these populations natural selection could act and generate genetic and phenotypic differences that I could follow over time.

Evolution in Experimental *Drosophila* Populations

Over the past fifteen years our understanding of the phenotypic changes that ensue while *D. melanogaster* adapts to crowded cultures has progressed dramatically. A summary of the fitness and phenotypic differences that have been documented from these studies are summarized in figure 17.1. The major effects of density-dependent natural selection on laboratory populations appear to be on phenotypic and physiological traits in larvae.

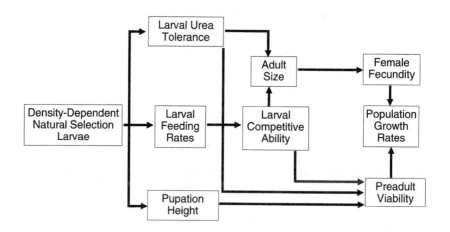

FIGURE 17.1.
The fitness and phenotypic consequences of density-dependent natural selection. For instance, density-dependent natural selection in the larvae will increase larval feeding rates in crowded cultures, and this leads to an increase in larval competitive ability, which in turn affects the ability of the larvae to survive and their final adult size. Adult size affects female fecundity (Mueller 1987), and both this trait and viability affect the population growth rates.

These effects result in changes in density-dependent rates of population growth in the manner predicted by the simple theories cited earlier (Mueller and Ayala 1981d; Mueller, Guo, and Ayala 1991).

A crucial and generally untested assumption of theories of density-dependent natural selection is that per-capita growth rates are useful measures of fitness. With *Drosophila* density-dependent selection affects traits that cause changes in preadult viability and female fecundity. Thus, for this model system, population growth rates do respond to selection. Below I summarize some of the detailed understanding we now have for the *Drosophila* model system.

When larvae are crowded, food is one resource that is in short supply. The ability of larvae to compete for limited food is genetically variable (Bakker 1961) and is therefore potentially amenable to change by natural selection. In food-limited environments better competitors are expected to both survive better and be larger as adults (Mueller 1988a). In *Drosophila*, as well as many other organisms, large body size is correlated with higher fertility in both males (Wilkinson 1987) and females (Mueller 1987). We have documented the increase in competitive ability of larvae kept under crowded conditions relative to low-density controls (Mueller 1988b). In addition we have shown that the differences in competitive ability may be attributed to differences in larval feeding rates (Joshi and Mueller 1988; Guo, Mueller, and Ayala 1991; Mueller, Graves, and Rose 1993; figure 17.2).

In addition to searching for food, larvae must find sites to pupate. In crowded cultures these sites may be limited, and some locations (on the surface of the food) may in fact increase the chances of death during the pupal stage (Joshi and Mueller 1993). Populations kept under crowded conditions show a preference to pupate higher on the side of the vials than populations kept at low larval densities (Mueller and Sweet 1986; Guo, Mueller, and Ayala 1991; Mueller, Graves, and Rose 1993; figure 17.3). Populations raised under low-density conditions and then moved to crowded environments will initially undergo directional selection for increased pupation height (Joshi and Mueller 1993). After some time in a crowded culture, selection is expected to become stabilizing and bring to a halt further increases in pupation height (Joshi and Mueller 1993).

Another aspect of crowded cultures is the rapid accumulation of waste products (Botella et al. 1985). One toxic waste product found in crowded cultures is urea. Larvae that are biting and swallowing food indiscriminately in crowded cultures are expected to ingest large quantities of these waste products. Urea slows development time, increases preadult mortality, reduces adult size, and decreases adult longevity and fecundity (Botella

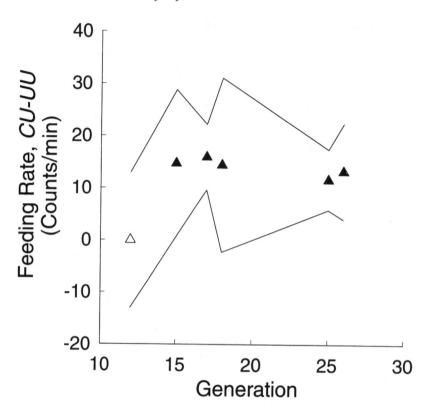

FIGURE 17.2.

Difference (*CU–UU*) in feeding rates of populations of D. *melanogaster* exposed to different levels of larval crowding. The *CU* populations are crowded as larvae and uncrowded as adults while the *UU* populations are uncrowded as larvae and adults (Mueller, Graves, and Rose 1993). The solid lines indicate the upper and lower 95% confidence interval computed at each generation. The solid triangles indicate statistically significant differences between the feeding rates of the two types of populations; the open triangles are not significantly different.

et al. 1985; Mueller, unpublished). However, populations of *Drosophila* that have evolved under crowded larval conditions show higher viability, increased adult size, and adult longevity relative to populations kept under uncrowded conditions (Mueller, unpublished).

An important issue in life history evolution has been the notion that the potential scope of possible life histories is constrained by physiological or energetic trade-offs. In our system there is some evidence that the larvae that have been selected under crowded conditions may be less efficient at utilizing food than corresponding controls (Mueller 1990). One potential hypothesis is that increasing a trait like feeding rates may increase overall metabolic rates, decrease time available for food digestion, and thus ultimately decrease the ability of the larva to extract energy from food. Detailed studies of the energy budgets of larvae would provide direct evidence to potentially support or refute this hypothesis.

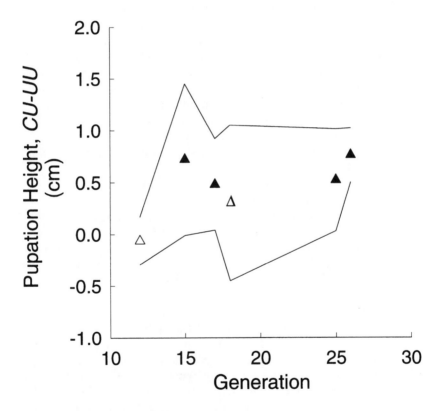

FIGURE 17.3.

Difference (CU–UU) in pupation height for populations of D. *melanogaster* exposed to different levels of larval crowding. Symbols follow figure 17.2.

Population Stability

A long-standing problem in population biology has been understanding the factors in the environment and properties of organisms that affect the stability of population dynamics. Fluctuations in population size around an equilibrium may be due to the manner in which life history characters respond to population density, or they may reflect fluctuations in qualities of the environment that affect the numbers of organisms that can be supported.

Theoretical models of the population dynamics of *Drosophila* (Mueller 1988a) suggest that population stability will be determined, to a great extent, by the response of female fecundity to increasing adult density. When adults are fed ad lib amounts of yeast the decline in female fecundity is sufficiently weak that populations may become destabilized (Mueller and Huynh 1994). To study this problem experimentally, we created 15 populations (5 replicates of 3 treatments) of *D. melanogaster* that were supplied with differing amounts of food as larvae (either high or low) and as adults (either high or low). The results of these experiments showed that population dynamics may become destabilized when adults are supplied high amounts of food and larvae are given low amounts of food (figure 17.4). The cycles that these environmental conditions generated (see curve LH figure 17.4) are reminiscent and perhaps caused by the same factors as the cycles exhibited in Nicholson's classical experiments with blowflies (Nicholson 1957).

An obvious question is whether natural selection on individual life history characters may affect population stability in some predictable way. Even the theory on this issue is unclear (Heckel and Roughgarden 1980; Turelli and Petry 1980; Mueller and Ayala 1981b; Hansen 1992; Gatto 1993), although an analysis of Nicholson's blowfly data suggests that the cycles exhibited in these populations may have become attenuated over time due to natural selection (Stokes et al. 1988).

In the experiments conducted by Mueller and Huynh, 15 populations were created from the same source populations (the $r \times r$ population described in Mueller, Guo, and Ayala 1991). We have measured the per-capita fecundity of these populations after they have been kept in their different environmental regimes for 9–12 generations. The per-capita fecundity under the low adult food conditions shows little differences between the populations (figure 17.5). However, the per-capita fecundity under high food levels appears to differ at the high adult densities (figure 17.6). This effect is confirmed by the significant population by food level interaction at densities 32 and 64 in the analysis of variance (table 17.1).

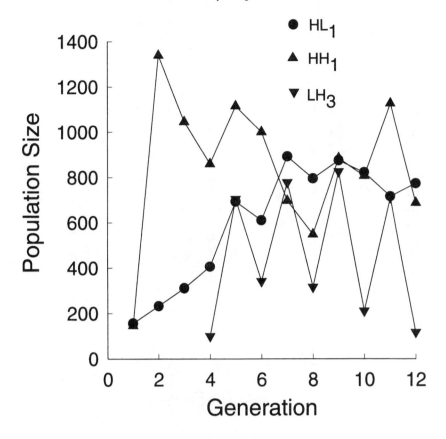

FIGURE 17.4.
Adult numbers in three populations of *D. melanogaster.* The *HL* population was maintained on high levels of larval food and low levels of adult food, *HH* on high adult and larval food, LH on low larval and high adult food levels. The subscript designates a particular population.

As discussed in Mueller and Huynh (1994), the stability of laboratory populations is critically dependent on the response of female fecundity (F) to adult density (N), which can be modeled by the relationship,

$$F = \frac{f}{1+aN}$$

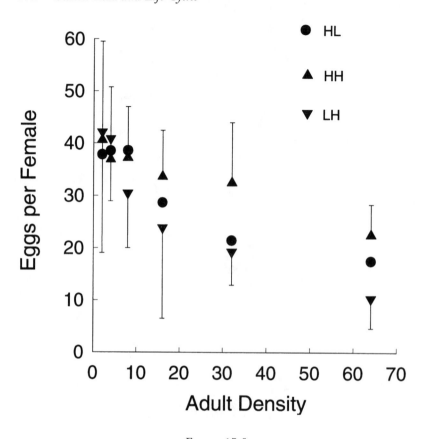

FIGURE 17.5.
Female fecundity as a function of adult density when adults from the *HL*, *HH*, and LH populations are cultured on low levels of food. Error bars are 95% confidence intervals.

When a is large, population stability is enhanced. If natural selection in fact enhances population stability, as suggested by Stokes et al. (1988) and Turelli and Petry (1980), then we would expect that under the high adult food conditions the populations with unstable dynamics (LH) may show an increase in a relative to the other populations.

We have estimated a and f for all 15 populations (table 17.2). Comparing the 5 values of a from the LH populations (table 17.2) to the 10 values of a from the *HL* and *HH* populations shows a significant difference (Mann-Whitney U test, $p = 0.04$). Thus, we have some preliminary evidence that natural selection may affect the stability of *Drosophila* populations. A number of features of this experiment should, however, be improved before definitive conclusions can be reached. The LH populations reached adult population sizes as low as 60 during this experiment. This means that perhaps the observed genetic differences were a result of drift during these bottlenecks rather than natural selection. These populations were also adapting to crowded larval cultures, and the results may reflect some differential adaptation to different levels of crowding rather than the details of the dynamic behavior of the population. Nevertheless, the results are suggestive and consistent with previous observations with blowflies.

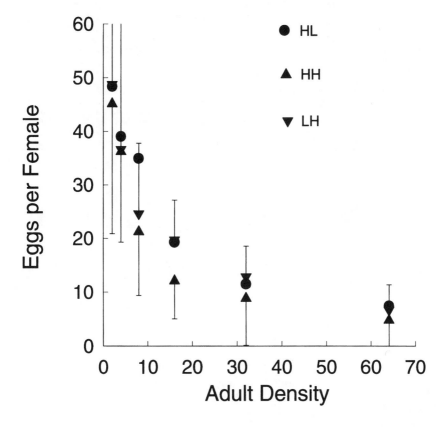

FIGURE 17.6.
Female fecundity as a function of adult density when adults from the *HL*, *HH*, and LH populations are cultured on high levels of food. Error bars are 95% confidence intervals.

TABLE 17.1.

*Analysis of Variance (ANOVA) of Fecundity as a Function of
Food Level and Population at Two Densities: 32 and 64 adults*

Source of Variation	Sum of Squares	Degrees of Freedom	Mean Square	F	p
		Density = 32 Adults			
Population	32,845	2	16,422	1.57	0.229
Food level	343,827	1	343,827	32.87	<0.001
Population × food	105,531	2	52,766	5.04	0.015
Error	251,064	24	10,461		

Source of Variation	Sum of Squares	Degrees of Freedom	Mean Square	F	p
		Density = 64 Adults			
Population	141,844	2	70,922	4.52	0.023
Food level	788,024	1	788,024	50.21	<0.001
Population × food	238,650	2	119,325	7.60	0.003
Error	345,276	22	15,694		

NOTE: The experimental protocols followed the methods described in Mueller and Huynh (1994). Briefly, females two generations removed from their selection environments were kept for three days in vials either with low amounts of live yeast or high amounts of live yeast. Adults were kept at six different densities: 2, 4, 8, 16, 32, and 64 adults with equal numbers of each sex. After the three-day period adults laid eggs in unyeasted vials for twenty-four hours and the total number of eggs were recorded. The interactions from ANOVAs at the four lowest densities were not significant. The mean values for fecundity for each population were used as the basic observation in this analysis.

TABLE 17.2.

The Values of a *and* f *from Equation (1) for Each of 15 Populations
Under the High Food Conditions*

Population	a	f
HL_1	0.0103	43.7
HL_2	0.118	60.3
HL_3	0.0237	51.2
HL_4	0.0219	34.1
HL_5	0.0340	35.5
HH_1	0.0141	44.5
HH_2	−0.002	27.7
HH_3	0.0113	53.6
HH_4	0.0243	41.6
HH_5	0.0163	38.0
LH_1	0.0392	60.7
LH_2	0.0662	35.4
LH_3	0.0550	42.5
LH_4	0.235	95.1
LH_5	0.0127	29.6

NOTE: Standard nonlinear regression techniques were used to obtain these estimates.

Advantages of the *Drosophila* System

The use of *Drosophila* to study density-dependent natural selection offers many advantages relative to the comparative approach, which was the primary method of analysis in the 1970s.

- Populations may be replicated. Thus, differences between experimental and control populations may be safely attributed to selection rather than drift or founding effects when differences arise multiple times.
- The ability to experimentally control the ecological conditions permits sound inference about the cause of the selection pressure. A severe problem with essentially all studies of natural populations is the inability to determine that population density is the only relevant difference between the populations being contrasted.
- The large body of research already carried out on *Drosophila*. For instance, in my own studies several prior studies of *D. melanogaster* have proved useful. My own research on the evolution of competition was greatly aided by the work of Bakker (1961), Sewell, Burnet, and Connolly (1975), Burnet, Sewell, and Bos (1977), and Nunney (1983). Likewise, the potential importance of urea in molding the adaptations of populations evolving in crowded populations was suggested by the work of Botella et al. (1985).

The assets of the experimental system described here will apply to a wide variety of problems in evolutionary biology. Another excellent example is the application of replicated populations of *Drosophila* to the study of aging (Rose 1984, 1991). With this approach, significant understanding of the physiological and genetic correlates of aging in *Drosophila* is now unraveling.

REFERENCES

Anderson, W. 1989. Selection in natural populations of *D. pseudoobscura*. *Genome* 31:239–45.

Bakker, K. 1961. An analysis of factors which determine success in competition for food among larvae in *Drosophila melanogaster. Arch. Neerl. Zool.* 14:200–81.

Beardmore, J. A., T. Dobzhansky, and O. A. Pavlovsky. 1960. An attempt to compare the fitness of polymorphic and monomorphic populations of *Drosophila pseudoobscura* at 16°. *Heredity* 14:19–33.

Birch, C. L. 1955. Selection in *Drosophila pseudoobscura* in relation to crowding. *Evolution* 9:389–99.

Botella, L. M., A. Moya, M. C. Gonzalez, and J. L. Mensua. 1985. Larval stop, delayed development and survival in overcrowded cultures of *Drosophila melanogaster:* Effect of urea and uric acid. *J. Insect Physiol.* 31:179–85.

Burnet, B., D. Sewell, and M. Bos. 1977. Genetic analysis of larval feeding behavior in *Drosophila melanogaster* II: Growth relations and competition between selected lines. *Genet. Res. Cambr.* 30:149–61.

Dobzhansky, Th. 1950. Evolution in the tropics. *Am. Scientist* 38:209–21.

———. 1964. How do the genetic loads affect the fitness of their carriers in *Drosophila* populations? *Am. Natur.* 98:151–66.

———. 1970. Evolutionary oscillations in *Drosophila pseudoobscura*. In: R. Creed, ed., *Ecological Genetics and Evolution*, pp. 109–33. Oxford: Blackwell.

Dobzhansky, Th., C. Krimbas, and M. G. Krimbas. 1960. Genetics of natural populations. XXX: Is genetic load in *D. pseudoobscura* a mutational or balanced load? *Genetics* 45:741–53.

Dobzhansky, Th., R. C. Lewontin, and O. Pavlovsky. 1964. The capacity for increase in chromosomally polymorphic and monomorphic populations of *Drosophila pseudoobscura. Heredity* 19:597–614.

Dobzhansky, Th. and O. A. Pavlovsky. 1961. A further study of fitness of chromosomally polymorphic and monomorphic populations of *Drosophila pseudoobscura. Heredity* 16:169–77.

Dobzhansky, Th. and B. Spassky. 1954. Genetics of natural populations. XXII: A comparison of the concealed variability in *Drosophila prosaltans* with that in other species. *Genetics* 39:472–87.

———. 1963. Genetics of natural populations. XXXIV: Adaptive norm, genetic load, and genetic elite in *Drosophila pseudoobscura. Genetics* 48:1467–85.

———. 1968. Genetics of natural populations. XXXL: Heterotic and deleterious effects of recessive lethals in populations of *Drosophila pseudoobscura. Genetics* 59:411–25.

Gatto, M. 1993. The evolutionary optimality of oscillatory and chaotic dynamics in simple population models. *Theor. Popul. Biol.* 43:310–36.

Guo, P. Z., L. D. Mueller, and F. J. Ayala. 1991. Evolution of behavior by density-dependent natural selection. *Proc. Natl. Acad. Sci.* 88:10905–6.

Hansen, T. F. 1992. Evolution of stability parameters in single-species population models: Stability or chaos? *Theor. Popul. Biol.* 42:199–217.

Heckel, D. G. and J. Roughgarden. 1980. A species near equilibrium size in a fluctuating environment can evolve a lower intrinsic rate of increase. *Proc. Natl. Acad. Sci.* 77:7497–500.

Joshi, A. and L. D. Mueller. 1988. Evolution of higher feeding rate in *Drosophila* due to density-dependent natural selection. *Evolution* 42:1090–93.

———. 1993. Directional and stabilizing density-dependent natural selection for pupation height in *Drosophila melanogaster. Evolution* 47:176–84.

Lewontin, R. C. 1974. *The Genetic Basis of Evolutionary Change.* New York: Columbia University Press.

Lotka, A. J. 1925. *Elements of Physical Biology.* Baltimore: Williams and Wilkins.

MacArthur, R. H. 1962. Some generalized theorems of natural selection. *Proc. Natl. Acad. Sci.* 48:1893–97.

MacArthur, R. H. and E. O. Wilson. 1967. *The Theory of Island Biogeography.* Princeton, N.J: Princeton University Press.

Mueller, L. D., 1987. Evolution of senescence in laboratory populations of *Drosophila. Proc. Natl. Acad. Sci.* 84:1974–77.

———. 1988a. Density-dependent population growth and natural selection in food limited environments: The *Drosophila* model. *Am. Natur.* 132:786–809.

———. 1988b. Evolution of competitive ability in *Drosophila* due to density-dependent natural selection. *Proc. Natl. Acad. Sci.* 85:4383–86.

———. 1990. Density-dependent natural selection does not increase efficiency. *Evol. Ecol.* 4:290–97.

Mueller, L. D. and F. J. Ayala. 1981a. Fitness and density dependent population growth in *Drosophila melanogaster. Genetics* 97:667–77.

———. 1981b. Dynamics of single species population growth: Stability or chaos? *Ecology* 62:1148–54.

———. 1981c. Dynamics of single species population growth: Experimental and statistical analysis. *Theor. Popul. Biol.* 20:101–17.

———. 1981d. Trade-off between r-selection and K-selection in *Drosophila* populations. *Proc. Natl. Acad. Sci.* 78:1303–05.

Mueller, L. D., J. L. Graves Jr., and M. R. Rose. 1993. Interactions between density-dependent and age-specific selection in *Drosophila melanogaster. Functional Ecol.* 7:469–79.

Mueller, L. D., P. Z. Guo, and F. J. Ayala. 1991. Density-dependent natural selection and trade-offs in life history traits. *Science* 253:433–35.

Mueller, L. D. and P. T. Huynh. 1994. Ecological determinants of stability in model populations. *Ecology* 75:430–37.

Mueller, L. D. and V. F. Sweet. 1986. Density-dependent natural selection in *Drosophila:* Evolution of pupation height. *Evolution* 40:1354–56.

Nicholson, A. J. 1957. The self adjustment of populations to change. *Cold Spring Harbor Symp. Quant. Biol.* 22:153–73.

Nunney, L. 1983. Sex differences in larval competition in *Drosophila melanogaster:* The testing of a competition model and its relevance to frequency-dependent selection. *Am. Natur.* 121:67–93.

Pianka, E. R. 1970. On r- and K-selection. *Am. Natur.* 104:592–96.

———. 1972. r- and K-selection or b and d selection? *Am. Natur.* 106:581–88.

Prout, T. 1980. Some relationships between density-independent selection and density-dependent population growth. *Evol. Biol.* 13:1–68.

Rose, M. R. 1984. Laboratory evolution of postponed senescence in *Drosophila melanogaster. Evolution* 38:1004–10.

————. 1991. *Evolutionary Biology of Aging.* New York: Oxford University Press.

Roughgarden, J. 1971. Density dependent natural selection. *Ecology* 52:453–68.

————. 1976. Resource partitioning among competing species—a coevolutionary approach. *Theor. Popul. Biol.* 9:388–424.

Sewell, D., B. Burnet, and K. Connolly. 1975. Genetic analysis of larval feeding behavior in *Drosophila melanogaster. Genet. Res. Cambr.* 24:163–73.

Stearns, S. 1976. Life history tactics: A review of the ideas. *Quart. Rev. Biol.* 51:3–47.

————. 1977. The evolution of life history traits: A critique of the theory and a review of the data. *Ann. Rev. Ecol. Syst.* 8:145–71.

Stokes, T. K., W. S. C. Gurney, R. M. Nisbet, and S. P. Blythe. 1988. Parameter evolution in a laboratory insect population. *Theor. Popul. Biol.* 34:248–65.

Turelli, M. and D. Petry. 1980. Density-dependent selection in a random environment: An evolutionary process that can maintain stable population dynamics. *Proc. Natl. Acad. Sci.* 77:7501–05.

Wilkinson, G. S. 1987. Equilibrium analysis of sexual selection in *Drosophila melanogaster. Evolution* 41:11–21.

PART FIVE

Speciation and Mating Behavior

18

Endosymbiotic Infectivity in
Drosophila paulistorum Semispecies

Lee Ehrman, Ira Perelle, and Jan R. Factor

To produce or reproduce species . . . is certainly difficult, because . . . numerous genetic differences would have to be combined; yet the task is not entirely hopeless. The superspecies Drosophila paulistorum *is composed of six incipient species. The incipient species seem to be identical morphologically, but they cross only with difficulty (that is, they are reproductively isolated), and among the hybrids produced, males are entirely sterile. . . . One strain, derived from a single female found in Llanos, Colombia, has become, after several years of cultivation in the laboratory, incapable of producing fertile male hybrids with the incipient species with which it was formerly fertile. By means of artificial selection for some fifty generations, this strain which became sexually isolated from strains with which it formerly crossed easily (Dobzhansky, Pavlovsky, and Powell 1976), may also be regarded as a new incipient species, though certainly still morphologically indistinguishable from its immediate antecedent. Yet the description and classification for species has traditionally been, and to a large extent continues to be, the province of systematists working in museums and herbaria. Sibling species cannot be distinguished, however, by classical museum techniques. For example, a pinned and dried female of* Drosophila paulistorum *can only be determined as belonging to the* willistoni *group of siblings. Species are, however, phenomena of nature that exist regardless of our ability to distinguish them. The techniques of biochemical tests are now used routinely to classify some microorganisms.*

—Dobzhansky 1970; Dobzhansky, Boesiger, and Wallace 1983.

In 1956, then Professor of Zoology at Columbia University, "Doby" (as Dobzhansky was affectionately called) returned from an extensive Central

Because he guided this project before his unexpected death at age sixty-two in April 1991, this article is dedicated to Professor Norman Somerson, microbiologist, Ohio State University, School of Medicine.

and South American trip with our initial *Drosophila paulistorum* strains. These were presented to Boris Spassky for crossability and other testing, resulting in the seminal paper, "*Drosophila paulistorum*, a cluster of species in *statu nascendi*" (Dobzhansky and Spassky 1959).

In those intervening years, my (Ehrman) initial scientific publication was devoted to a description of a *D. paulistorum* sibling, namely, *D. insularis*, a narrow island endemic (Dobzhansky, Ehrman, and Pavlovsky 1957), and I comfortably anticipated a doctoral thesis on the behavior genetics of the sharply delineated, discrete, and workable *species nova* (Ehrman and Parsons 1981). My husband and I were both in graduate/professional schools and profoundly depauperate.

One late 1957 morning on the crowded and happily messy eighth floor of the Schermerhorn Hall Annex, Doby stopped by the unairconditioned office-laboratory I shared with Leigh and Phoebe Van Valen (into which drainage from the ancient, wooden constant-temperature rooms emptied) to suggest a choice of doctoral problems: *Drosophila paulistorum* interstrain hybrid sterility or sexual isolation. He pointed out that elucidation of the genetic architecture (his very words) of either reproductive isolating mechanism would suffice more than adequately and that since only hybrid males were sterile, if I undertook the sterility genetics, I would have only one sex to worry about; with the manifest sexual isolation, there would be two. Doby then suggested I think these matters over and proceeded home to lunch with Natasha on Claremont Avenue, as he routinely did. That very afternoon, with no commitment of any kind yet on my part, Boris Spassky delivered my many *D. paulistorum* strains. Currently, some thirty-five years later, our work continues.

As part of a symposium, Ehrman, Somerson, and Kocka (1990) and all references therein, reviewed the technical history of the induction of hybrid male sterility in this unique species complex, *Drosophila paulistorum*. Sterility in hybrids between *D. paulistorum* semispecies is caused by a cell-wall-deficient microorganism, a streptococcal Group D L-form, located inside the testes of hybrid and nonhybrid males. These are infectious agents that rapidly proliferate only in the testes of hybrid males with the concomitant breakdown of spermatogenesis. Adult females do not seem to suffer from foreign L-forms. No untreated *D. paulistorum*, male or female, has ever been observed to be free of L-forms, and the flies do not survive long after their L-form is removed.

Each *D. paulistorum* semispecies has its own highly specific endosymbiont without which it cannot thrive. These wall-less streptococci appear to provide some essential factor or service, perhaps the synthesis of vitamins, although the exact nature of the factor(s) or service(s) remains wholly obscure. Abundant evidence is available of cases of maternal cytoplasmic inheritance of a microorganism that may be essential to the survival of its host (Ehrman, Somerson, and Kocka 1990 and references therein; Somerson et al. 1984).

> The genetic status of the superspecies *Drosophila paulistorum* is as complex as it is interesting. The Centro-American, Orinocan, Amazonian, Andean-Brazilian, and [Interior] groups of populations[1] have diverged genetically, and have evolved reproductive isolations, safe enough for them to coexist sympatrically without gene exchange. . . . However, the nature and the origin of the genetic differences between the incipient species need elucidation. . . . Our data favors the hypothesis that the hybrid sterility between incipient species *D. paulistorum* is the primary isolating mechanism while the ethological isolation has developed later by natural selection in response to the challenge of hybridization.
>
> (Dobzhansky and Pavlovsky 1967)

Isolates cultured from different semispecies are streptococcal L-forms but do not necessarily all belong to Group D. Laboratory-induced infections of foreign L-forms produce the same fertility patterns as hybridization. Extracts of testicular L-forms from semispecies "A" injected into a female of semispecies "B," which is subsequently fertilized by a male of semispecies "B," results in fertile daughters and sterile sons, indicating that a hybrid genotype is not absolutely necessary for the manifestation of this phenomenon. When L-forms isolated from one semispecies were passed into another, there was a reduction in the number of offspring also but no change in the sex ratio of offspring. At this point we were sufficiently well informed about requisite procedures and results of assorted injections into females (e.g., those into males are not productive of the induced sterility phenomenon, and neither are those containing only sterile sucrose into females) to realize that we were obligated to initiate massive "tissue from every semispecies into every other semispecies" injection experiments.

---

[1]The transitional semispecies, unlisted here by Dobzhansky, is likely to be the relict ancestor. The Interior semispecies was described later (see Dobzhansky and Pavlovsky 1975)

Methodology

Inocula and Crosses

Inocula for these studies are prepared according to the protocol utilized by Ehrman et al. (Ehrman, Somerson, and Gottlieb 1986; Ehrman et al. 1989), which should be consulted for methodological details. Inocula are freshly prepared before injection from material live just before extraction; i.e., we eschewed aging cultures and frozen samples. Each inoculum is a homogenate derived from one donor nonhybrid male. The homogenate is diluted in a solution of 15% sterile sucrose with the supernatant diluted to 10^{-5} after twenty minutes of 13,000 rpm centrifugation at 0–4° C; 0.25 μL of homogenate per 0.1 mL of solution. Briefly, starting with 20 living intact males, 0.25 μL of inoculum is injected into each twenty-four-hour-old recipient female (with at least 45 [15×3] females per semispecies by 6 semispecies, or 270 such recipient females). Their postinjection sons are assayed for induced sterility, which we define operationally as a range of absolute reproductive failure from no or immotile spermatozoa to male inviability.

Microinjection

Our investigations into the reproductive effects of an unusual strain of the microbial endosymbiont *Streptococcus faecalis* have made use of a variety of microinjection techniques to transfer tissue preparations and cultured microbes into larval and adult drosophilids. At various times, a variety of needles have been used, including steel, hand-sharpened steel, microcapillary glass, and pulled-glass needles. These have evolved into a technique that allows carefully controlled injections of submicroliter volumes into larval and adult drosophilids with minimal damage to the integrity of the individual insect (Factor, Ehrman, and Inocencio 1991). This technique has been used successfully in the *Drosophila paulistorum* studies.

Microinjection apparatus: A fluid-filled 10 mL syringe (Hamilton #701, Hamilton Co., PO Box 10030, Reno, Nev. 89510) is fitted with a 30-cm length of silicon tubing (Cole Parmer #6411-6, ID 1/32 in, OD # 3/32 in, Cole Parmer Instrument Co., 7425 N. Oak Park Ave., Chicago, Ill. 60648). The syringe is positioned in a repeating dispenser (Hamilton #PB600) and placed onto a Hamilton Electric Thumb (Model EPB-600, Hamilton

#76710) equipped with a foot-pedal switch. Glass needles are produced manually, in the standard way, by pulling 50 μL Accu-fill glass microcapillary pipettes (Clay Adams Division of Becton Dickinson and Co., Parsippany, N.J. 07054) over a Touch-O-Matic gas burner (Hanau Engineering Co., Buffalo, N.Y.). The resulting needles are checked with a dissecting microscope and selected for uniformity (examples of needle tips are illustrated in fig. 18.1). Manual needle pulling is quicker than using an automatic needle puller and routinely yields excellent results. The pulled-glass injection needle is inserted into the end of the silicon tubing of the apparatus.

Microinjection technique: The 10 mL syringe is filled with sterile mineral oil, and a tuberculin syringe is used to fill the silicon tubing with mineral oil. The tip of the pulled-glass injection needle is placed directly in inoculum and filled by capillary action; it is then inserted into the free end of the silicon tubing. Sterile technique is observed throughout.

Adult flies are lightly etherized and held steady against the tip of a camel hair brush. The injection needle is shallowly inserted into the abdomen at a narrow angle pointing anteriorly. In adults, it can be placed under the fold between the third and fourth abdominal sternites, but it sometimes pierces the cuticle in the third segment. Micromanipulators can be used, but injection by hand is faster and simpler and yields excellent results after some practice. Delivery of the preset volume is initiated by pressing the foot pedal switch controlling the Electric Thumb. Neither antibiotics nor other special procedures to inhibit infection of the wound seem necessary.

The volume of inoculum injected is controlled by the repeating dispenser, with 0.20–0.25 μL of inoculum routinely injected in adults, and 0.01 μL routinely into larvae. When injected, the *Drosophila paulistorum* proboscis becomes extended and contracted, an indication that as much volume as can be tolerated has already been delivered. Following injection, flies are immediately transferred to a clean, dry vial; held until they recover from etherization (three to five minutes); and then transferred to a fresh food vial or culture.

Results: The results of this technique are excellent in adults and no discernible mortality occurs owing to injection. Larvae often pupate immediately after injection, regardless of the instar injected, triggering early pupation that often results in death. Additional specimens must be injected to carry out a study, but the technique remains useful for larvae. See Gottlieb et al. (1977) for details concerning larval recipients. In each experiment, a sterile insect saline solution can be injected into a control group of flies to assess any possible effects of the injection procedure.

Histology and Ultrastructure

Reproductive tissues of *Drosophila paulistorum* can be prepared mechanically for microscopy in three ways: (1) animals can be fixed whole and intact (except wings are removed); (2) entire abdomens can be severed from the thorax and perforated at the posterior tip before fixing; and (3) gonads can be dissected from the abdomen in a drop of saline and then immediately transferred to fixative.

Tissues for transmission electron microscopy are fixed in 3% glutaraldehyde in 0.15 M sodium cacodylate buffer at pH 7.4. Fixation is carried out at 6°C for twelve to sixteen hours and is followed by four fifteen-minute buffer rinses. Tissues are initially postfixed in 1% OsO_4 in buffer for at least five hours; additional postfixation is then carried out for twelve to sixteen hours in fresh osmium solution. Four fifteen-minute buffer washes and three distilled water washes follow after fixation. Specimens are then treated with 2% uranyl acetate for twelve to sixteen hours and dehydrated in an ethanol series. Propylene oxide is used to replace the ethanol and as a solvent for infiltration with plastic. Specimens are embedded in "Embed-812" (Electron Microscopy Sciences, Ft. Washington, Pa.), a modification of the Mollenhauer (1963) mixture of Epon 812 and Araldite 502. Thin sections are stained with uranyl acetate and lead citrate and observed with a transmission electron microscope.

Tissues prepared for light microscopy are fixed and embedded as for transmission electron microscopy. Thick plastic sections (0.5 to 1.0 µm) are cut, stained with methylene blue, and examined with the light microscope.

Colonies of the microbial endosymbiont *Streptococcus faecalis*, cultured on solid media, are spooned out with surrounding agar (to avoid mechanical disruption) and placed in vials of fixative. The fixation and embedding procedure for microbial colonies is the same as that described above for reproductive tissues.

Scanning Electron Microscopy

Flies intended for examination with scanning electron microscopy are fixed in glutaraldehyde (as above) or in 70% ethanol, dehydrated in an ethanol series of increasing concentrations, transferred to acetone, and dried in a critical-point drier with CO_2 to replace the acetone. Dried specimens are mounted with double-coated tape to expose the features of interest. Specimens are then coated with gold/palladium in a sputter coater and observed with a scanning electron microscope.

Research Results

The results of these experiments are reported here in a six-by-six matrix (table 18.1). This table represents 90 injections and 1,620 (45×36) dissections, since we examined 15 sons from each of 3 broods per cross in our six-by-six design. As a result of the analysis of the 1990 data cited and explained above, we selected broods seven, eight, and nine (days postinjection) as most

TABLE 18.1.

Infectivity of All the Drosophila paulistorum *Semispecies*:*
Number of Sterile Sons of 15 Sons Examined
from Each of Broods 7, 8, and 9

			Female Recipient					
Male Donor:	AM	AN	CA	IN	OR	TR	Total	χ^2 ***
AM	** 1	17	25	32	26	40	141	
	0.7	12.1	17.7	22.7	18.4	28.4	100.0	38.36
	0.5	9.1	13.2	17.4	14.3	20.5	12.6	
AN	42	23	45	10	30	36	186	
	22.6	12.4	24.2	5.4	16.1	19.4	100.0	27.35
	23.1	12.4	23.7	5.4	16.5	18.5	16.6	
CA	23	27	5	44	31	34	164	
	14.0	16.5	3.0	26.8	18.9	20.7	100.0	31.22
	12.6	14.5	2.6	23.9	17.0	17.4	14.7	
IN	42	31	33	16	37	43	202	
	20.8	15.3	16.3	7.9	18.3	21.3	100.0	14.48
	23.1	14.5	17.4	8.7	20.3	22.1	18.1	
OR	33	43	43	37	14	28	198	
	16.7	21.7	21.7	18.7	7.1	14.1	100.0	18.24
	18.1	23.1	22.6	20.1	7.7	14.4	17.7	
TR	41	45	39	45	44	14	228	
	18.0	19.7	17.1	19.7	19.3	6.1	100.0	18.95
	22.5	24.2	20.5	24.5	24.2	7.2	20.4	
Total	182	186	190	184	182	195	1119	
	16.3	16.6	17.0	16.4	16.3	17.4	100.0	0.76
	100.0	100.0	100.0	100.0	100.0	100.0	100.0	
χ^2 ***	43.10	20.91	35.28	34.80	17.05	16.72	25.05	

* Semispecies: AM = Amazonian; AN = Andean; CA = Centro- American; IN = Interior; OR = Orinocan; TR = Transitional.

**Cell data: Frequency, Percent of Row, Percent of Column

***All row and column χ^2: $p < 0.001$ *except* column total row (0.76).

informative. These include neither the earliest nor the final broods generated, which are often unreliable in number and contain few sons.

The data in each table 18.1 cell represent the results of dissection, in physiologic saline where possible, of 45 sons of 15 injected females (the females were, immediately after injection, crossed to males of their own strains so that no genetic hybridity was involved).

Each cell in table 18.1 presents:

1. the number of sterile male sons of the 45 male sons dissected (fifteen male sons from each of three broods)
2. the row percentage (i.e., the proportion of sterile male sons each cell represents of the total sons across females of the various semispecies, within each male semispecific donor of inoculated material)
3. the column percentage (i.e., the proportion of sterile male sons each cell represents of the total sons within females of the various semispecies, across male semispecific donor of inoculated material)

Chi-square statistics for each row and column are also provided.

The chi-square goodness-of-fit test (Sokal and Rohlf 1981) was chosen for this analysis for several reasons. We did not want to make any assumptions about the underlying normality of the distribution of sterility of *D. paulistorum* sons or about the equality of variance of the various crossings. We also believe the goodness-of-fit test to be the most appropriate since our null hypotheses state:

1. There is no difference in sterility across females of the various semispecies within each injected inocula strain.
2. There is no difference in sterility across injected inocula strain within females of the various semispecies.

Expected frequencies for the chi-square statistics were calculated by the more conservative method of summing the numbers of sterile sons in each row and column and dividing by the number of cells in each row and column (6). The more liberal method of using as the expected frequencies the number of sterile sons produced by each semispecies of female, when injected with inoculum prepared from males of her own semispecies, is certainly justified by the nature of this experiment. However, the resulting chi squares would have been so large as to have been meaningless, e.g., the chi square for row 1 would have been 2,971.

All row and column chi squares are significant at alpha = 0.001, which indicates that the null hypotheses must be rejected for all rows and columns. An examination of row 1 will serve as a model for most, but not all, the rows and columns. It can be seen that the minimum number of sterile sons, row 1 (cell 1,1), occurred when inoculum prepared from Amazonian males was injected into Amazonian females. Amazonian inoculum injected into females of all other semispecies profoundly increased the number of sterile sons hatched to those females (minimum =17, maximum=40). Similar results can be seen in examining column 1, where cell 1,1 is also the base cell. Amazonian females injected with inoculum prepared from any other semispecies male produced many more sterile sons than Amazonian females injected with Amazonian inoculum did.

The balance of table 18.1 can be examined in a similar fashion, where the negative diagonal provides the base cells for each row and column (cells 2,2; 3,3; 4,4; 5,5; and 6,6). In almost every cross, the base cell, the one showing a female injected with inoculum from a male of her own semispecies, records a smaller number of sterile sons than any other cell in the row or column does. There are only two exceptions to this pattern. Andean females produced fewer sterile sons when injected with inoculum prepared from Amazonian males than with inoculum prepared from Andean males, and Interior females produced fewer sterile sons when injected with inoculum produced from Andean males than with inoculum produced from Interior males.

Amazonian and Central American females seemed to be the most sensitive to inoculum injections, producing very few sterile sons when injected with inoculum prepared from males of their *own semispecies* but many sterile sons when injected with inoculum prepared from other semispecies. With Orinocan and Transitional semispecies females, the differences were much less pronounced. Amazonian inoculum appeared to produce the fewest sterile sons (141) and Transitional inoculum appeared to produce the largest number of sterile sons (228) across all semispecies females.

Transitional is believed to be the relict ancestral *D. paulistorum* semispecies, though no one can be sure. Currently, Amazonian and Centro American exhibit the greatest remoteness from the *D. paulistorum* cluster as measured by reproductive isolating mechanisms. (See Ehrman and Powell 1982; Ehrman and Wasserman 1987 for discussions of these points.)

The row total and column total chi squares provide an indication of the female and male effect. Note that the column total chi square (0.76, $p=0.95$) shows virtually no difference in the total number of sterile sons produced across females of all semispecies. Chi squares across row totals, however, show a highly significant difference in the number of sterile sons across male inoculum donors. This seems to indicate that although there are differences within each semispecies, overall, females produced approximately the same number of sterile sons while male inocula, across semispecies, were significantly different in effects on reproduction.

By now we are sure that each semispecific endosymbiont differs from each of the five others in a variety of biochemical (e.g., antibiotic sensitivities) and physiological ways. Their precise characterizations and normal *intra*semispecific functions represent future goals.

We have employed extensive transmission electron microscopy to visualize this endosymbiotic microbe in culture and within reproductive tissues of larval and adult *D. paulistorum*. Figure 18.1 depicts our injection procedure and its sequelae. Figure 18.2 (in vitro growth) shows the contents of our inocula when Mesitas males donate; see row 2 of table 18.1. Figure 18.3 represents nonhybrid but nonetheless infected tissue; see column 2, row 3 in table 18.1. Figure 18.4 shows sections from a symbiont-free *D. paulistorum* sibling species, *D. pavlovskiana*. We offer these electron micrographs as controls. Figures 18.5 and 18.6 depict the results of crosses between these two hybridizable and closely related, but microbiologically different, sibling species. Figures 18.7 and 18.8 were taken of larval tissue from *D. pavlovskiana* (fig. 18.7) and *D. paulistorum* (fig. 18.8) respectively. These provide prereproductive evidence of symbiont presence during development. We are now about to transfer egg cytoplasm between semispecies (into very young embryonated eggs), stimulated by the fine work of Louis and Nigro (1989).

And what would the mentor of all *D. paulistorum* research (surely not an exaggeration) have made of all of this? He would have repeatedly noted that

> Organic evolution presents an essential peculiarity, not found in other sciences, which sets limits of fact but not of principle. Each evolutionary step is a historical event that has never occurred before and will never be repeated. Experiments cannot be repeated in the laboratory until the underlying mechanisms are understood . . . . Indeed, the study of evolutionary successions is more difficult than the study of a chemical reaction, which can be endlessly reproduced in the laboratory in seemingly identical fashion. Because each event is unique and because large numbers of factors intervene, extremely complex interactions arise. Is that sufficient reason, however, for fixing limits to research?

(Published posthumously in 1983: Dobzhansky, Boesiger, and Wallace 1983)

Then he would have planned at least several collection trips to fill in the manifest geographic gaps in our stock collection representing all the semi-

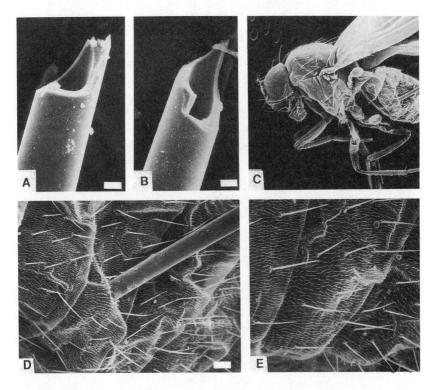

FIGURE 18.1

Scanning electron micrographs illustrating the microinjection technique for *Drosophila paulistorum*. A and B. Examples of the tips of typical injections needles; diameters near tips are approximately 40 μm. C. An adult *Drosophila paulistorum* female with injection needle still in place. The fly is approximately 2.1 mm long. D. Abdomen of the same adult female in C, with injection needle still in place in the third abdominal sternite. E. A wound in the third abdominal sternite of an adult *Drosophila paulistorum* female one hour after injection. Scale bars for A, B, C, D, and E represent 4.0 μm, 4.0 μm, 250 μm, 40.0 μm, and 40.0 μm respectively. From Factor Ehrman, and Inocencio 1991.

species. These trips would not have included Lee Ehrman unless her husband, his dentist, also accompanied us. His prudishness prohibited sole female accompaniment. But, rather than admit this, he would have gleefully pointed out Lee's uselessness on such trips, owing to the absence of Spanish.

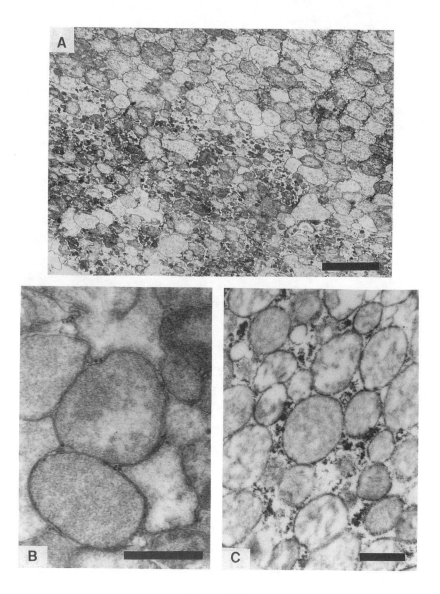

FIGURE 18.2

Sections of a colony (from culture) of the microorganism isolated from testes of *Drosophila paulistorum* (Mesitas, Colombia strain, Andean semispecies), identified as Group D *Streptococcus faecalis* L-forms (often referred to as "FM1"). Note that they are cell-wall deficient. Some cells are senescent, have lost intracellular differentiation, and appear almost empty of cytoplasmic content. A. Scale bar represents 5.0 μm.

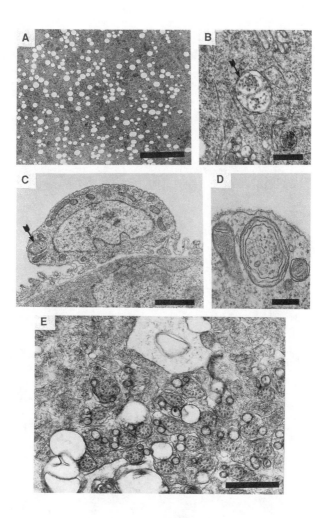

FIGURE 18.3

Drosophila paulistorum (Mesitas, Colombia strain, Andean semispecies). A–C. Fertile, nonhybrid, naturally infected female gonadal tissue. A. Egg cytoplasm containing electron-transparent yolk granules, mitochondria, and intracellular microbial symbionts of the type illustrated in B (arrow). C. Polar body, shed by the egg cytoplasm, containing the same microbial symbionts (arrow). D. The intracellular symbiont from C showing several layers of surrounding membranes; the innermost membrane belongs to the symbiont, while the outer membranes are contributed by the host cell; the symbiont is flanked by two mitochondria. E. Fertile, nonhybrid, naturally infected male gonadal tissue: a portion of a single cyst illustrating the motillar apparatus and mitochondria of spermatid tails; note areas devoid of healthy, maturing sperm, occupied instead by microorganisms (upper left, upper right). Scale bars for A, B, C, D, and E represent 5.0 μm, 0.5 μm, 1.0 μm, 0.2 μm, and 1.0 μm, respectively.

(From Ehrman et al. 1989. © 1989 by the University of Chicago.)

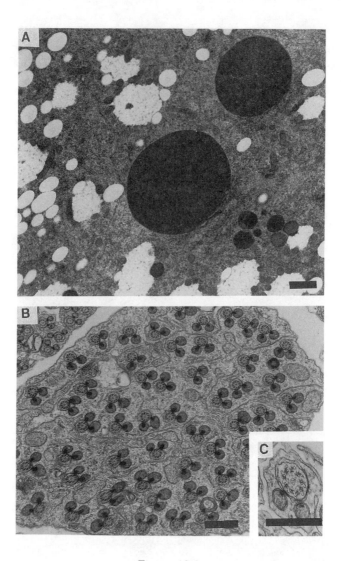

FIGURE 18.4

Drosophila pavlovskiana. A Fertile, nonhybrid, naturally uninfected female gonadal tissue: egg cytoplasm, including electron-dense lipid granules and electron-transparent yolk granules; no microorganisms are present in this tissue. B, C. Fertile, nonhybrid, naturally uninfected male gonadal tissue. B. Apparently healthy spermatids illustrating normal motillar apparatus and the two mitochondria that eventually fuse to form the nebenkern; transverse section. C. A single spermatid showing microtubular details of motillar apparatus; transverse section. No microorganisms are present in this tissue. Scale bars for A, B, an C represent 1.0 µm, 1.0 µm, and 0.5 µm, respectively.

(From Ehrman et al. 1989. © Copyright © 1989 The University of Chicago.)

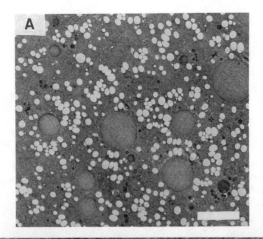

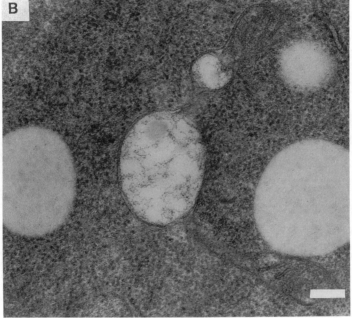

FIGURE 18.5

Fertile, hybrid gonadal tissue of F_1 progeny resulting from a cross between a *Drosophila paulistorum* (Mesitas) female and a *Drosophila pavlovskiana* male. A. Fertile, infected egg cytoplasm (apparently normal in morphology) containing electron-dense lipid granules, electron-transparent yolk granules. B. A portion of the cytoplasm containing a membrane-bound microorganism (center), flanked by two yolk granules of about the same size. Scale bars for A and B represent 5.0 μm and 0.2 μm, respectively.

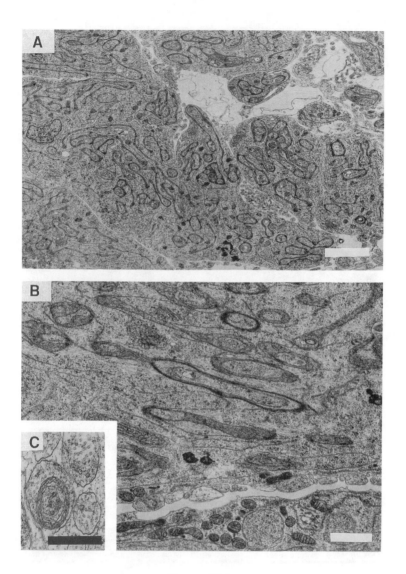

FIGURE 18.6

Sterile, hybrid F_1 male gonadal tissue from a cross between a *Drosophila paulistorum* (Mesitas) female and a *Drosophila pavlovskiana* male. A. Sterile, infected, thoroughly degenerated testis. B. Note the complete failure of spermatogenesis and the abundant presence of pleomorphic microorganisms; apparently normal mitochondria are visible in the testicular sheath (bottom). C. A microorganism surrounded by several layers of membranes; the innermost layer belongs to the microorganism, while the outer membranes are contributed by the host cell. Scale bars for A, B, and C represent 5.0 μm, 1.0 μm, and 0.5 μm, respectively.

(From Ehrman et al. 1989. Copyright © 1989 The University of Chicago.)

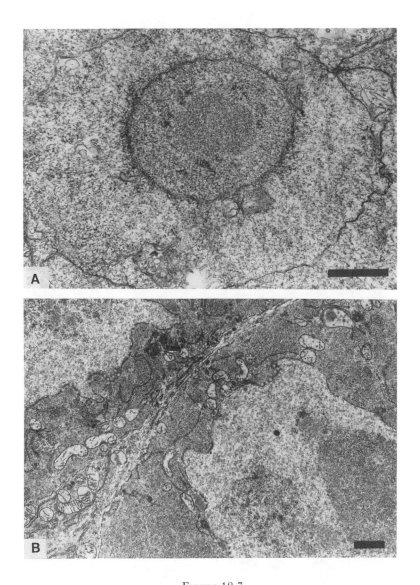

FIGURE 18.7

Third larval instar of *Drosophila pavlovskiana*. A. Female, gonadal primordium. B. Male, gonadal primordium. Note absence of cytoplasmic endosymbiont from both male and female tissues. Scale bars for A and B represent 1.0 μm and 10.0 μm, respectively.

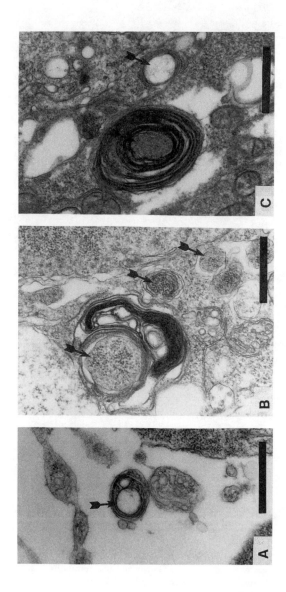

FIGURE 18.8

Third larval instar of *Drosophila paulistorum*, Mesitas strain. A. Female, gonadal primordium; note cytoplasmic endosymbiont (arrow) in a cell process. B. Male, gonadal primordium; note cytoplasmic endosymbionts (arrows), some of which are surrounded by concentric host-cell membranes. C. Male, gonadal primordium; note cytoplasmic endosymbionts (arrows), surrounded by concentric host-cell membranes. Scale bars for A–C represent 1.0 μm.

Then he would have suggested that Dr. Richard Ehrman go with him alone because this would have meant horseback riding at some point.

Finally, when we announced plans to proceed with polymerase chain reaction (PCR) studies of endosymbiontic RNAs, for each semispecies, Doby would have muttered something or other about trendiness and he would have been wrong. In preliminary experiments (conducted by Dr. S. Miller of the Agricultural Research Service, U.S. Department of Agriculture at Gainesville, Florida, because of the untimely death of Dr. Somerson), cloned 16s ribosomal RNAs (rRNAs) of bacterial origin were obtained from *D. paulistorum* by means of PCR. DNA prepared from Santa Marta and Mesitas testes (representing the Transitional and Andean semispecies, respectively) served as template, and conserved, "universal" eubacterial PCR primers were used in the amplification reactions.

Owing to the relatively constant rate of nucleotide substitution over evolutionary time within 16s rRNA genes and the large number of sequences available in gene catalogues for comparative analyses, this type of analysis has provided determinative information regarding phylogenetic relationships among bacteria and other organisms (Woese 1987; Weisburg et al. 1991).

From the sequence analysis of cloned PCR products, only two types of sequences were obtained for both Santa Marta and Mesitas, each of which differed by approximately 5% from its homologue in these two semispecies. Gene bank searches revealed that both of the bacteria identified were bacilli. One of these, a single bacterium, proved to be readily cultivated from extracts of a number of adult tissues by use of either the SP4 formulation utilized earlier by Somerson (Somerson et al. 1984) or Luria-Bertani medium, used for the routine cultivation of *E. coli*.

Since the spermicidal bacterium had previously been reported as cultivatable (in spite of its inherent pathogenicity) in lepidopteran larvae (Gottlieb et al. 1977, 1981), transinfections into a variety of moth species were conducted in order to provide confirmatory information. Indeed, inocula prepared from either testes extracts or from reconstituted colonies on nutrient agar plates proved to be pathogenic when injected into at least seven species of noctuid and pyralid moths. Furthermore, cultivation of infected hemolymph from live, but moribund, larvae yielded a monoculture of the original bacterium isolated from *D. paulistorum*. Sequence analysis of 16s rRNA PCR products obtained from DNA isolated from individual colonies in all cases gave identical results, namely, that the cultivatable and infectious bacterium was the *Proteus vulgaris*-like microbe that was identified in the preliminary screen. It was not a streptococcal variant, and therefore the *D. paulistorum* male sterility agent still remains at large and further work will be required to fully control it.

Renewed efforts at developing culture conditions that inhibit the prolific growth of the *P. vulgaris*-like bacterium, while permitting the growth of an apparently fastidious and certainly cryptic coccus, will likely be productive. It may also prove that the use of streptococcal 16s rRNA PCR primers in amplification reactions may likewise be useful for preliminary diagnoses. In any case, the availability of probes (either DNA or antisera) would permit tracking of the temporal and spatial distribution of the microorganism in both naturally infected and transinfected insects. In addition, transinfections of this microorganism (or a related organism obtained from more distantly related insects) into economically important insects may prove to generate male sterility in the field fostering area-wide suppression.

REFERENCES

Dobzhansky, Th. 1970. *Genetics of the Evolutionary Process.* New York: Columbia University Press.

Dobzhansky, Th., E. Boesiger, and B. Wallace. 1983. *Human Culture: A Moment in Evolution.* New York: Columbia University Press.

Dobzhansky, Th., L. Ehrman, and O. Pavlovsky. 1957. *Drosophila insularis:* New sibling species of the *willistoni* group. University Texas Publ. 5721:39–47.

Dobzhansky, Th. and O. Pavlovsky. 1967. Experiments on the incipent species of the *Drosophila paulistorum* complex. *Genetics* 55:141–56.

———. 1975. Unstable intermediates between Orinocan and Interior semispecies of *Drosophila paulistorum. Evolution* 29:242–48.

Dobzhansky, Th., O. Pavlovsky, and J. Powell. 1976. Partially successful attempt to enhance reproductive isolation between semispecies of *Drosophila paulistorum. Evolution* 30:201–12.

Dobzhansky, Th. and B. Spassky. 1959. *Drosophila paulistorum,* a cluster of species in *statu nascendi. Proc. Natl. Acad. Sci.* 45:419–28.

Ehrman, L. and J. R. Factor. 1992. The *Drosophila paulistorum* endosymbiont in larval gonads. *Chromatin* 1:125–34.

Ehrman, L., J. R. Factor, N. Somerson, and P. Manzo. 1989. The *Drosophila paulistorum* endosymbiont in an alternative species. *Am. Natur.* 134:890–96.

Ehrman, L. and P. Parsons. 1981. *Behavior Genetics and Evolution.* New York: McGraw-Hill.

Ehrman, L. and J. Powell. 1982. The *Drosophila willistoni* species group. In M. Ashburner and J. Thompson, eds., *The Genetics and Biology of Drosophila.* Vol. 3b, pp. 193–225. New York: Academic Press.

Ehrman, L., N. Somerson, and F. Gottlieb. 1986. Reproductive isolation in a neotropical insect: Behavior and microbiology. In M. Huettel, ed., *Evolutionary Genetics of Invertebrate Behavior,* pp. 97–108. New York: Plenum.

Ehrman, L., N. Somerson, and J. Kocka. 1990. Induced hybrid sterility by injection of streptococcal L-forms into *Drosophila paulistorum:* Dynamics of infection. *Can. J. Zool.* 68:1735–40.

Ehrman, L. and M. Wasserman. 1987. Asymmetric evolution. *Evol. Biol.* 21:1–20 (and comments).

Factor, J., L. Ehrman, and B. Inocencio. 1991. A microinjection technique for drosophilids. *Drosophila Info. Serv.* 70:242–44.

Gottlieb, F., R. Goitein, L. Ehrman, and B. Inocencio. 1977. Interorder transfer of mycoplasma-like microorganisms between *Drosophila paulistorum* and *Ephestia kuehniella:* Tissues, dosages, and effects. *J. Inverteb. Pathol.* 30:140–50.

Gottlieb, F., G. Simmons, L. Ehrman, B. Inocencio, J. Kocka, and N. Somerson. 1981. Characteristics of the *Drosophila paulistorum* male sterility agent in a secondary host, *Ephestia kuehniella. Appl. Environ. Micro.* 42:838–42.

Louis, C. and L. Nigro. 1989. Ultrastructural evidence of *Wolbachia rickettsiales* in *Drosophila simulans* and their relationships with unidirectional cross-incompatibility. *J. Inverteb. Pathol.* 54:39–44.

Mollenhauer, H. 1963. Plastic embedding mixtures for use in electron microscopy. *Stain Technol.* 39:111–14.

Sokol, R. and F. J. Rohlf. 1981. *Biometry.* 2d ed. San Francisco: Freeman.

Somerson, N., L. Ehrman, J. Kocka, and F. Gottlieb. 1984. Streptococcal L-forms isolated from *Drosophila paulistorum* semispecies cause sterility in male progeny. *Proc. Natl. Acad. Sci.* 81:282–85.

Weisburg, W., S. Barns, D. Pelletier, and D. Lane. 1991. 16s ribosomal DNA amplification for phylogenetic study. *J. Bact.* 173:697–703.

Woese, C. 1987. Bacterial evolution. *Microbiol. Rev.* 51:221–71.

19

Levels of Evolutionary Divergence of *Drosophila willistoni* Sibling Species

Antonio R. Cordeiro and Helga Winge

One of the most successful tactics used by Dobzhansky to stimulate our imaginations and to induce us to hard work was to raise problems on the nature of cryptic species and their origin. Some of these problems were clearly and masterfully presented in two of his famous books (Dobzhansky 1951, 1970). All the work in the *D. willistoni* siblings was accomplished owing to the personal involvement, the keen interest, and masterly competence of Dobzhansky in both field and laboratory research. He was not only the friendly, attentive mentor but also the dynamic hard worker enjoying all activities and inspiring us with his example. Brazil was his greatest beneficiary, since several research centers in population genetics and evolution that are now well-developed departments were started by his former students.

A group of six species, many subspecies, and semispecies closely resembling *Drosophila willistoni* have been discovered and intensively studied over the vast territory of Central and South America by Dobzhansky and his students, mainly from 1943 to 1975. This master work is now completing fifty years in Brazil, and significant investigations are still in progress, some of which are reported in this paper.

The authors are grateful to the long-standing support of the Rockefeller Foundation and the Conselho Nacional de Pesquisas for laboratory and fieldwork. We are indebted to our technicians and colleagues for many years of friendly cooperation in laboratory work and collecting excursions.

These sibling species can hardly be characterized by usual taxonomic traits, and that constitutes the major difficulty of this work. This project began with a taxonomic survey of *Drosophila* by Dobzhansky and Pavan (1943), although there had been an earlier discovery of the first cryptic form of *D. willistoni* Sturtevant, which was later described as *Drosophila paulistorum* Dobzhansky and Pavan (in Burla et al. 1949). In that paper, a new sibling, *D. tropicalis* Burla and da Cunha, was described together with a study of the sexual isolation, morphologic details, and polytene salivary gland chromosomes of the four known siblings, including *D. equinoxialis* Dobzhansky (1946). This cryptic species had been collected in the Amazon and studied by Dobzhansky and Mayr (1944).

Another interesting form was added by Townsend (1954), who described as a subspecies of *D. tropicalis*, *D. t. cubana*. The reproductive isolation between these forms was found to be at the same level as that of the other sibling species by Winge (1965), who proposed the species status, as *Drosophila cubana* Townsend, to *D. t. cubana*, and found that *D. cubana* presents more affinity with several geographic races of the Amazonian region than *D. tropicalis* does. Furthermore, Townsend (1963) suggested the existence of six subspecies for *D. tropicalis* complex. More intensive studies may disclose two new groups of subspecies.

The biggest and the darkest among the siblings, a new species restricted to the Lesser Antilles, *D. insularis*, was found and described by Dobzhansky (in Dobzhansky, Ehrman, and Pavlovsky 1957). The careful study of Spassky (1957) allowed the discrimination of living male individuals of the sibling species by microscopic examination of external genitalia.

Cordeiro (1952) found geographic variations in *D. paulistorum* external male genitalia and showed that the shape of hypandria lobes is controlled by a single autosomal locus. A few years later Dobzhansky and coworkers discovered that *D. paulistorum* is in fact a superspecies composed of five semispecies or species in *statu nascendi*, plus a Transitional population, living mainly in the northern region of South America (Dobzhansky et al. 1964 and the review of Ehrman and Powell 1982).

Finally, *D. pavlovskiana*, the former Guiana race of *D. paulistorum* complex, was described by Kastritsis and Dobzhansky (1967).

A series of contributions from Ehrman and coworkers elucidated most aspects of the role of mating behavior, sexual isolation, and infective agents on reproductive isolation of *D. paulistorum* semispecies (Ehrman, Somerson, and Kocka 1990). The *P* elements found in all siblings of *D. willistoni* by Daniels et al. (1990) may also have an effect in reproductive isolation. Some kind of infection was related to the spontaneous origin of a new "incipient species" of *D. paulistorum*, which originated in laboratory cultures (Dobzhansky and Pavlovsky 1966).

The presumed remainder of the primitive species, *D. willistoni*, is apparently unified in a great part of its immense territory, with certain exceptions. For example, it was found that females of *D. willistoni* from Lima, Peru, produced F_1 sterile males when crossed with males of other populations (for review, see Winge 1971). In addition, a northern race from Florida, Mexico, and the Caribbean, with a high degree of isolation, when tested with populations from Brazil, was found by Winge (1971). In 1973, Ayala described the Lima race as *D. willistoni quechua*. In the same paper he also named a subspecies of *D. equinoxialis*, *D. e. caribbensis* from Puerto Rico, Hispaniola, and Costa Rica. Dobzhansky (1975) added a genetic study of male sterility and sexual behavior, finding evidences "that very weak, but statistically assured, preference of homogamic matings exists between *D. w. quechua* and both strains of *D. w. willistoni* with which it has been tested."

A detailed geographic distribution of *D. willistoni* siblings and of the semispecies of *D. paulistorum* was presented by Spassky et al. (1971) and updated by Dobzhansky and Powell (1975). The geographic distribution of *D. willistoni* is the largest among its siblings (Fig. 19.1.A, B) extending from Florida (Townsend 1952), Bahamas (Sturtevant 1921), to Montes Tordillos near the town of General Conesa, on Parallel 40, 300 km south of Buenos Aires, Argentina (Cordeiro and Townsend, 1955 expedition). Carson, Val, and Wheeler (1983) refer to the occurrence of *D. willistoni quechua* in a 1977 collection from three major Galapagos Islands. The distribution of *D. paulistorum* is included within the range of *D willistoni*, and the territories of the other species are reduced in the order: *D. equinoxialis* >*D. tropicalis* = *D. cubana* > *D. insularis* (Fig. 19.1.A, B).

Chromosome inversion polymorphism is an important class of genetic variability because of its adaptive role. Consequently, it would be expected that inversion polymorphism plays some role in speciation. The evidence shows that *D. willistoni* inversion polymorphism is very high in central Brazil, averaging 9.36 +/−0.26 heterozygous inversions per individual female, in Monjolinho, Goyaz, Central Brazil; and decreasing in all directions toward the edges of its distribution (Fig. 19.1.A and B) as in: Florida, USA; Pitanga, Bahia, and Tainhas, Rio Grande do Sul, Brazil; and Delta de la Plata; Argentina, with 1.2–1.9 (da Cunha and Dobzhansky 1954; Townsend 1952, 1958). *Drosophila willistoni* populations of the highland region of Tainhas, at 100 km north of Porto Alegre, Rio Grande do Sul, and the populations of the central valley 30 km south of that city, studied

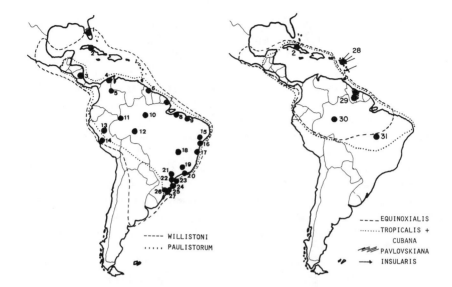

----- WILLISTONI
..... PAULISTORUM

----- EQUINOXIALIS
.... TROPICALIS +
          CUBANA
≈≈≈≈ PAVLOVSKIANA
  ⟶ INSULARIS

FIGURE 19.1.A.

Geographic distribution and the collecting sites of *Drosophila willistoni* and *D. paulistorum*. Population samples used in this work: (1) Placid Lake, Florida, USA; (2) Cuba; (3) Costa Rica; (4) Santa Marta, (5) Bucaramanga, Colombia; (6) Georgetown, Guiana; (7) Serra do Navio, (8) Belem, Pará, (9) Sacaven, Maranhão, (10) Manaus, (11) Tabatinga, (12) Porto Velho, Amazonas, Brazil; (13) Tingo Maria, (14) Lima, Peru; (15) Cassarongongo, (16) Pitanga, (17) Pedras de Una, Bahia, (18) Brasília, (19) Itatiaia, (20) Angra dos Reis, Rio de Janeiro, (21) Paranaí and Apiaí, (22) Praia do Leste, (23) Ilha das Cobras, Paraná, (24) Florianopolis + Ilha São Francisco + Tubarão, Santa Catarina, (25) Itapeva + São Pedro, (26) Taínhas, (27) Capão da Canoa, Rio Grande do Sul, Brazil.

FIGURE 19.1.B.

Geographic distribution and the collecting sites of *Drosophila willistoni* siblings. Populations samples used in this work: (2) Cuba; (28) St. Kitts; (29) Apoteri, Guiana; (30) Tefé, Amazonas, (31) Palma, Tocantins, Brazil.

by Cordeiro, Salzano, and Marques (1960) and Cordeiro (1961), were shown to be nearly devoid of inversions in the X-chromosomes, confirming that the southern, as well as the northern (Townsend 1952), marginal regions are ecologically stringent for this species. This paper gives some evidence that these marginal populations produce more interspecific experimental hybrids with *D. paulistorum* than the central ones do. Yet, even in these regions, individuals completely devoid of inversions rarely occur, and some new "southern" inversions have been found by Valente and Morales (1985) and Valente and Araújo (1986). The *D. willistoni* chromosomal polymorphism is also reduced in proportion to the degree of urbanization in the city of Porto Alegre, RS, Brazil, as reported by Valente, Ruszczyk, and Santos (1993). Inversion frequencies of IIL changed more than the IIR and III chromosomes, which are relatively insensitive to environmental changes. More surprising is that *D. paulistorum*, formerly limited to 100 km east of Porto Alegre, has invaded this city. The city population was found to contain 18 inversions, distributed in all five arms, and this may represent the first step in the urbanization of this species.

Allelic substitutions measured from the allozyme variability allowed Ayala and Powell (1972) and Ayala et al. (1972) to calculate the genetic distances among the cryptic species of *D. willistoni*. With an average of 23 substitutions per 100 loci, these siblings are comparable to many other good species in this respect. These extensive studies by Ayala (1975) culminated with the construction of a very interesting dendrogram of the phylogenetic relationships of the seven species, their semispecies and subspecies, based in 36 enzyme coding genes. This dendrogram coincides with the proposed phylogenetic tree for the siblings of *D. willistoni*, based on the many different traits used by Winge (1971).

Experimental Hybrids Between
Drosophila willistoni and *D. paulistorum*

Materials and Methods

Samples of these species from several regions of Brazil were studied by the usual method of crossings without choice with Virgin females of *D. paulistorum*, strains #12 and #14, of São Pedro (locality #25, fig. 19.1.A), isolated every four hours and aged four to five days, were mated to equally aged virgin males from each of 857 isofemale strains of *D. willistoni*, recently collected in 21 localities (table 19.1.; figure 19.1.A) of Brazil. Species identification was made by microscopic inspection of external male genitalia, spermathecae, and/or polytene chromosomes slides, as well as by crosses to "tester" strains of each species. Ten pairs of flies were transferred every three to four days to new vials with an agarless food (Marques et al. 1966). The reciprocal crosses: *D. w. willistoni* São Pedro, strain #18, females × males of *D. paulistorum* from 108 isofemales strains, recently collected in 8 Brazilian localities, (fig. 19.1.A) were performed as described above.

Results and Discussion

Tables 19.1 and 19.2 show that once sexual isolation is overcome, as signaled by the appearance of larvae (starting ten to twenty days after the crosses), more than half of the cultures with larvae produced adults. In the first experiment, 8570, and in the second, 1080, crossing pairs were replicated every third or fourth day for forty or more days. The production of hybrids was very low, females usually being fertile and males mostly sterile. In the first experiment the average number of adult F_1 was 3.2 and, in the

TABLE 19.1.

Results of Crosses Between Females of D. paulistorum *Tester Strain #12, São Pedro, Locality #25, Rio Grande do Sul, and* D. willistoni *Males (of Isofemale Strains), from 21 Localities in Brazil*

willistoni Localities #	Number of Strains	(1) Range Average	(2) Range Average
Rio Grande Sul #25a, b, 26, 27	201	0.04–0.61 0.25	0.45–1.0 0.70
Santa Catarina #23, 24a, b, c	150	0.10–0.50 0.26	0–0.88 0.43
Parana #21a, 21b, 22	172	0.06–0.20 0.13	0.57–1.0 0.86
Rio and Brasilia #18, 19, 20	125	0–0.12 0.06	0–1.0 0.33
Northeast Brazil #15, 16, 17	155	0–0.11 0.06	0–1.0 0.44
Amazonas #8, 10, 11, 12	54	0–0.43 0.13	0–1.0 0.42
All localities	857	0–0.61 0.11	0–1.0 0.63

(1) Localities, range of averages, and pooled average of crosses that produced larvae; (2) ranges and averages of cultures with larvae that produced adults. Localities are indicated in figure 19.1A.

TABLE 19.2.

Results of Crosses Between Males of Drosophila paulistorum *from Localities
in Brazil and Females of* D. willistoni *Tester Strain #18 from São Pedro,
Locality #25, Rio Grande do Sul, Brazil*

D. paulistorum Localities #	Number of Strains	(1) Range Average	(2) Range Average
Rio Grande Sul #25, 27	22	0.06–1.0 0.60	0–1.0 0.50
Santa Catarina #24a, b, c	35	0–0.67 0.38	0–1.0 0.63
Parana #22, 23	38	0.10–0.11 0.10	0–1.0 0.50
Rio de Janeiro #19, 20	13	0.25–1.0 0.63	1.0 1.0
All localities	108	0.00–1.0 0.44	0.00–1.0 0.75

(1) Localities' ranges, and averages of crosses that produced larvae; (2) ranges and averages of cultures with larvae that produced adults. Localities are indicated in figure 19.1.A.

second, 4.5 individuals per parental couple. The hybrid flies are frequently abnormal with taxi wings and many other anomalies.

Tables 19.1 and 19.2 show that the localities closer to the tester strains' origin, in Rio Grande do Sul, Santa Catarina, and Parana States (fig. 19.1.A) present a higher average of crosses producing larvae and adults. This appears contrary to the expected stronger sexual isolation of geographically closer populations. Yet, the crosses were forced for many days, under no choice of mating, reducing sexual isolation. However, as the other strains were submitted to the same conditions, the hypothesis that some marginal populations' strains have more facility to generate interspecific hybrids appears to be supported.

Another series of crossings, of males *D. paulistorum* S. Pedro strain #12 × females *D. willistoni* São Pedro, strain #18, (locality #25) were made. The flies were transferred every three or four days to fresh food vials for about sixty days. With that amount of pair × time, the 660 couples produced 150 hybrids and four times more larvae (table 19.3). In homospecific crosses, thousands of fertile flies were produced. See Winge and Cordeiro (1963) for the first account on those hybrids and more details on techniques. These experiments have been designed, not only to study reproductive isolation, but also to produce interspecific hybrids and study their chromosomes. For this reason, our results differ from those reported by other authors (see Ehrman and Powell 1982).

TABLE 19.3.

Interspecific Hybrid Production in Crosses Between Males of
D. paulistorum *Strain #12 and Females of* D. willistoni *Tester Strain #18*
from Locality #25 São Pedro, Rio Grande do Sul, Brazil

No. Couples per Culture	No. of Cultures	Total No. of Pairs	No. Cultures with F_1	Total No. Hybrids
20	20	400	8	46
30	6	180	6	69
40	2	80	2	35
Totals	28	660	16	150

Crossings Between Semispecies of *D. paulistorum* with *D. w. willistoni* and *D. w. quechua*

The same methods were used as in the previous experiments, except that each bottle received 40 interspecific couples, of virgin females and males. Most strains from outside Brazil have been sent to us by Dobzhansky (Fig. 19.1A, locals #: *D. willistoni:* Lima #14; *D. paulistorum:* Costa Rica #3, Santa Marta #4, Georgetown #6, Bucaramanga #5, Tingo Maria #13) and by J. I. Townsend (*D. willistoni:* Placid Lake #1, Florida; and Cuba #2). The strains from Brazil were collected by us (table 19.4; fig. 19.1.A).

Results and Discussion

Table 19.4 shows that fertile hybrid males and females were produced by the crosses between *D. paulistorum* Transitional, Santa Marta Colombia females × males *D. w. quechua* Lima, Peru, and *D. w. willistoni*, Pitanga, Bahia. Several other crosses produced fertile females but sterile males: *D. paulistorum* St. Marta males with *D. willistoni* females from Florida, Pitanga, and São Pedro, and *D. paulistorum* Centro American males from Costa Rica with *D. willistoni* females from Florida. The Orinocan *D. paulistorum* males also produced fertile females with *D. willistoni* females from Pitanga. As can be seen in table 19.4, *D. willistoni* Pitanga males produced fertile female with *D. paulistorum* females from São Pedro, the remaining crosses produced sterile hybrids or not a single adult fly, and two produced only pupae. The contributions of Dobzhansky (1941, 1951, 1970) on hybrid sterility remain classic.

TABLE 19.4.

Results of Interspecific Crosses Between Drosophila w. willistoni *and*
Drosophila w. quechua (*) *with Semispecies of* Drosophila paulistorum

paul will	CA #3	TR #4	OR #6	AM #8	AM #5	ABR #13	ABR #18	ABR #25
Florida	0	S	S	0	Ff	0	0	S
m: #1		1 f	1 f		5 f			1 f
Cuba	S	0	P	0	0	S	0	S
m: #2	1 m, 1 f		3			1 m		1 m, 2 f
Lima *	0	F	P	0	0	S	0	S
m: #14		3 m, 5 f	2			3 m, 1 f		3 m
Pitanga	0	F	0	0	S	0	S	Ff
m: #16		1 m, 2 f			1 f		1 m	1 m, 1 f
Brasilia	0	0	S	0	0	0	0	0
m: #18			3 m, 1 f					
São Pedro	0	0	S	0	0	S	S	S
m: #25			3 m, 8 f			3 m, 2 f	1 m	5 m, 7 f
Florida	Ff	Ff	0	S	0	0*	0	S
f: #1	118 m, 95 f	36 m, 40 f		4 m, 6 f				1 m
Cuba	0	0	S	0	0	0	0	0
f: #2		1 m						
Lima *	S	S	0	S	0	0	0	0
f: #14	26 m, 14 f	17 m, 10 m		1 f				
Pitanga	S	Ff	Ff	S	0	0	0	0
f: #16	12 m, 18 f	2 m, 12 f	1 m, 5 f	8 m, 9 f				
Brasilia	S	S	S	S	0	0	S	0
f: #18	1 f	2 m, 4 f	5 m, 5 f	3 m, 5 f			1 f	
São Pedro	0	Ff	S	S	0	0	0	Ff
f: #25		4 m, 5 f	2 m, 2 f	1 f				1 m, 2 f

CA=Centro American; TR=Transitional; OR=Orinocan; AM=Amazonian; ABR=Andean Brazilian. Localities numbers are indicated in figure 19.1.A.

F= F_1 fertile, Ff = only females fertile; S = sterile; P = pupae only; O = no F_1. Numbers of hybrid males: m and females: f, obtained from 96 combinations of 40 crossing pairs, cultured during forty days

It can be seen that marginal populations of *D. w. willistoni*, from Florida (USA); Pitanga; São Pedro (Brazil); and *D. w. quechua* are the ones that produced more hybrids, and this supports our observation about the relative ease of marginal populations to hybridize. Here again, cultures with many larvae usually produced adults, despite the apparent great mortality in several larvae instars. Hybrids are frequently phenotypically abnormal.

The progenies of the two fertile crosses, of females *D. paulistorum* Santa Marta with males *D. w. quechua* Lima and also with males *D. w. willistoni*, Pitanga, Bahia, are being cultivated until now in mass cultures for hundreds of generations. However, the hybrid nature of these cultures was apparently lost. The allozymes of *D. paulistorum* for, esterases, alkaline phosphatases, and leucine aminopeptidases, have been completely lost since the fifteenth generation. However, polytenic chromosomes maintained some signs of their hybrid nature for a longer period of time, and the chromosome sets gradually became better paired, losing the *D. paulistorum* pattern (figs. 19.2 and 19.3). It is apparent that *D. willistoni* is better adapted to culture bottles; this can be observed in culturing these species, in our standard conditions (25–27°C, in Marques et al. 1966 culture media). The possibility that hybrids can be "purified" by differential selection is an interesting point. This may be favoring *D. willistoni* to preserve its integrity. If any interspecific cross has occurred in nature, it has never been observed in our collections.

The first species shown to produce hybrids with the other siblings was *D. insularis* (Dobzhansky 1957; Dobzhansky, Ehrman, and Pavlovsky 1957). These authors found that females of this species mate more easily with *D. tropicalis* than with *D. willistoni* and the other siblings. No hybrids were obtained in reciprocal crosses. Winge (1965) confirmed these obser-

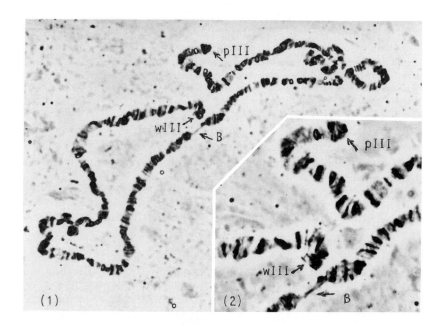

FIGURE 19.2

F_1 hybrid of male *D. paulistorum* Santa Marta × female *D. willistoni* Pitanga unpaired III chromosomes: P III and W III are the tips of these chromosomes: (1) at lower and (2) at greater enlargement. (B) the basal end.

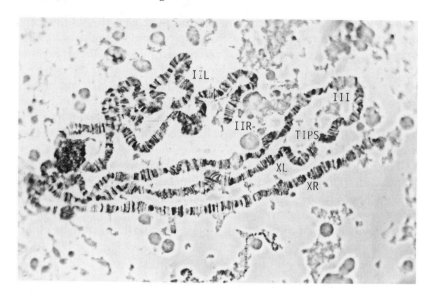

FIGURE 19.3

Hybrid at seventeenth generation of the cross: males *D. willistoni* Pitanga × female *D. paulistorum* Santa Marta. Notice the complete pairing (not observed in all cells) and the likeness to *D. willistoni* banding pattern.

vations and showed that an average of five hybrids per P_1 couple were produced. In several crosses the sex ratio was greatly disturbed. The F_1 males were sterile, their reproductive organs exhibiting gross morphological abnormalities.

Sexual and Reproductive Isolation Among the Siblings

Materials and Methods

Strains used (fig. 19.1.A, B): *D. willistoni*, São Pedro #25, RS, (Brazil), Lima #14 (Peru) (TD); *D. paulistorum*, São Pedro #25, RS, Florianopolis #24, SC (Brazil); *D. equinoxialis*, Tefé #30, AM, (Brazil); *D. tropicalis*, Palma #31, Tocantins, (Brazil); *D. cubana*, Cuba #2 (JIT); *D. pavlovskiana*, Apoteri #29 (Guiana) (TD); *D. insularis*, St. Kitts #28 (TD) (sent by: [TD], T. Dobzhansky, [JIT], J. I. Townsend). We followed the techniques of crossings without male choice, with 10 virgin females and 10 males per vial, at 24–26°C, 60–70% RH, and twelve hours day/night period. Making 15 to

20 replicate experiments, 150 to 200 mating pairs were examined for each cross. Ten days after crossing, the females were dissected and insemination scores obtained. For brevity, only the maximal indexes are reported here. The formulae of Levene (1949) or Merrell (1950) were used.

Figure 19.4 shows that the sexual affinity between *D. paulistorum* and *D. tropicalis* is highest, followed by the pairs: *D. cubana* males × females of *D. tropicalis*, *D. w. willistoni*, and *D. paulistorum*. The other pairs present higher sexual isolation. Figure 19.5 summarizes the results of our experiments on reproductive isolation. The pairs with higher reproductive affinity are: (males × females), *D. paulistorum* × *D. equinoxialis* and *D. paulistorum* × *D. willistoni*, followed by *D. paulistorum* × *D. pavlovskiana*. More technical details and descriptions of hybrid phenotypes were presented by Winge (1965), which include part of these data. There are some inverse correlations, namely, the species with higher sexual attraction are the ones with lower production of hybrids.

Degrees of Chromosome Pairing in Hybrids

Materials and Methods

The hybrid larvae obtained in experiments I and II were used for this work; besides that, the strains used in experiment II were crossed to produce more larvae for chromosome studies. Despite the great number of larvae, especially between *D. paulistorum* × *D. equinoxialis*, many were unsuitable for chromosome analysis. Slides are prepared by dissecting well-fed third-instar larvae in a Ringer's solution and quickly placing the salivary glands into a 3:1 solution of 98% ethanol and 60% acetic acid, for two to three minutes, and then to 2% orcein in 45% acetic acid. The salivary glands are then transferred to a glass slide with drops of 0.5% orcein + 0.1% Janus Green B, in 1:1 solution of 40% lactic acid and 60% acetic acid. A cover slip is applied and gently tapered with a small glass rod until the chromosomes are spread. The preparation is placed between filter paper pieces and squashed. The lactic acid increases chromosome spreading, and the Janus Green B, the durability and contrast. The preparation can be sealed and kept in refrigeration at +5°C, for several months.

Results and Discussion

Chromosome pairing is usually related to the degree of genetic affinity. However, each interspecific cross exhibited great variability in this respect.

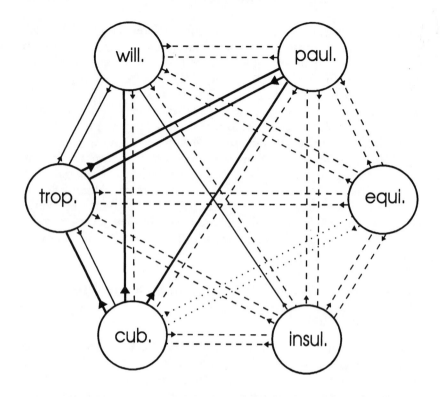

FIGURE 19.4

Maximal insemination frequencies between six species of the *D. willistoni* cryptic group: . . . .
0.0; ----- 0.1–25.0; —— 25.1–50.0; ▬▬ 50.1–99.0. The arrows show the direction of
crosses: male/female. Strains used are indicated in the text and in Figure 19.1.A, B.

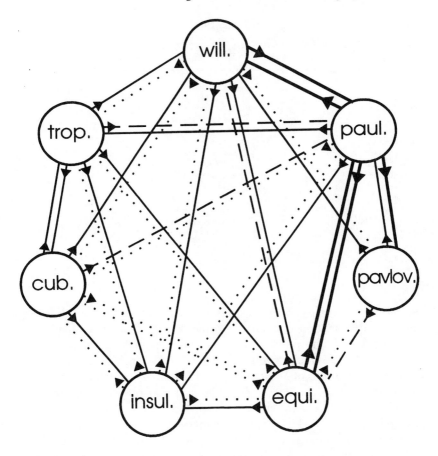

FIGURE 19.5

Maximal reproductive affinities between seven species of the *D. willistoni* cryptic group: . . . . no offsprings; ------ hybrid inviability; ——— sterile F_1 hybrids; ▬▬ fertile F_1 hybrids. The arrows show the direction of crosses: male/female. Strains used are indicated in the text and in Figure 19.1.A, B.

Even the same strains of two species that showed complete paired chromosome sets in one slide may present unpaired regions in other slides. Table 19.5 shows that *D. paulistorum* has the highest degree of chromosome pairing with the greatest number of species, while *D. willistoni* presents the least affinity. The pairwise crosses in table 19.5 are presented in the decreasing order of affinity, in each level. Pairing is absent or limited to few and very small segments in *D. paulistorum* × *D. tropicalis* and *D.*

TABLE 19.5.

Maximum Degree of Pairing Between Polytenic Chromosomes of Interspecific Hybrids

Maximum Degree	Hybrids
Level 1: Extensive pairing of 3, 4, or 5 chromosomes	*paulistorum* × *willistoni*
	paulistorum × *equinoxialis*
	paulistorum × *insularis*
	cubana × *tropicalis*
Level 2: Extensive pairing of 1 or 2 chromosomes	*paulistorum* × *pavlovskiana*
	paulistorum × *cubana*
	cubana × *insularis*
Level 3: Small segments paired consistently	*willistoni* × *equinoxialis*
	willistoni × *tropicalis*
	tropicalis × *insularis*
Level 4: Little paired segments, not consistent, or no pairing	*willistoni* × *cubana*
	willistoni × *tropicalis*
	willistoni × *insularis*
	paulistorum × *tropicalis*
	equinoxialis × *tropicalis*

NOTE: The origin of strains is indicated in the text and in figure 19.1.A, B.

equinoxialis × *D. tropicalis.* These observations over hundreds of slides are limited to the mentioned strains. Dobzhansky (1957), studying hybrids of *D. insularis* with the other siblings, found the polytene chromosomes unpaired. However, fairly good chromosome pairings occur in crosses between some strains of *D. paulistorum* and *D. cubana*, as table 19.5 shows. Yet, other crosses confirm Dobzhansky's (1957) observations. In male hybrid larvae, the X-chromosome with its greatly enlarged width is reported by Dobzhansky (1957) who wrote: "this is a microscopically visible counterpart of the genetic phenomenon of dosage compensation."

We would like to suggest that this excess of enlargement may be the sign of an overdosage compensation that could be related to male sterility and inviability. This phenomenon occurs in most if not all male hybrids of these cryptic species (fig. 19.6). Most hybrids also show an anomalous enlarged nucleolus (Cordeiro 1968). The unexpected observation that the hybrid chromosome sets, when extensively or completely paired, show few or no inversions may have a simple but interesting interpretation. It is conceivable that the homosequential hybrid larvae are more viable and develop to the late third instar, bearing salivary glands good enough to be studied. We are tempted to suggest that such flies, with homosequential genomes, are relics of primitive populations.

This sibling group is in an active process of speciation, and all levels of this process can be observed. The first step, of incipient isolation, is repre-

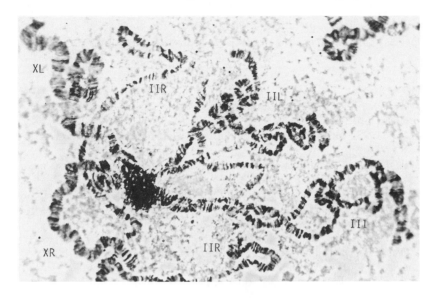

FIGURE 19.6

Male larvae chromosome set of F_1 hybrid between males *D. cubana* Trinidad × females *D. tropicalis* Falma. The III chromosomes are partially paired with the tips unpaired as usual. The chromosomes XL XR with greatly enlarged width. The IIL and IIR are unpaired.

sented by subspecies like *D. w. quechua* and *D. e. caribbensis* (Ayala 1973), the northern race of *D. willistoni* (Winge 1971), the proposed geographic races of *D. cubana* (Winge 1965), and subspecies of *D. tropicalis* (Townsend 1963). As Lewontin wrote (1974, ch. 4, p. 183) referring to the *D. paulistorum* cluster:

> If the semispecies can be regarded as a model for the second step in speciation, which, in the past, led to the formation of the present-day sibling species, then we see that most of the differentiation of the sibling species has occurred since speciation was completed. The differences we observe among the sibling species of the *willistoni* group are best explained as the result of phyletic evolution after successful speciation, whereas the speciation process itself resulted in very little differentiation.

We conclude remembering that, even in the third level, the sibling species of *D. willistoni* shows several levels of differentiation.

REFERENCES

Ayala, F. J. 1973. Two new subspecies of *Drosophila willistoni* Group. *Pan-Pacific Entomol.* 49:273–79.

———. 1975. Genetic differentiation during speciation process. *Evol. Biol.* 8:1–77.

Ayala, F. J. and J. R. Powell. 1972. Allozymes as diagnostic characters of sibling species of *Drosophila. Proc. Natl. Acad. Sci.* 69:1094–96.

Ayala, F. J., J. R. Tracey, C. Mourao, and S. Perez-Salas. 1972. Enzyme variability in *Drosophila willistoni* Group, IV: Genetic variation in natural populations of *Drosophila willistoni. Genetics* 70:113–39.

Burla, H., A. B. da Cunha, A. R. Cordeiro, T. Dobzhansky, and C. Pavan. 1949. The *willistoni* group of sibling species of *Drosophila. Evolution* 4:300–14.

Carson, H. L., F. C. Val, and M. R. Wheeler. 1983. *Drosophilidae* of the Galapagos Islands, with descriptions of two new species. *Int. J. Entomol.* 25:239–48.

Cordeiro, A. R. 1952. Inheritance of variations in the male genitalia of *Drosophila paulistorum. Am. Natur.* 86:185–88.

———. 1961. Chromosomal polymorphism decrease due to *gamma*-radiation on natural populations of *Drosophila willistoni. Experientia* 17:405.

———. 1968. Chromosomal pairing variability of interspecific hybrids of *Drosophila willistoni* Cryptic Group. *Proc. Int. Congr. Genetics, Tokyo* 1:191.

Cordeiro, A. R., F. M. Salzano, and V. B. Marques. 1960. An interracial hybridization experiment in natural populations of *Drosophila willistoni. Heredity* 15:35–45.

Da Cunha, A. B. and T. Dobzhansky. 1954. A further study of chromosomal polymorphism of *Drosophila willistoni* in its relation to the environment. *Evolution* 8:119–34.

Daniels, S. B., K. R. Peterson, L. D. Strausbaugh, M. G. Kidwell, and A. Chovnick. 1990. Evidence for horizontal transmission of the **P** transposable element between *Drosophila* species. *Genetics* 124:339–55.

Dobzhansky, Th. 1941. *Genetics and the Origin of Species.* 2d ed. New York: Columbia University Press.

———. 1946. Complete reproductive isolation between two morphologically similar species of *Drosophila. Ecology* 27:205–11.

———. 1951. *Genetics and the Origin of Species.* 3d ed. New York: Columbia University Press.

———. 1957. Genetics of Natural Populations. XXVI: Chromosomal variability in islands and continental populations of *Drosophila willistoni* from Central America and West Indies. *Evolution* 11:280–93.

———. 1970. *Genetics of the Evolutionary Process*. New York: Columbia University Press.

———. 1975. Analysis of incipient reproductive isolation within a species of *Drosophila*. *Proc. Natl. Acad. Sci.* 72:3638–41.

Dobzhansky, Th., L. Ehrman, and O. Pavlovsky. 1957. Genetics of *Drosophila*. *IX: Drosophila insularis*, a new sibling species of *willistoni* Group. University Texas Publ. 5721:39–47.

Dobzhansky, Th. and E. Mayr. 1944. Experiments on sexual isolation in *Drosophila*. I: Geographic strains of *Drosophila willistoni*. *Proc. Natl. Acad. Sci.* 30:238–44.

Dobzhansky, Th., L. Ehrman, O. Pavlovsky, and B. Spassky. 1964. The superspecies of *Drosophila paulistorum*. *Proc. Natl. Acad. Sci.* 51:3–9.

Dobzhansky, Th. and C. Pavan. 1943. Studies on Brazilian species of *Drosophila*. *Bol. Faculd. Fil. Cien. Letr. Univ. S. Paulo* 36:7–72.

Dobzhansky, Th. and O. Pavlovsky. 1966. Spontaneous origin of incipient species in the *Drosophila paulistorum* complex. *Proc. Natl. Acad. Sci.* 55:727–33.

Dobzhansky, Th. and J. R. Powell. 1975. The *willistoni* group of the sibling species of *Drosophila*. In R. C. King, ed., *Handbook of Genetics* 3:589–622. New York: Plenum.

Ehrman, L. and J. R. Powell. 1982. The *Drosophila willistoni* species group. In M. Ashburner, H. L. Carson, and J. N. Thompson Jr., eds., *The Genetics and Biology of Drosophila*. vol. 3b, pp. 193–225. New York: Academic Press.

Ehrman, L., N. L. Somerson, and J. P. Kocka. 1990. Induced hybrid sterility by injection of streptococcal L-forms into *Drosophila paulistorum*: Dynamics of infection. *Can. J. Zool.* 68:1735–40.

Kastritsis, C. D. and Th. Dobzhansky. 1967. *Drosophila pavlovskiana*, a race or a species? *Am. Midl. Natur.* 78:244–48.

Levene, H. 1949. A new measure of sexual isolation. *Evolution* 3:315–21.

Lewontin, R. C. 1974. *The Genetic Basis of Evolutionary Change*. New York: Columbia University Press.

Marques, E. K., M. Napp, H. Winge, and A. R. Cordeiro. 1966. A cornmeal, soybean flour, wheat germ medium for *Drosophila*. *Drosophila Info. Serv.* 41:187.

Merrell, D. J. 1950. Measurement of sexual isolation and selective mating. *Evolution* 4:326–31.

Spassky, B. 1957. Morphological differences between sibling species of *Drosophila*. University of Texas Publ. 5721:48–61.

Spassky, B., R. C. Richmond, S. Perez-Salas, O. Pavlovsky, C. A. Mourao, A. S. Hunter, H. Hoenigsberg, T. Dobzhansky, and F. J. Ayala. 1971. Geography of sibling *willistoni* species and of the semispecies of the *Drosophila paulistorum* complex. *Evolution* 25:129–43.

Sturtevant, A. H. 1921. The North American species of *Drosophila*. Carnegie Inst. Wash. Publ. 301:1–150.

Townsend, J. I. 1952. Genetics of marginal populations of *Drosophila willistoni*. *Evolution* 6:428–42.

———. 1954. Cryptic subspeciation in *Drosophila* belonging to the subgenus *Sophophora*. *Am. Natur.* 88:339–51.

———. 1958. Chromosomal polymorphism in Caribbian Island population of *Drosophila willistoni*, *Proc. Natl. Acad. Sci.* 44:38–42.

———. 1963. A preliminary investigation of complex incipient speciation in *Drosophila tropicalis*. *Va. J. Sci.* 14:212.

Valente, V. L. S., A. M. Araújo. 1986. Chromosomal polymorphism, climatic factors and variation in population size of *Drosophila willistoni* in Southern Brazil. *Heredity* 57:149–59.

Valente, V. L. S. and N. B. Morales. 1985. New inversions and qualitative description of inversion heterozygotes in natural populations of *Drosophila willistoni* inhabiting two different regions in the State of Rio Grande do Sul Brazil. *Rev. Brasil Genet.* 8:167–73.

Valente, V. L. S., A. Ruszczyk, and R. Santos. 1993. Chromosomal polymorphism in urban *Drosophila willistoni*. *Rev. Brasil Genet.* 16:10–23.

Winge, H. 1965. Interspecific hybridization between the six cryptic species of *Drosophila willistoni* Group. *Heredity* 20:9–19.

———. 1971. Niveis de divergencia evolutiva no grupo criptico da *Drosophila willistoni*. Thesis, 371 pp., univ. Federal do Rio Grande do Sul, Porto Alegre, RS Brazil.

Winge, H. and A. R. Cordeiro. 1963. Experimental hybrids between *Drosophila willistoni* Sturtevant and *Drosophila paulistorum* Dobzhansky and Pavan from southern marginal populations. *Heredity* 18:215–22.

20

The Founder Effect in Speciation: *Drosophila pseudoobscura* as a Model Case

Agustí Galiana, Andrés Moya, and Francisco J. Ayala

> *As the situation appears to be now, there is one consideration which seems unlikely to be changed: there is not a single kind but there are several kinds of species and of processes of speciation in Drosophila and, of course, even more in the living world at large.*
>
> —Theodosius Dobzhansky (1972)

Dobzhansky's Ideas on Speciation

No other author has contributed more to understanding the process of speciation from a genetic perspective than Theodosius Dobzhansky. He was the first to articulate the biological species concept, he postulated a general mechanism and explored alternative models of speciation, he was the first to perform experiments for the genetic analysis of reproductive isolating mechanisms (RIMs) between species, and he even tried to produce new species experimentally by selecting for reproductive isolation between laboratory populations.

Dobzhansky proposed a general model of speciation essentially consisting of two main stages. First, genetic divergence occurs between geographically separated populations as a consequence of their adaptation to local conditions, as well as of drift. There is no direct selection for reproductive isolation during this first stage, but postmating RIMs may arise as a consequence of genetic divergence. The second stage occurs when genetically diverged populations come into geographic contact where the opportunity for intermating exists. Natural selection against the hybrids directly promotes the evolution of premating RIMs and, thus, the completion of species differentiation. This second stage, often called "reinforcement" after Blair (1955), is identical to the "Wallace effect" (Wallace 1889). The reality of this second stage has been challenged by some biologists as theoretically unlikely (Darwin 1872, 1903; Templeton 1981; Patterson 1982; Butlin 1989). Nevertheless, Dobzhansky's two-stage model of speciation counts with much evidential support and is still the most generally accepted model (Ayala 1991).

Dobzhansky (1940) explored two difficulties encountered by his model of speciation by reinforcement. First, how would barriers built in the sympatric area extend to the whole species range? Second, how would reproductive isolation evolve in species inhabiting different oceanic islands or otherwise completely separate populations so that no hybridization between incipient species can ever take place? One answer to this last question is that reproductive isolation may eventually become complete because "physiological" incompatibility will arise as a by-product of adaptation to the distinct local ecological conditions of the island, whenever the geographic isolation would long persist. In other words, "reinforcement" might not always be part of the process.

Carson (1971, see also Carson 1975, 1978, 1987; Carson and Templeton 1984) proposed a different explanation for speciation in oceanic islands, based on his investigations of *Drosophila* endemic in the Hawaiian archipelago. Carson's model assumes that colonization is often by only one pair or very few individuals and that population expansion occurs unhindered by competition for limited resources and, hence, without natural selection eliminating poorly adapted genotypes. Selection comes eventually into play when the environmental resources become saturated and yields genotypes quite divergent from those of the founders, owing to extensive opportunities for recombination during the noncompetitive part of the process. Genetic drift plays a major role in Carson's speciation model, not only because of the vagaries of the colonization process ("founders"), but also because the rapid expansion of the population ("flush") is not regulated by selection promoting adaptation to local conditions.

Dobzhansky (1972) noted that this relaxation of natural selection in Carson's model was free of any adaptive purpose, but he was quite receptive to the possibility of this mode (as well as other) of speciation mechanisms. He wrote, for instance:

In several brilliantly argued contributions, Carson advances a novel and unorthodox view—speciation may occur rapidly, and a neospecies of *Drosophila* may, without prior adaptive divergence, emerge within relatively few generations. The idea stems from

Mayr's founder principle, which is in turn a special case of Wright's random genetic drift.

This is a radical departure from the orthodox view. What is the biological function of speciation? The most reasonable interpretation seemed to be that speciation makes the adaptive divergence of evolving populations irreversible. . . . According to Carson's scheme, reproductive isolation and speciation precede differential adaptedness. If so, speciation would seem devoid of biological function, until a differential adaptedness arises following the speciation.

(Dobzhansky 1972)

Founder-Flush Speciation

As Carson (1971) himself suggested, his founder-effect speciation model can be tested in the laboratory, by means of repeated colonizations of cultures by a few individuals. This has been done at least in four independent experiments, those of Powell (1989), Ringo et al. (1985), Meffert and Bryant (1991), and by us (Galiana, Moya, and Ayala 1993).

There are other colonization experiments, for instance, those carried out with isofemale lines, among which those of Sperlich, Karlik, and Pfriem (1982) and Terzian and Biémont (1988) are most interesting. Dobzhansky himself tested the founder effect with populations of *Drosophila pseudoobscura* (Dobzhansky and Pavlovsky 1953), observing "a greater variety of outcomes than in comparable populations descended from numerous founders," and also an increase in intrapopulation variability.

Carson derived his model from the study of Hawaiian *Drosophila* (Carson et al. 1970), and all founder-flush-crash experiments have used flies as materials: *D. pseudoobscura* (Powell 1989; Galiana, Ayala, and Moya, 1989; Galiana, Moya, and Ayala 1993), *D. simulans* (Ringo et al. 1985), and *Musca domestica* (Meffert and Bryant 1991).

Test of Founder-Flush Theory

We have described earlier our experimental design and reported our extensive results comprising 7 founder-flush-crash cycles (Galiana, Moya, and Ayala 1993). We now extend our results to a total of 13 cycles spanning seven years of an experiment with several score laboratory populations and hundreds of assortative mating tests between them. The tests performed during the 6 additional cycles included in the present paper show that no case of assortative mating persists over time.

The laboratory populations are derived from 2 natural populations of *Drosophila pseudoobscura*, collected in Bryce Canyon National Park, Utah (abbreviated as BC), and Lake Zirahuén, Mexico (M). The experiment consists of 45 populations subject to the full experimental protocol but with different numbers of founders ("Flush-Crash" populations) and 14 control populations, consisting of 2 populations maintained by mass culture ("Ancestral" populations) and two sets of 6 populations each, subject to two levels of inbreeding ("Prima" and "Endogamic" populations). These populations are briefly described here. For additional details, see Galiana, Moya, and Ayala (1993).

Ancestral Populations

Two ancestral populations of *Drosophila pseudoobscura*, BCA, from Bryce Canyon National Park, Utah, and MA, from Lake Zirahuén, México, were established in June 1984 (from strains generously provided by Professor Wyatt W. Anderson). The Bryce Canyon strains were chromosomally monomorphic; the Lake Zirahuén strains were extremely polymorphic for inversion arrangements in the third chromosome.

The BCA and MA populations were established by combining a number of isofemale lines and mass cultures and maintained throughout the experiment by serial transfer: every week the adults recovered from four culture bottles were placed in a new culture bottle and the oldest one in the series was discarded. The minimum number of adult flies in a population was about 1000 individuals.

Founder-Flush Populations

In December 1984, 27 populations were derived from BC and 18 from M (see table 20.1; slightly modified from Galiana, Moya, and Ayala 1993, table 1). These populations were maintained according to a founder-flush-crash protocol. Each population was founded with n virgin pairs, with $n = 1, 3, 5, 7,$ or 9. These populations were first allowed to grow "exponentially" for a few generations (flush phase) as follows. The n pairs laid eggs for a week in one culture bottle with standard *Drosophila* medium and for another week in a second bottle. Fifty progeny flies were collected from each of these two bottles and evenly distributed among five new culture

TABLE 20.1.

The 59 Experimental Populations of Drosophila pseudoobscura;
n *Is the Number of Pairs in Each Bottleneck*

n	BC	m	N_e
Founder-flush-crash (derived) populations			
1	BC2, BC4, BC32, BC33	M3, M5, M37,	9.79
3	BC7, BC9, BC10, BC11, BC12, BC34	M7, M8, M10, M11, M12, M36	29.06
5	BC13, BC14, BC15, BC17, BC18	M13, M14, M15	47.45
7	BC19, BC20, BC21, BC22, BC23, BC24	M19, M20, M21	66.66
9	BC25, BC26, BC27, BC28, BC29, BC30	M25, M26, M27	84.56
Prima populations			
1	BC2', BC4', BC32'	M3', M5', M37'	3.99
Endogamic populations			
1	BAC1, BAC2, BAC3	MAC1, MAC2, MAC3	2.75
Ancestral populations			
	BCA	MA	~100

NOTE: N_e is the estimated effective population size. BC stands for Bryce Canyon (Utah) and M for Mexico (Lake Zirahuén).

bottles, which thus had 20 flies each. After one week the 20 flies in each bottle were transferred to another bottle in a second set of five cultures, so that each population consisted of ten cultures each started with 10 pairs of adult flies from the previous generation. The progenies (F_2 generation) were again distributed among five new bottles but in a particular way: 2 flies were sampled from each of the ten previous bottles of each set and these 20 flies were placed in a fresh culture. This was repeated five times, so that once again we had five cultures each with 20 flies that were transferred to five new bottles after one week. Additional generations of flush were prepared in a like manner. The density of 10 pairs of flies per culture was maintained throughout the flush phase, so that competition for limited resources would be slight or absent, as required by Carson's model.

After a certain number of flush generations (ranging from 4 to 7 in different cycles, 6 on average), a crash was induced as follows. All flies (usually several hundred) emerging in any 2 nonconsecutive bottles (i.e., in two bottles that had different parents) were combined into a single culture bottle, left there for a week, and then transferred to a second bottle for another week. This was done for all 10 bottles, so that all progeny flies from the last flush generation (typically, several thousand) were collected to become the parents of the crash generation. From the emerging progenies of each crash-generation culture, n virgin pairs were randomly chosen in order to start another founder-flush-crash cycle. The value of n was kept constant throughout all cycles for any particular population.

Thus, each cycle starts with a founder event, where the number of founder pairs, n, ranges 1–9. During each of the following flush generations, only 10 pairs of adult egg-laying flies are present in any one culture, so that competition for food and other resources is slight or absent. This allows recombinant genotypes, even some with low fitness, to survive along with nonrecombinants, so that they may in turn generate new recombinant genotypes through the various flush generations. In the crash generation, resource competition is fierce, since several hundred parental flies are placed together in each culture bottle. This population crash is an essential feature of the model, since Carson postulates that it is the crash phase that selects new adaptive genotypes.

Prima Populations

The 6 "prima" populations were subjected to founder-flush-crash cycles similar to those used for the derived populations, except that the crash bottleneck consisted of only 1 pair and lasted 3 generations rather than 1. All flush cycles following the first one were accordingly 2 generations shorter in the "prima" than in all other populations.

Endogamic Populations

The 6 "endogamic" populations were derived 3 from each ancestral population, BCA and MA, and each kept for 8 generations by brother x sister mating (1 pair per generation) and thereafter by mass serial transfer.

Assortative Mating Tests

We performed multiple-choice mating tests by placing together in a mating chamber 48 flies (12 virgin pairs from each of 2 populations) and recording the matings by visual observation for forty-five minutes. The virgin flies were kept in small vials with food for five to six days before being introduced into the mating chambers. At least four replications were done for each test.

The mating choice data were analyzed by means of various assortative mating indices that give, on the whole, similar outcomes. We report here the results obtained by using the statistic Y (Ringo 1987) defined by

$$Y = \frac{\sqrt{(AD/BC)} - 1}{\sqrt{(AD/BC)} + 1}$$

where A and D are the numbers of homogamic matings and B and C the numbers of heterogamic matings. The significance of Y is tested by the statistic $\chi^2(Y)$, which is chi square distributed with one degree of freedom (Fienberg 1977):

$$\chi^2 (Y) = \frac{(\ln AD/BC)^2}{(1/A) + (1/B) + (1/C) + (1/D)}$$

The index Y is based on the cross-product ratio AD/BC and is a margin-free association measure (Bishop, Fienberg, and Holland 1975); i.e., it reflects only assortative mating and not differences in mating propensities. Y can range from -1 to $+1$, so that negative and positive values stand for negative and positive assortative mating, respectively, and zero for random mating. (Neither B nor C was zero in any of our experiments, which would have made it necessary to give them arbitrary low numbers to avoid zero in the denominator.)

Experimental Results

The large scale of the experiment we are summarizing is apparent by noting that 59 populations were maintained for some 80 generations (seven years) by use of a complex protocol calling for elaborate handling of the populations. Assortative mating tests between pairs of populations were performed throughout the experiment to ascertain the development of sexual isolation through time. A total of 678 assortative mating tests were done, each replicated four or more times (i.e., each test involving matings between 200 virgin flies, approximately).

The results are summarized in figure 20.1, where the black component of the histograms represents the assortative mating tests yielding significant ($p < 0.05$) Y values. There is a modest number of significant cases: 49 (10.1%) of 486 tests between founder-flush populations, and another 10 (7.0%) of 142 between a founder-flush and an ancestral population (table 20.2). That this result cannot be fully accounted for as a chance outcome is

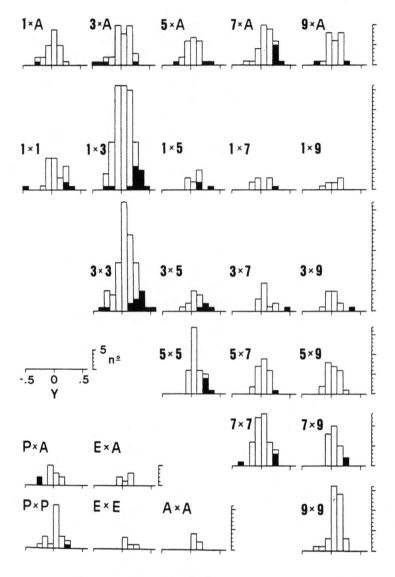

FIGURE 20.1.

Histograms showing the number of tests (ordinate) that give particular Y values (abscissa). The numbers refer to the numbers of founders (1, 3, 5, 7, or 9) in the derived (founder-flush-crash) populations. A = ancestral, P = prima, E = endogamic populations. Statistically significant tests ($p < 0.05$) are shown in black.

TABLE 20.2.

Summary of the Results for the Various Cycles

Cycle	Mating Combination	Number	Positive Cases (%)	Negative Cases (%)
4	A × D	20	0 (0.0)	1 (5.0)
	D × D	45	7 (15.6)	0 (0.0)
5	A × D	38	0 (0.0)	2 (5.3)
	D × D	171	19 (11.1)	3 (1.8)
7	A × D	60	8 (13.3)	2 (3.3)
	D × D	165	10 (6.4)	1 (0.6)
11	A × D	13	0 (0.0)	0 (0.0)
	D × D	66	7 (10.6)	1 (1.5)
12	A × D	8	1 (12.5)	1 (12.5)
	D × D	38	6 (15.9)	0 (0.0)
13	A × D	3	1 (33.3)	0 (0.0)
	D × D	12	0 (0.0)	1 (6.7)
Total	A × D	142	10 (7.0)	6 (4.2)
	D × D	486	49 (10.1)	6 (1.2)
Grand total		628	59 (9.4)	12 (1.9)
Control crosses				
	A × A	6	0 (0.0)	0 (0.0)
	A × E	6	0 (0.0)	0 (0.0)
	A × P	6	0 (0.0)	2 (33.3)
	E × E	6	0 (0.0)	0 (0.0)
	P × P	21	1 (4.7)	0 (0.0)
Total		50	1 (2.0)	2 (4.0)

NOTE: The number of significant cases ($p < 0.05$) of assortative mating are shown. The symbols for the populations are D=Derived (or founder-flush-crash), A=Ancestral, E=Endogamic, and P=Prima.

apparent by three considerations. First is that the number of significant negative Y values is only 12 (1.9%) of the 628 tests between founder-flush populations or between them and the ancestral populations. Second is that there is only 1 case (2.0%) of positive assortative mating between control populations (and 2 of negative assortative mating). Third is that the proportion (10.1%) of positive mating tests that are significant among founder-flush populations is greater than the 5 percent that would be expected by chance; moreover, two positive Y values are significant even by the very severe Bonferroni sequential test (see Galiana, Moya, and Ayala 1993).

Enthusiasm for the success of the founder-flush-crash protocol in generating divergent mating preferences must, however, be tempered. The incidence of significant cases is low, as noted. Moreover, this incidence does not increase over time in spite of the increasing number of founder-flush-crash cycles. As shown in table 20.2, the largest numbers of tests were made after cycles 5, 7, and 11; the corresponding percents of positive significant tests are 11.1, 6.4, and 10.6, roughly similar to each other and to the 10.1% of cases observed for all cycles combined. The lack of progress through time would not seem attributable simply to loss of genetic variance, since there is no discernable pattern that would indicate less differentiation among populations with fewer founders. (Indeed, the opposite appears to be the case; the percent of positive assortative mating combinations are 10.7, 13.8, 14.3, 7.1, and 0.0 for populations with $n=1, 3, 5, 7$, and 9 respectively.)

The conclusion that the divergence between populations does not increase with the number of cycles is true for particular populations and not only for the averages. Table 20.3 displays the Y values between pairs of populations that exhibited significant positive assortment in at least one test and, therefore, were selected for additional testing. There is no evidence that mating preferences persist, let alone increase, through time. One particular combination that seemed promising is the pair M3 × BC7. These 2 populations exhibited significant positive assortative mating in three successive tests and thus had been highlighted by Galiana, Moya, and Ayala (1993). However, none of the three more recent tests (after 11, 12, and 13 cycles) is statistically significant (see table 20.3).

TABLE 20.3.

Value of Y *(Index of Sexual Isolation) for Mating Combinations That Yield at Least One Case of Significant Assortative Mating*

Type	Mating	4	5	7	11	12	13
1× A	BC2 × MA	−.278*	.023	−.030			
3 × A	BC7 × BCA	.042	−.238*	.040	.084		
	BC12 × BCA		−.307*				
	M7 × MA	−.017	−.081	.213*	−.091	−.451*	
5 × A	BC18 × BCA			−.232*			
	BC18 × MA			.219*		−.072	.361***
7 × A	BC21 × BCA			.272*		.209	.108
	BC21 × MA			.271*		.319*	.029
	BC22 × BCA			.287**	.072	−.037	
	BC22 × MA			.280**	.069	.182	
	M20 × BCA			.222*		.049	
9 × A	BC26 × MA			.218*		.119	
	M25 × MA	.121		−.237	−.044		
1 × 1	BC32 × BC2		.293*			−.029	−.500*
	BC33 × BC32		.390***			.068	
	M5 × M3		.277*			.215	

(continued)

TABLE 20.3. (*continued*)

Type	Mating	4	5	7	11	12	13
1 × 3	BC4 × M8		.305*			−.095	
	BC32 × M7		.324***		−.031	.136	
	BC33 × M11		−.256*				
	M3 × BC7	.293*	.363***	.161*	.103	−.014	.124
	M3 × BC9		.280*			.065	.000
	M3 × BC10		−.003		.301***	.050	
	M3 × BC11		.412*			.024	
	M3 × M7	.284*	.011	−.078	−.157*	−.011	
	M3 × M11		.335**			.134	
	M5 × BC7		.235*			.209*	
	M5 × M11		.238*				
1 × 5	BC32 × M14				.193*	.137	
	M3 × BC13	.330**				−.084	
	M3 × M14				.166*	.002	
1 × 7	M3 × BC19	.252*				.000	
3 × 3	BC12 × BC11		.238*			−.009	.101
	BC34 × BC7		.284*			.352***	.081
	BC34 × BC9		−.244*				
	M7 × BC7	.088	−.068	−.184	.156*	−.057	.209
	M7 × BC10		.527***		.050	.225	.029
	M8 × BC9		.314*			−.050	
	M10 × BC7		.351***			.451*	
	M11 × M10		−.307*				
	M12 × BC7		.323**		.098	.197*	
	M12 × BC12		.374***			.229	
	M36 × BC12		.288*				−.268
3 × 5	BC7 × M14				.201*	.224*	.139
	M7 × BC13	.329**					
	M7 × M14				.173*	.042	
3 × 7	BC7 × M19	.414***			−.028	.268	.220
3 × 9	M7 × BC25	.352***				−.046	.129
5 × 5	M13 × BC13			.250*		.029	
	M14 × BC13			.277*		.225	
	M14 × BC14			.218*	.005	.159	
	M15 × BC14			.357***		.065	
	M14 × BC15			.296*		.027	
5 × 7	M14 × M19			.282*	.000	−.084	
7 × 7	BC21 × BC19			−.372***		.214*	
	BC22 × BC19			.263*		.046	
	BC22 × BC20			.256*		−.113	.000
9 × 7	BC26 × BC20			.275*			
	BC30 × M19				.212*	−.071	
P × P	M3′ × BC32′			.243*		−.113	

* *p*<.05   ** *p*<.01   ***p<.001.
NOTE: The numbers on top (4, 5, 7, 11, 12, 13) refer to the number of founder-flush-crash cycles after which the tests were made.

Discussion and Conclusions

Carson (1968, 1971) and Carson and Templeton (1984) have proposed a theory according to which the process of speciation is triggered by stochastic genetic events associated with colonization by very few individuals. This founder-flush-crash model of speciation postulates that during the flush phase of the cycle there are opportunities for extensive recombination in the absence of strong resource competition and thus opportunities for creation of novel genotypes. According to the model, the crash phase sorts out the resulting genotypes, selecting those well fit to succeed in the competition for resources, many of which may be novel genotypes. The sequence of genetic events thought to occur during a founder-flush-crash cycle is thought to enhance the divergence of a colonizing population relative to its ancestral population or to other colonizing populations (Carson 1975; Carson and Templeton 1984).

Carson (1971 and later) has pointed out that his founder- flush-crash model lends itself to experimental testing by means of suitably designed laboratory systems; he outlined an experimental protocol for the purpose. Several experimental tests have been carried out that follow Carson's protocol, with *Drosophila pseudoobscura* (Powell 1978; Dodd and Powell 1985), *D. simulans* (Ringo et al. 1985), and the housefly (Meffert and Bryant 1991). In the present paper, we have reported the results of a large experiment designed to test Carson's founder-flush-crash model of speciation by mimicking in the laboratory to the extent possible the essential features of the model. The experimental design is largely similar to those of Powell (1978) and Ringo et al. (1985), although it differs from them in important features. Our experimental organism is *D. pseudoobscura*, the same as in Powell's (1978; Dodd and Powell 1985), but our results are substantially different.

The most significant differences concern the frequency and the magnitude of cases of positive assortative mating, which are much larger in Powell's (1978) than in our experiment, and their persistence over time, which is conspicuous in the previous experiment (Powell 1978; Dodd and Powell 1985) but absent in ours. Powell (1978, table 4) tested 36 two-population combinations involving 8 derived and 1 "original" (O) population and observed 31% significant chi squares for positive (and none for negative) assortative mating; if we exclude the tests involving the 0 population (1 significant case out of 8 tests), the incidence of significant positive assortment is 36%. In contrast, we have observed only 10.1% cases of positive assortative mating between pairs of derived (founder-flush-crash) populations and

7.0% between one of them and one ancestral population (table 20.2); and we have also observed significant cases of negative assortative mating in both kinds of test. When Powell (1978, table 6) again tested pairs of populations that had earlier exhibited positive assortative mating, he observed 4 significant cases in 5 tests. The values of Y (which we have calculated for his data in table 4) range 0.28–0.74. In contrast, we have observed no instance of assortative mating that would persist over the whole experiment and our Y values are lower. Some pairs of Powell's populations still exhibited positive assortative mating several years later, although they had been maintained by mass culture in the intervening years (Dodd and Powell 1985).

We have no obvious explanation for these differences between Powell's and our experiments. The experimental design was different in some important respects, but we attempted to meet the postulates of Carson's model more accurately. Perhaps the relevant difference is the origin of the founding populations. Powell started with a polyhybrid population (derived by intercrossing 4 geographically distant natural populations), whereas our experimental populations each derive from only 1 natural population (one or other of 2 populations). If this is the explanation for the different results, ours would be more relevant to Carson's model, since a polyhybrid origin is not appropriate for Carson's founder-flush-crash model, as pointed out by Charlesworth, Lande, and Slatkin (1982), Barton and Charlesworth (1984), and Barton (1989).

Our results are quantitatively more similar to those of Ringo et al. (1985; Ringo 1987) than to Powell's. As in the present experiment, that of Ringo et al. yielded little and erratic assortative mating. Sexual isolation occurred in 12 of 216 tests (Ringo et al. 1985, table 4), of which 2/48 were between 1 ancestral and 1 derived population and 8/168 between 2 derived populations. The significance of this low incidence of assortative mating has been discounted on the grounds that *D. simulans* is a cosmopolitan species and, hence, not a good choice to test the founder-flush-crash model (Templeton 1980a, b). This reservation does not apply to our experiment.

Positive assortative mating appears also as a rare event in a founder-flush experiment with another cosmopolitan species, *Musca domestica* (Meffert and Bryant 1991). Six founder-flush populations were studied, 2 for each of three bottleneck sizes ($n = 1, 4, 16$). Tests after 5 founder-flush cycles gave 2 significant cases of positive and 1 of negative assortative mating. Fifteen population pairs were tested, 9 between derived populations and 6 between 1 derived and 1 ancestral population. The experiments yielding positive assortative mating were two separate tests involving the same 2 populations (1 with $n = 1$, the other $n = 4$), which suggests that assortative mating was not a chance event in this case (Meffert and Bryant 1991). The assortative index between the 2 populations was $Y=0.21$, $p<0.01$ (calculated by us from table 2 in Meffert and Bryant 1991 with the data for the two tests combined).

In our experiment, a majority of the tests yielding significant positive assortment are between pairs of derived populations rather than between ancestral and derived populations. This was also the case for Powell (1978). Barton and Charlesworth (1984) have suggested that this finding is more consistent with Wills's (1977) model—where the critical parameter is inbreeding rather than founder effects as such—than with Carson's model. If this were correct, ethological isolation would also be expected between inbred populations, such as the "prima" and "endogamic" populations in our experiment, or the ones in Powell (1978) and Powell and Morton (1979), which is not the case.

The number of individuals at the bottleneck might be an important parameter that was not tested in earlier *Drosophila* experiments. In our experiment, it seems that populations with a larger number of founders ($n \geq 5$ pairs) are less likely to evolve mating preferences under the founder-flush-crash protocol. However, the instances of assortative mating are too few, in spite of the large scale of the experiment, for drawing any definitive conclusion. The matter would deserve additional testing, but the enormous amount of work involved makes it very difficult to carry out experiments of the needed magnitude. The one general conclusion of our experiment is that the founder-flush-crash protocol yields sexual isolation only as a rare (and erratic) event, which makes it difficult to quantify the contribution of the different variables involved.

The results of our large-scale experiment, which attempts to model Carson's founder-flush-crash theory as accurately as feasible, may then be seen as supporting Carson's model to the extent that they show that the conditions of the model may lead to the evolution of ethological isolation. However, significant ethological isolation between individual populations is observed in only a few of the very many cases tested, although a few more than is expected by chance, and ethological isolation between particular populations does not persist through the various cycles nor does the average of all populations increase over time. Thus, our results substantiate Barton and Charlesworth's (1984) position that "although founder effects may cause speciation under sufficiently stringent conditions" (p. 133), "[t]here are no empirical . . . grounds for supposing that rapid evolutionary divergence usually takes place in extremely small populations" (p. 157; see also Charlesworth and Smith 1982).

We conclude that although founder events may occasionally lead to the evolution of assortative mating and hence to speciation, our results do not support the claim that the founder-flush-crash model identifies conditions very likely to result in speciation events.

The process of speciation was a subject of great interest to Theodosius Dobzhansky. He was particularly intrigued by Carson's founder-flush-crash model as a speciation process that may particularly apply to island and other colonizations. We have conducted a large, seven-year experiment with *Drosophila pseudoobscura* designed to meet the essential postulates of the model and to test separately some of the postulates. Forty-five experimental and 14 control populations have been studied during 13 successive founder-flush-crash cycles, or about 80 generations. Sexual isolation tests yielded significantly positive assortative mating only in about 10% of tests between pairs of experimental populations. Sexual isolation did not increase or even persist over time. Populations that exhibited significant assortative mating at a point in time did not retain the isolation over the remaining time of the experiment. The frequency of cases of assortative mating between experimental populations did not augment as the number of cycles increased from 4 through 13. No significant assortative mating occurred between control populations, including highly inbred ones. In conclusion, our results do not support the claim that the founder-flush-crash model identifies conditions very likely to result in speciation events.

REFERENCES

Ayala, F. J. 1991. On the evolution of reproductive isolation and the origin of species. In S. Osawa and T. Honjo, eds., *Evolution of Life*, pp. 253–70. Tokyo: Springer-Verlag.

Barton, N. H. 1989. Founder effect speciation. In D. Otte and J. A. Endler, eds., *Speciation and Its Consequences*, pp. 229–56. Sunderland, Mass.: Sinauer.

Barton, N. H. and B. Charlesworth. 1984. Founder effects, genetic revolutions, and speciation. *Ann. Rev. Ecol. Syst.* 15:133–64.

Bishop, Y. M. M., S. E. Fienberg, and P. W. Holland. 1975. *Discrete Multivariate Analysis*. Cambridge, Mass.: MIT Press.

Blair, W. F. 1955. Mating call and stage of speciation in the *Microphyla Olivacea-M. carolinensis* complex. *Evolution* 9:469–80.

Butlin, R. K. 1989. Reinforcement of premating isolation. In D. Otte and J. A. Endler, eds., *Speciation and Its Consequences*, pp. 158–79. Sunderland, Mass.: Sinauer.

Carson, H. L. 1968. The population flush and its genetic consequences. In R. C. Lewontin, ed., *Population Biology and Evolution*, pp. 123–37. Syracuse, N.Y.: Syracuse University Press.

———. 1971. Speciation and the founder principle. *Stadler Genet. Symp.* 3:51–70.

———. 1975. The genetics of speciation at the diploid level. *Am. Natur.* 109:83–92.

———. 1978. Speciation and sexual selection in Hawaiian *Drosophila*. In P. F. Brussard, ed., *Ecological Genetics: The Interface*, pp. 93–107. New York: Springer-Verlag.

———. 1987. The genetic system, the deme and the origin of species. *Ann Rev. Genet.* 21:405–23.

Carson, H. L., D. E. Hardy, H. T. Spieth, and W. S. Stone. 1970. The evolutionary biology of the Hawaiian Drosophilidae. In M. K. Hecht and W. C. Steere, eds., *Essays in Evolution and Genetics in Honor of Th. Dobzhansky*, pp. 437–543. New York: Appleton-Century-Crofts.

Carson, H. L. and A. R. Templeton. 1984. Genetic revolutions in relation to speciation phenomena: The founding of new populations. *Ann. Rev. Ecol. Syst.* 15:97–131.

Charlesworth, B., R. Lande, and M. Slatkin. 1982. A neo-Darwinian commentary on macroevolution. *Evolution* 36:474–98.

Charlesworth, B., and D. B. Smith. 1982. A computer model of founder effect speciation. *Genet. Res. Cambr.* 39:227–36.

Darwin, C. R. 1872. *On the Origin of Species by Means of Natural Selection.* 6th ed. London: John Murray. Re-edition: Oxford University Press.

———. 1903. In F. Darwin, ed., *More Letters of Charles Darwin*. London: John Murray.

Dobzhansky, Th. 1940. Speciation as a stage in evolutionary divergence. *Am. Natur.* 74:312–21.

———. 1972. Species of *Drosophila*. *Science* 177:664–69.

Dobzhansky, Th. and O. Pavlovsky 1953. Indeterminate outcome of certain experiments on *Drosophila* populations. *Evolution* 7:198–210.

Dodd, D. M. B. and J. R. Powell. 1985. Founder-effect speciation: An update of experimental results with *Drosophila*. *Evolution* 39:1388–92.

Fienberg, S. E. 1977. *The Analysis of Cross-Classified Categorical Data.* Cambridge, Mass.: MIT Press.

Galiana, A., A. Moya, and F. J. Ayala. 1993. Founder-flush speciation in *Drosophila pseudoobscura*: A large-scale experiment. *Evolution* 47:432–44.

Galiana, A., F. J. Ayala, and A. Moya. 1989. Flush-crash experiments in *Drosophila*. In A. Fontdevila, ed., *Evolutionary Biology of Transient Unstable Populations*, pp. 58–73. Berlin: Springer-Verlag.

Meffert, L. M. and E. H. Bryant. 1991. Mating propensity and courtship behavior in serially bottlenecked lines of the housefly. *Evolution* 45:293–306.

Patterson, H. E. H. 1982. Perspective on speciation by reinforcement. *South Afr. J. Sci.* 78:53–57.

Powell, J. R. 1978. The founder-flush speciation theory: An experimental approach. *Evolution* 32:465–74.

———. 1989. The effects of founder-flush cycles on ethological isolation in laboratory populations of *Drosophila*. In L. V. Giddings, K. Y. Kaneshiro, and W. W. Anderson, eds., *Genetics, Speciation and the Founder Principle*, pp. 239–51. Oxford: Oxford University Press.

Powell, J. R. and L. Morton. 1979. Inbreeding and mating patterns in *Drosophila pseudoobscura*. *Behav. Genet.* 9:425–29.

Ringo, J. M. 1987. The effect of successive founder events on mating propensity of *Drosophila*. In M. Huettel, ed., *Evolutionary Genetics of Invertebrate Behaviour*, pp. 79–88. New York: Plenum.

Ringo, J. M., D. Wood, R. Rockwell, and H. Dowse. 1985. An experiment testing two hypothesis of speciation. *Am. Natur.* 126:642–61.

Sperlich, D., A. Karlik, and P. Pfriem. 1982. Genetic properties of experimental founder populations of *Drosophila melanogaster*. *Biol. Zentralbl.* 101:395–411.

Templeton, A. R. 1980a. The theory of speciation via the founder principle. *Genetics* 94:1011–38.

———. 1980b. Modes of speciation and inferences based on genetic distances. *Evolution* 34:719–29.

———. 1981. Mechanisms of speciation: A population genetic approach. *Ann. Rev. Ecol. Syst.* 12:23–48.

Terzian, C. and C. Biémont. 1988. The founder effect theory: Quantitative variation and MDG-1 mobile element polymorphism in experimental populations of *Drosophila melanogaster*. *Genetica* 76:53–63.

Wallace, A. R. 1889. *Darwinism*. London: Macmillan.

Wills, C. J. 1977. A mechanism for rapid allopatric speciation. *Am. Natur.* 111:603–5.

21

Courtship Behavior of *Drosophila obscura* Group Flies

Seppo Lakovaara and Jaana O. Liimatainen

This paper is dedicated to the memory of Theodosius Dobzhansky. One of us (S. L.) met him in Helsinki, Finland, in the spring of 1970. Dobzhansky visited the University of Helsinki and gave a lecture on genetic load. During his visit Dobzhansky became interested in the *Drosophila* research done in the physiological zoology department of the University of Helsinki by me and Anssi Saura. Some months later he invited us to his laboratory in Rockefeller University, New York City, where I spent one year. Dobzhansky and the research in his laboratory inspired me so much that I am still working with *Drosophila* and the processes of evolution.

Studies on the behavior of *Drosophila* have concentrated on courtship and mating behavior. This is natural, for these behaviors are complicated and in general species-specific. Courtship behavior is also believed to be of paramount importance as an isolating mechanism between species. As such it is a major factor in evolution.

Observations on the courtship behavior of *Drosophila* were published first by Sturtevant (1915, 1921). He described some courtship traits in the *Drosophila virilis* group species. Following publications of these two papers the courtship and mating behavior of *Drosophila* was studied by many others.

The *Drosophila obscura* group, the favorite group of Theodosius Dobzhansky (1970), is 1 of the 7 species groups of the subgenus *Sophophora* in the family Drosophilidae. The *Drosophila obscura* group can be divided into 4 subgroups: *obscura, affinis, pseudoobscura,* and *microlabis* (Lakovaara and Saura 1982; Cariou et al. 1988).

The behavior of at least 17 species of the approximately 30 species of the *D. obscura* group has been recorded. The best known species are the American *D. pseudoobscura* and the European and now also American *D. subobscura*. If we try to rank the papers published during the last three decades on the behavior of the *D. obscura* group flies, two papers of Brown (1964, 1965) stand out in importance.

In his first paper Brown (1964) describes and analyzes the courtship of *D. pseudoobscura*, using the quantitative techniques developed by Bastock and Manning (1955) for *D. melanogaster.* He splits the courtship of *D. pseudoobscura* males into three main phases, namely, *orientation*, i.e., the male turns to face the female, taps her with his forelegs, and follows her whenever she moves; *vibration*, i.e., he extends the wing nearest the female's head to 90° and vibrates it; and then *jumps* and attempts genital contact.

Brown (1964) also described three additional phases of courtship, thereby indicating that the male may also stand beside the female with wings raised (wing posture) or, very occasionally, face her and rotate his middle legs (rowing). He may also perform the first three main phases, together with some others (countersignaling) directed to other males. The courted female may run away, fend, kick, or slash at the male with her legs, flutter her wings, curl her abdomen under her or to the side away from the male, spread her vaginal plates, or extrude her ovipositor, but she does so only in response to the male's courtship.

Brown also analyzed the effect of the female's behavior on the male. According to him all the female's movements, except spreading, inhibit the male. Extrusion, performed only by a fertilized female, is particularly effective in making the male turn away. Spreading is performed by virgin females and appears to assist genital contact after the jump but is not a positively stimulating acceptance response. In his courtship, the *D. pseudoobscura* male does not depend on fresh, specific stimulation from the female to pass from one courtship element to the next. The only thing needed is the presence of a conspecific female.

In the paper published one year later Brown (1965) reported on a comparative behavior study. He described and compared the courtship behavior of 3 *D. pseudoobscura* subgroup species, namely, *pseudoobscura*, *persimilis*, and *miranda*; 6 *D. obscura* subgroup species, *obscura*, *subsilvestris* (*silvestris* in the paper), *tristis*, *bifasciata*, *ambigua*, and *subobscura*; and 2 *D. affinis* subgroup species, *affinis* and *helvetica*. These species, though differing in many details, are more like each other than any is to members of other groups. The main variation is in the form of wing vibration, and Brown suggested that this serves to maintain species isolation.

D. pseudoobscura, *D. persimilis*, and *D. miranda* are 3 closely related American species; their behavior differs in quantitative details only. The European species, *D. obscura*, *D. subsilvestris*, *D. tristis*, *D. bifasciata*, and *D. ambigua*, follow the basic courtship pattern of the 3 American species but differ in more frequent rowing, and in addition, they have a frontal, flatter wing posture.

D. subobscura differs from all the others in vibrating rarely; instead, the normal prelude to copulation is wing posture, combined with a side-to-side wing dance. According to Brown (1965) this species relies more on visual stimuli than the others, and it is suggested that the predominance of wing posture might be linked to this. Brown concluded that in *D. subobscura* there is a continuous interaction of fresh external stimuli between male and female during courtship, in contrast to the situation in *D. pseudoobscura* and probably the other members of the *D. obscura* group.

Twenty-one years before Brown's paper, Philip et al. (1944) showed that the courtship of *D. subobscura* is extremely light dependent; *D. subobscura* males are not able to inseminate females in complete darkness. The importance of light for general activity in *D. subobscura* was shown by Wallace and Dobzhansky (1946), who observed highly reduced locomotion of the flies in red light above 610 nm. Kekić and Marinković (1974) demonstrated the dependence of mating success on the intensity of light. More details about the role of visual signals were provided by Grossfield (1968), who pointed out that the wings of *D. subobscura* are used for optical stimulation during courtship. The decisive importance of visual courtship elements was demonstrated by Pinsker and Doschek (1979). They found a clear-cut correlation between mating success and contrast discrimination ability.

Springer (1973) selected a laboratory strain of *D. subobscura* for light-independent mating by reducing the light intensity gradually over 14 generations. This strain, called *lin*, can copulate in complete darkness. The males recognize the females by tapping and receive thereby tactile and/or chemical stimuli and copulate with them without a preceding courtship.

An additional detail in the courtship behavior of *D. subobscura* is the courtship feeding of the female by the male. Steele (1986) showed that the male *D. subobscura* provides the female with a drop of regurgitated food during courtship and that females feeding on this drop of food lay more eggs on a low-nutrient medium than those females not receiving a drop.

The two species of the *D. affinis* subgroup, *affinis* and *helvetica*, court in a way very similar to the other *D. obscura* group species but differ particularly in circling the female more actively and in the rarity or absence of wing posture and rowing (Brown 1965).

All these studies have been made in laboratories. The difficulty of this approach is that one never knows whether the observed behavior has any bearing on what happens out in the field. Flies may, in fact, behave quite differently outdoors. In honoring Theodosius Dobzhansky, we must remember that he loved outdoor life. It was an irony that the ecology and habitat of his beloved *D. pseudoobscura* was to him, and still is to us, very much of a mystery.

One may also look at the role of the female in courtship. As we mentioned before, Brown (1965) concluded that in *D. subobscura* there is a continuous interaction of stimuli between male and female during courtship, whereas the males of other members of the group court relatively independently of female stimulation. The male takes an active and the female a passive part in courtship. So, the active role of the *D. subobscura* female in courtship represents a rare exception in the *D. obscura* species group.

In this study we look at females in the wild and see whether the coy and passive role that they assume indoors and under observation occurs as well in nature. Further, we also look at the behavior of the female in the laboratory.

Behavior in the Wild

We have recorded different aspects of behavior of *D. alpina* Burla and *D. obscura* Fallén in the field. Both species belong to the *D. obscura* subgroup. Hundreds of flies have been under scrutiny—exactly how many we do not know, for flying and running flies are difficult to count. Field observations were made in Kuusamo, Finland (66° 22′N, 29° 21′E) and in Kuopio, Finland (62° 55′N, 27° 46′E), where both species are common.

Equipment

Recording were made by JVC KY-210B (Victor Company of Japan, Ltd., Japan) three-tube color video camera with phase alternate line (PAL) signals. The camera was placed on a suitable site, where the flies were expected to fly and come in contact with other flies. Flies were attracted to the vicinity of the camera by fermenting malt. The distance between the lens and the flies varied from 5 to 10 cm. The camera was connected to indoor studio equipment with remote control panel, video monitor, and Sony VO-6800PS (Sony Corporation, Japan) portable U-matic low-band video casette recorder. With this equipment it was possible to turn the camera, to zoom and focus the lens, to see what was happening under the lens, and, of course, to record the movements of flies. After recordings, the videotapes were played back for analysis of the behavior of flies.

The Male and His Territory

In the wild, *D. alpina* and *D. obscura* males display territorial behavior; i.e., the male marks, patrols, and defends a small definite area. The male marks his territory by touching or dragging the tip of his abdomen against the substrate. The marking secretion is unknown, but most probably it is a normal anal fluid. On his territory, which may be food (malt bait), a leaf, or any limited object in the wild, the male takes a position from where he can easily supervise his territory. At times the male patrols the territory waving and scissoring his wings. If the male observes another fly or any moving object in his territory, he runs toward the intruder, scissoring his wings. This scissoring is much faster and more frequent in *D. alpina* than in *D. obscura*. When the male meets an intruder, he first lifts and straightens his foreleg and taps the intruder. Tapping enables the male to recognize the species and sex of the intruder. If the intruder is a conspecific male, the resident starts to repel the intruder by wing scissoring and pushing. Again *D. alpina* is more active than *D. obscura*. In general a *D. alpina* resident male starts to fight by jumping on the intruder. If the intruder belongs to some other species, the resident disregards it completely.

If the resident meets a conspecific female, he commences the courtship by orienting and then vibrating or fluttering his wings. These courtship elements differ between *D. obscura* and *D. alpina*. A *D. obscura* male stands facing the side of the female's abdomen, extends the wing nearest the female to an angle of 120°, and vibrates it for about one second. The direction of beat is up and down but also to and from the female. A *D. alpina* male stands behind the female, extends his wing again nearest the female to an angle of 45° to 90°, and flutters or vibrates it for one to five seconds.

Female Holds the Key

As Brown (1964, 1965) already stated, a courted female may perform a number of movements directed toward the male. Interestingly, the most effective repellent signal is undescribed thus far. It is the abdomen curling or turning toward the male. The ventral side of the abdomen of the *D. obscura* group female is white. This color seems to have an essential role in the courtship behavior. It serves as an inhibitory and probably also as an attraction signal for the male, depending on the female's reproductive state. While repelling a courting male the *D. obscura* female turns her white belly toward the male, to the right or to the left, depending on the approaching direction of the male. She may also extrude her ovipositor. This female behavior is triggered by the mere presence of a male in the close vicinity of the female, the male's touch and his wing vibration.

This female signal is astonishingly effective. In most cases it breaks the courtship immediately. It may occasionally lead to the wing posture behavior of the male; i.e., the male stands in front of and face to face with the female, his wings opened to an angle of about 45° in the horizontal plane. Often one wing is opened more than the other. If the female abdomen turning and ovipositor extrusion leads to the wing posture of the male, the courtship may go on and on. We have seen unrequited courtships lasting sixteen minutes without any break.

Mature *D. obscura* females shine under ultraviolet (UV) light (366 nm). The reflection derives mainly from the ventral side of the abdomen and further from the yolk of her eggs. The more mature the eggs in the ovaries, the brighter the reflection. Immature and old females shine just faintly. A male may see the fluorescence and can, accordingly, distinguish mature inseminated females from potentially receptive virgins.

When repelling courting males, the *D. alpina* female lifts the tip of her abdomen and turns her white belly always backward. She may also extrude her ovipositor. The activity is directed toward the approaching and courting *D. alpina* male. Interestingly, one can see why a *D. alpina* female behaves this way. One factor is the approaching direction of the courting male. A *D. alpina* female shines in UV light, too, but in a fashion different from *D. obscura*. Immature eggs shine, but mature ones do not reflect at all. Many *D. alpina* females may flash a fluorescent sign ventrally at the tip of the abdomen. A female may secrete droplets of transparent liquid into her ovipositor. These droplets fluoresce strongly. Evidently this sign can be switched on and off. If *on*, it tells the courting male that the female needs rest. A female with the sign *off* may well be receptive.

A *D. alpina* female directs, in addition, some other movements toward the male, but not as many times as a *D. obscura* female does. The *D. alpina* female may run or fly away from the male at his approach. We never saw slashing or kicking of the legs or flicking, fluttering, or quivering of the wings. If the repelling mechanisms of the *D. alpina* female do not work, the male jumps immediately on the female and attempts genital contact. Again *D. alpina* males are faster and more active in jumping and copulating than

D. obscura males. Several *D. alpina* males may also try copulating with a single female at the same time.

Courtship in the Laboratory

Many studies suggest that a successful courtship and mating in *Drosophila* requires an interplay between the male and female (e.g., Ahearn 1980; Welbergen, van Dijken, and Scharloo 1987; Welbergen, Spruijt, and van Dijken 1992; Liimatainen et al. 1992). Does this hold in the *D. obscura* group? To find out, we have also studied the behavior of flies in the laboratory.

We collected flies by traps baited with fermenting malt from wild populations of *D. obscura* and *D. subobscura* Collin in Kuopio, Finland (62° 55′N, 27° 46′E). As mentioned above, Brown (1965) observed that in *D. subobscura* there is a clear interaction between male and female during courtship. This contrasts to the other members of the group, including *D. obscura*.

Flies were reared in the laboratory on malt medium in culture bottles at 22–23°C and in continuous light for 3 to 4 generations. Freshly emerged flies were collected in separate vials containing the same malt medium, each fly in a vial of its own. For experiments the flies were used at the age of five to seven days.

Camera, Recorder, and Computer

Single-pair courtships of *D. obscura* and *D. subobscura* were recorded at 22–23°C. A male and a female at the same age were placed without anesthesia on different sides of a circular plastic chamber (diameter 40 mm, height 8 mm) divided into two compartments by a removable wall. The chamber was placed under the lens of a three-tube color video camera with PAL signals (JVC KY-1900E, Victor Company of Japan, Ltd., Japan). The divider in the chamber was then removed and the behavior of the flies was recorded by a Panasonic NV-FS1 (Matsushita Electric Industrial Co., Ltd., Japan) S-VHS video cassette recorder connected to the camera until they copulated or until thirty minutes had elapsed, whichever came first. After each experiment the chamber was washed.

By this technique we studied the behavior of 58 pairs of virgin *D. obscura* flies and 65 pairs of *D. subobscura* flies. Four pairs of *D. obscura* and 14 pairs of *D. subobscura* did not show any courtship behavior. In 32 pairs of *D. obscura* and in 26 pairs of *D. subobscura* the courtship ended in copulation.

We reviewed the videotapes and recorded the successful courtships using an MS-DOS adapted version of a behavioral analysis program developed by Welbergen, van Dijken, and Scharloo (1987). Male behavior was classified into the following eight elements: no courtship, orienting, following, touching, wing vibration, wing posture (in *D. obscura*) or wing dance (in *D. subobscura* including wing posture), rowing, and attempting to copulate. The eight female behavior elements were: no courtship, wing fluttering, abdomen curling, decamping, walking, standing, wing dance (in *D. subobscura* only), and copulation. Changes in the behavior of flies were recorded on a continuous time scale in a computer.

Analysis of behavioral transitions was based on the assumption that the probability of a given act depends only on the identity of the act immediately preceding that act (first-order Markov chain). We first calculated the frequencies of changes from each behavioral element to any other element within and between individuals of each courting pair. Then we generated preceding-following behavior matrices for each courtship. Finally we added the matrices within each species. The total transition matrix was divided into male-male, male-female, female-male, and female-female matrices. The actual and expected values of transitions between behavioral acts in a matrix were compared with a chi-square test. To identify the categories responsible for a significant chi-square value we calculated for each cell in a matrix an adjusted residual (after Everitt 1977). Any adjusted residual higher than 1.96 (standard normal deviate) indicates that the given transition had occurred significantly more often than expected ($p < 0.05$) assuming a random model. The matrix contained at least $5 \times n \times m$ transitions, where n is the number of preceding behavioral elements and m the number of following ones in the matrix.

The frequencies of transitions were calculated as a conditional probability (Wood, Ringo, and Johnson 1980). Transitions with a conditional probability of at least 15% and an occurrence of at least five times in the courtship of the experimental group were used to set up pictograms, where intraindividual and interindividual transitions are drawn on the same figure. Because the method filters out the interindividual transitions in the intraindividual transitions and vice versa, the behavioral changes within individuals and interactions between sexes must be studied separately. The female's standing and moving during other elements were detected from the continuous string of courtship elements.

Courtship Elements

In a successful laboratory courtship the mean length of active courtship was 112 ± 24 seconds for *D. obscura* and 158 ± 27 seconds for *D. subobscura*

(0.01 < p < 0.05, Mann-Whitney U-test). In *D. obscura*, courtship breaks comprised 51% of the time elapsed from the first courtship act of the male to the copulation. In *D. subobscura* the proportion of courtship breaks was 50%. The average duration of copulation was 22 ± 2 seconds for *D. obscura* and 403 ± 30 seconds for *D. subobscura* (p < 0.001, Mann-Whitney U-test).

D. obscura males changed their behavior from one element to another 14.1 times per minute during the courtship. *D. subobscura* males were significantly more passive. They changed their behavior 8.4 times per minute only (0.001 < p < 0.01, Mann-Whitney U-test). In both species the most common male behavioral element was orientation (table 21.1). The next most common behavioral elements in *D. obscura* males were following the female and wing vibration and, in *D. subobscura* males, touching and following. *D. obscura* males vibrated their wings more frequently than *D. subobscura* males did. In *D. obscura* 87% of wing vibrations were directed to standing females and the remaining 13% to walking females. For *D. subobscura* these percentages were 83% and 17%, respectively. Rowing and the wing dance were recorded in *D. subobscura* only. Courtship feeding was not detected at all in our experiments.

TABLE 21.1.

Means and Standard Errors of the Duration of Given Courtship Elements in Single-Pair Courtships Between Virgin Females and Males of Drosophila obscura *and* D. subobscura. *Means Are Given as Percentages of Active Courtship Time; n=Number of Courtships*

Courtship Elements	*D. obscura* *n=32*	*D. subobscura* *n=26*
Male		
Orienting	42.4 ± 3.9	50.6 ± 4.5
Following	23.8 ± 3.6	15.7 ± 3.2
Touching	11.2 ± 3.6	18.0 ± 18.0
Wing vibration	16.9 ± 2.4	3.6 ± 3.6***
Wing posture	4.3 ± 2.6	—
Wing dance	—	11.6 ± 2.7
Rowing	0.0 ± 0.0	0.1 ± 0.1
Attempted copulation	1.5 ± 0.5	0.3 ± 0.2*
Female		
Wing fluttering	5.8 ± 1.5	1.7 ± 0.8*
Abdomen curling	1.7 ± 0.8	0.6 ± 0.4
Decamping	1.2 ± 0.5	1.0 ± 0.4
Walking	26.3 ± 3.4	17.6 ± 3.4
Standing	65.0 ± 3.5	71.0 ± 4.0
Wing dance	—	8.0 ± 2.3

*0.01 < p < 0.05, ***p < 0.001, Mann-Whitney U-test.

D. obscura females changed their behavior from one element to the other 10.8 times per minute. They were more active than the females of *D. subobscura*, which changed behavior 6.3 times per minute ($0.001 < p < 0.01$, Mann-Whitney U-test). During active courtship the females of both species were mainly standing (table 21.1). However, they also walked around the chamber quite frequently. In addition, *D. obscura* females fluttered their wings more frequently than *D. subobscura* females did.

Courtship and Interaction

D. obscura males began the courtship by either orienting toward the standing female or following her, if she was walking in the chamber (fig. 21.1). After orientation the male may touch the female and/or vibrate his wings. Following also led to touching and, further, to the vibration of wings. Vibration was the last behavioral element preceding a copulation attempt. An eventual male wing posture is in general followed by orientation. The *D. obscura* male can also repeat the behavioral elements he already displayed.

D. subobscura males commenced courting by orientation only (fig. 21.2). Orientation led to touching, and touching to the wing dance or back to the orientation again. The next step from the dance was the attempt to copulate. As in *D. obscura* the courtship in *D. subobscura* did not always proceed straight to copulation, but a male often repeated his behavioral elements.

D. obscura females had a distinct effect on the flow of courtship. The standing of a female led to the orientation and/or touching behavior of a male, and a walking female was followed by the male. If the male touched the female, she fluttered her wings. Quite often during the courtship the female curled or turned her abdomen away from the male. This direction is diametrically opposite to the inhibiting behavior in the field. She was in general standing when performing these responses to the courting male. At the end of courtship, when the male wing postured or attempted copulation, the female was mainly standing (100% and 96%, respectively). If the female decamps, the male stops courting.

The interaction between male and female of *D. subobscura* in courtship is at the beginning very similar to that in *D. obscura*. A standing female activated the male to orientation, and a walking female, to following behavior. If the male touched the female, she fluttered her wings and curled her abdomen away from the male. A very striking interaction between male and female is the wing dance performed by male and female together. The male positions himself face to face with the female. If the female accepts him, both flies dance; i.e., they step side to side while remaining face to

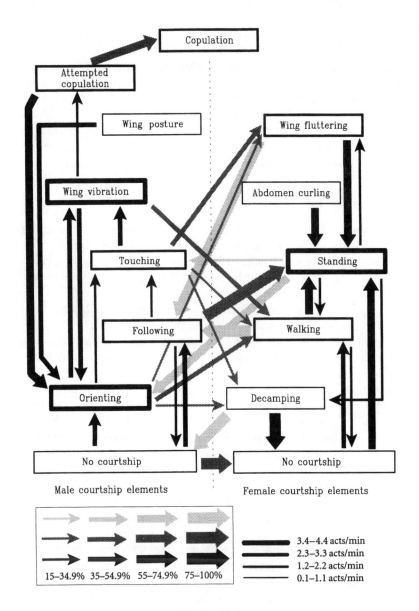

FIGURE 21.1

Pictograms of successful courtships in *Drosophila obscura*. Within-individual (black arrows) and between-individual (gray arrows, dark from male to female, light from female to male) transitions must be studied separately, because the within-individual transitions have been filtered from between-individual transitions and vice versa. The thickness of frame in a behavioral element box describes the frequency of occurrence of that particular element per minute. The thickness of arrow describes the conditional probability of that transition.

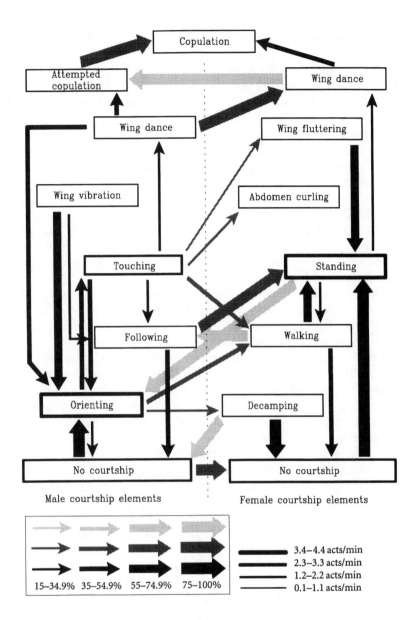

Pictograms of successful courtships in *Drosophila subobscura*. See figure 21.1 legend.

face. During the dance the male opens his wings and finally raises them. The dance, in our experiments, was similar to that described by Maynard Smith (1956) and Brown (1965). After dancing, the male attempts genital contact with the female. The male never attempted copulation before the wing dance, so the dance is a prerequisite for copulation in *D. subobscura*. Before the wing dance and the copulation attempt the female was always (100%) standing, even though the latter cannot be seen in the behavioral transitions.

In the field, males of *D. obscura* and *D. alpina* established territories, which they marked and defended. There are good reasons to suppose that many other species of the *D. obscura* group also show territoriality in the wild. The behavior is very similar to that observed in the Hawaiian *Drosophila* species (e.g., Spieth 1974).

Territoriality was never found in our laboratory experiments. The conditions in the laboratory are, of course, different and simpler than those in the wild. The placing of a pair of flies in a plastic chamber does not provide the variety of stimuli for the flies that is found in nature.

D. melanogaster males defend areas of food against other males (Hoffmann 1987). Hoffmann and Cacoyianni (1990) examined the factors affecting the establishment of territories and the incidence of territorial defense in *D. melanogaster* males in connection with the relative mating success of territorial and nonterritorial males. The study was made in population cages in the laboratory. They suggested that the territorial behavior of males is flexible, changing in response to variation in environmental conditions and the presence of conspecifics. Under field conditions territoriality is a conditional strategy that varies with changes in male behavior. Even though we did not examine the reasons underlying the territorial behavior of our flies, the explanation given by Hoffmann and Cacoyianni for *D. melanogaster* sounds reasonable and might be extended to cover territorial behavior of *D. obscura* group flies.

Laboratory courtship behavior of males in *D. obscura* and *D. subobscura* had a distinct sequential order. A *D. obscura* male first oriented to or followed the female, touched her, vibrated his wing, and finally attempted genital contact. A *D. subobscura* male oriented first and then touched the female. At the end of courtship the male danced with the female before attempting to copulate.

The most common behavioral elements in females of both *D. obscura* and *D. subobscura* were standing and walking. In particular the last phases of the male's courtship, i.e., the wing posture and copulation attempt in *D. obscura* and the wing dance and copulation attempt in *D. subobscura*, were directed almost totally toward the standing female. The females of *D. melanogaster* (Cook 1979, 1980; Markow and Hanson 1981; Tompkins et al. 1982), *D. simulans* (Cobb, Burnet, and Connolly 1986), *D. sechellia* (Cobb et al. 1989), and *D. montana* (Liimatainen et al. 1992) are often in motion at the beginning of the courtship but become stationary as courtship progresses, much as the *D. obscura* group females did. The standing female seems to be a stimulating signal for the male to continue the courtship sequence.

When the male touched the female, she fluttered her wings. The fluttering may produce air currents that transport pheromones that stimulate the courting male. This idea has been presented by many authors (e.g., Spieth 1974), but it is very difficult to prove.

In the laboratory virgin females curled or turned the white ventral side of their abdomens away from the courting male. In the field we saw that the turning of the white female underbelly toward the male is an extremely effective inhibiting signal, which turns the male off. Brown (1964) described how the *D. pseudoobscura* male will court most other members of the group, but not pale *D. obscura* females. Brown (1965) also stated that *D. obscura* males will not actively court dark females of *D. subobscura, D. ambigua, D. bifasciata,* or *D. tristis.* Brown suggested that the color of the cuticle may convey some nonvisual stimulus. Brown found also that *D. obscura* males will not actively court pale females. They never went beyond tapping one-day conspecific females. We suppose that the white ventral side of the female's abdomen, enhanced by fluorescence, is a very important signal. A receptive female hides it, but a nonreceptive one displays it.

In summary, the territoriality of *D. obscura* flies makes their behavior in the wild different from that in the laboratory, but the factors underlying the territorial behavior are as yet unknown. *D. obscura* group females have an active role in courtship. The female is the one that upholds the integrity of the species through sexual isolation. To mate or not to mate is up to her to decide.

We hope this paper gives an answer to a question about possible differences in life in artificial and in natural conditions. Dobzhansky was committed to knowing everything about the differences between laboratory conditions and the wild. We want to follow him.

REFERENCES

Ahearn, J. N. 1980. Evolution of behavioral reproductive isolation in a laboratory stock of *Drosophila silvestris. Experientia* 36:63–64.

Bastock, M. and A. Manning. 1955. The courtship of *Drosophila melanogaster.* *Behaviour* 8:86–111.

Brown, R. G. B. 1964. Courtship behaviour in the *Drosophila obscura* group. I: *D. pseudoobscura. Behaviour* 23:61–106.

———. 1965. Courtship behaviour in the *Drosophila obscura* group. Part II: Comparative studies. *Behaviour* 25:281–323.

Cariou, M. L., D. Lachaise, L. Tsacas, J. Sourdis, C. Krimbas, and M. Ashburner. 1988. New African species in the *Drosophila obscura* species group: Genetic variation, differentiation and evolution. *Heredity* 61:73–84.

Cobb, M., B. Burnet, and K. Connolly. 1986. The structure of courtship in the *Drosophila melanogaster* species subgroup. *Behaviour* 97:182–211.

Cobb, M., B. Burnet, R. Blizard, and J.-M. Jallon. 1989. Courtship in *Drosophila sechellia:* Its structure, functional aspects and relationship to those of other members of the *Drosophila melanogaster* species subgroup. *J. Insect Behav.* 2:63–89.

Cook, R. M. 1979. The courtship tracking in *Drosophila melanogaster. Biol. Cybernetics* 34:91–106.

———. 1980. The extent of visual control in the courtship tracking of *Drosophila melanogaster. Biol. Cybernetics* 37:41–51.

Dobzhansky, Th. 1970. *Genetics of the Evolutionary Process.* New York: Columbia University Press.

Everitt, B. S. 1977. *The Analysis of Contingency Tables.* New York: Wiley.

Grossfield, J. 1968. The relative importance of wing utilization in light dependent courtship in *Drosophila.* In M. R. Wheeler, ed., *Studies in Genetics IV,* pp. 147–56. Austin: University of Texas Publ. 6818.

Hoffmann, A. A. 1987. A laboratory study of male territoriality in the sibling species *Drosophila melanogaster* and *D. simulans. Animal Behav.* 35:807–18.

Hoffmann, A. A. and Z. Cacoyianni. 1990. Territoriality in *Drosophila melanogaster* as a conditional strategy. *Animal Behav.* 40:526–37.

Kekić, V. and D. Marinković. 1974. Multiple-choice selection for light preference in *Drosophila subobscura. Behav. Genet.* 4:285–300.

Lakovaara, S. and A. Saura. 1982. Evolution and speciation in the *Drosophila obscura* group. In M. Ashburner, H. L. Carson, and J. N. Thompson, Jr., eds, *The Genetics and Biology of Drosophila,* pp. 1–59. London: Academic Press.

Liimatainen, J., A. Hoikkala, J. Aspi, and Ph. Welbergen. 1992. Courtship in *Drosophila montana:* The effects of male auditory signals on the behaviour of flies. *Animal Behav.* 43:35–48.

Markow, T. A. and S. J. Hanson. 1981. Multivariate analysis of *Drosophila* courtship. *Proc. Natl. Acad. Sci.* 78:430–34.

Maynard Smith, J. 1956. Fertility, mating behaviour and sexual selection in *Drosophila subobscura. J. Genet.* 54:261–79.

Philip, U., J. M. Rendel, H. Spurway, and J. B. S. Haldane. 1944. Genetics and karyology of *Drosophila subobscura. Nature* 154:260–62.

Pinsker, W. and E. Doschek. 1979. On the role of light in the mating behavior of *Drosophila subobscura. Z. Naturforsch.* 34c:1253–60.

Spieth, H. T. 1974. Courtship behavior in *Drosophila. Ann. Rev. Entomol.* 19:385–405.

Springer, R. 1973. Light-independent mating, probably a dominant character of behaviour in *Drosophila subobscura. Drosophila Info. Serv.* 50:133.

Steele, R. H. 1986. Courtship feeding in *Drosophila subobscura.* I: The nutritional significance of courtship feeding. *Animal Behav.* 34:1087–98.

Sturtevant, A. H. 1915. Experiments on sex recognition and the problem of sexual selection in *Drosophila. Animal Behav.* 5:351–66.

———. 1921. The North American species of *Drosophila.* Carnegie Inst. Wash. Publ. 301:1–150.

Tompkins, L., A. C. Gross, J. C. Hall, D. A. Gailey, and R. W. Siegel. 1982. The role of female movement in the sexual behaviour of *Drosophila melanogaster. Behav. Genet.* 12:295–307.

Wallace, B. and Th. Dobzhansky. 1946. Experiments on sexual isolation in *Drosophila.* VIII: Influence of light on the mating behavior of *Drosophila subobscura, Drosophila persimilis* and *Drosophila pseudoobscura. Proc. Natl. Acad. Sci.* 32:226–34.

Welbergen, Ph., F. R. van Dijken, and W. Scharloo. 1987. Collation of the courtship behaviour of the sympatric species *Drosophila melanogaster* and *Drosophila simulans. Behaviour* 101:253–74.

Welbergen, Ph., B. M. Spruijt, and F. R. van Dijken. 1992. Mating speed and the interplay between female and male courtship responses in *Drosophila melanogaster. J. Insect Behav.* 5:229–44.

Wood, D., J. M. Ringo, and L. L. Johnson. 1980. Analysis of courtship sequences of the hybrids between *Drosophila melanogaster* and *Drosophila simulans. Behav. Genet.* 10:459–66.

22

Adult Influence on Reproductive Behavior of *Drosophila*

Dragoslav Marinković and Snežana Stanić

Behavioral genetics is one of the important fields of biology in which Dobzhansky applied his ingenious ideas about evolutionary adaptations, using *Drosophila* as a model organism. According to Dobzhansky (1937b, 1951b), changes in mating behavior are the initial stages of the speciation process in animals, involving the development of reproductive isolating mechanisms (RIMs) (Dobzhansky and Pavlovsky 1967; Dobzhansky, Ehrman, and Kastritsis 1968). Consequently, it is of interest to know whether these behaviors, as the major components of fitness, are strictly inherited, or whether they are, at least in part, modified under the influence of a specific environment.

How much intrinsic and acquired factors influence the development of behavioral characters is one of the basic questions in the science of biology and genetics. Acquired behaviors are frequently described among vertebrates, but not so often in lower classes of animals, including such well-studied groups as insects.

On the effects of previous experiences on reproductive behavior, we can refer to Pruzan and Ehrman (1974), who have demonstrated that previous copulation experience and exposure to other males and females can alter succeeding mating patterns of *D. pseudoobscura* females. Kyriacou and Hall (1980) demonstrated that in *D. melanogaster* previous experience of males

The generous help of Drs. Lee Ehrman and Seppo Lakovaara in reading the manuscript is very much appreciated.

may modify their courtship behavior and that the exposure of females to "rhythmical songs" of males enhances their sexual receptivity. Siegel and Hall (1979) and Tompkins et al. (1983) suggest that in such cases a conditional learning should be taken into account.

In two preliminary reports (Stanić and Marinković 1990, 1992), we published initial results of a series of experiments showing that the newly eclosed *Drosophila* flies, when surrounded by older individuals, are able to notice their reproductive behavior visually, by olfactory, and/or by vibratory stimuli, which affects their own subsequent mating efficiency.

Materials and Methods

Our experiments were conducted with *D. melanogaster* flies, F_2 or F_3 progenies of a few hundred individuals collected at Slankamen, 60 km north of Belgrade. Fly cultures were maintained under optimal conditions, in 250 mL bottles with cornmeal medium, at 25°C and 60 percent relative humidity (RH), and in uncrowded conditions (i.e., fewer than 100 individuals raised per bottle).

In the first set of experiments, newly eclosed flies were split into two groups. In one group, females and males (separately, 10 flies transferred to a clean glass test tube with food medium) were surrounded for three days by a large number of older individuals in a jar, thus being able to observe the behavior of the older flies, including their courtships, matings, etc. In the other group, the test tubes with 10 females or males, were not surrounded by older flies, thus having no opportunity to experience the courtship and mating behavior of the older flies. When young flies were three to three and a half days old, their mating success was measured by placing 10 pairs from the same group together. Such tests were made synchronously, in 10–100 replicas, and were repeated from zero hour to one hundred sixty-eight hours after the removal of parental flies. In a number of additional experiments, the influence of surrounding flies was compared in three different ways: (1) The test tubes with young flies were covered by a dark paper. (2) Their entrance was tightly closed by a cork. (3) They were uncovered and closed with light cotton at the mouth of the test tube. This allowed an eventual receiving of specific cues from surrounding flies (1) by vibration and olfactory means, (2) only visually, or (3) by all stimuli.

In a second set of experiments, two groups of males were compared in regard to their mating experience. In one group, individual males were allowed to have one mating when two days old, and in another group the males of the same age did not have such an experience. The mating success with virgin females from a control was compared after three to one hundred twenty hours in these two sets of males, split into 5–6 experimental groups.

All experiments were conducted under equal conditions, and wholly without anesthesia of flies.

Results

Table 22.1 displays the comparable results from twenty synchronous experiments with groups of *Drosophila melanogaster* (previously surrounded or nonsurrounded by older flies), in which the number of mated pairs, out of 10, are listed. It turned out that females and males that had been surrounded by older individuals were subsequently reproductively more successful than those that did not have such a surrounding group at the onset of their adult lives. The mating success was compared in *Drosophila* flies among whom (A) both sexes, (B,C) one of the sexes or (D) none of them were surrounded by older couples during the first three days of their postpupal lives. It appears that the biological surrounding of females in the early phases of their adult lives was of prime importance for later reproductive success. That is, the number of matings was significantly greater among flies in which both sexes, or only females, were previously surrounded by older couples, in

TABLE 22.1.

Number of Matings Among Ten D. melanogaster *Pairs.*

No. Matings	Combinations of Postimaginal Contacts			
	A	B	C	D
0	—	—	2	0
1	—	—	3	2
2	2	0	4	1
3	0	2	1	2
4	0	1	4	4
5	4	4	0	4
6	4	7	3	6
7	8	4	1	1
8	1	2	2	0
9	1	—	—	—
10	—	—	—	—
$\overline{X} \pm$ SE	6.1	5.8	3.6	4.5
	± .3	± .3	± .5	± .3

NOTE: (A) both sexes previously surrounded by flies; (B) females only previously surrounded; (C) males only previously surrounded; (D) none of them surrounded by older couples during their postpupal ontogenesis (n = 4 × 20 replicas). tA/C, A/D, B/C, B/D =3.1−4.3 (p<.01) tA/B, C/D n.s.

Comparisons of D. melanogaster *Flies Grown During*
First Three Days of Their Postpupal Development

	$\overline{X}_A$	$\overline{X}_B$	n	A/B Index $\overline{X} \pm SE$
No. matings in 10 pairs/90 minutes	Exp. I = 7.0 Exp. II = 7.4	5.5 5.7	20	1.9 ± .5 *
No. matings in 10 pairs/90 minutes (flies grown in competitive conditions)	Exp. I = 6.0 Exp. II = 7.6	4.5 6.4	20	1.3 ± .1 ***c
Premating time in 10 pairs	Exp. I = 19 min 57 sec Exp. II = 14 min 42 sec	26 min 13 sec 17 min 52 sec	20	1.4 ± .2 **
No. progeny from eggs laid/48 hours	228	97	10	2.7 ± 1 *

* $p < 0.1$; ** $p < 0.05$; *** $p < 0.01$.
c More than 200 individuals grown in 100 cc bottle.
NOTE: (A) surrounded, (B) nonsurrounded by older individuals (n = 2 × 10 replicas in each experiment).

comparison with the pairs whose males had been surrounded or where none of the sexes experienced such a surrounding.

In table 22.2 are given the results of two additional series of experiments, in which, in two groups of tested flies, not only the mating success but also the duration of premating time and the number of produced progenies were compared in 10 synchronous replicates. It clearly shows that for individuals that have had the experience of being surrounded by flies, the premating time was, on average, 20–25% shorter, the mating success was 25–30% greater, and the number of progenies was more than doubled, in comparison with the parents without such an experience.

In our next experiment the mating success of three-day-old males that have (A) or have not (B) been surrounded previously by older flies was compared. This time, however, they were allowed to compete for virgin females being placed together (10A + 10B + 10 females). In 20 such experiments, the males from the A or B group were marked alternatively by micronized dusts, and after a copulation with 1 of 10 females it was possible to determine to which group they belonged. Again, males that had previously been surrounded by older individuals were more successful than those that had not experienced such a natural surrounding ($\overline{X}_A$ = 5.4 ± 0.3; $\overline{X}_B$ = 3.6 ± 0.3 matings; t = 3.76; $p < 0.01$).

The comparisons described so far were repeated after some time, in order to determine how permanent such behavioral differences are between flies that had a different biological environmental. The environmental conditions were equalized in both groups; i.e., three-day-old flies were kept without being surrounded by older individuals. The tests were conducted zero to one hundred sixty-eight hours after the removal of these surrounding flies from one of the comparable groups. The results for 15 synchronous replicas are presented in table 22.3. The proportion of mated pairs was found to be 6–15% greater among individuals previously surrounded by older flies, but significantly so in groups tested zero hour, twenty-four hours, one hundred twenty hours after termination of contact with older flies.

The results of other sets of experiments include comparisons of males that had previous mating experience with those without such an experience. The tested virgin females were of the same age in both groups. In table 22.4 the premating time has been compared (I) between males that three to one hundred twenty hours earlier had experienced a copulation with a female and (K) those that did not have such an experience. The premating time was shorter by 16–50% in all tested I groups, but significantly so in those experiments where males had a copulation forty-eight, seventy-two or ninety-six hours before. In table 22.5 the mating success of males that had a mating experience three to seventy-two hours earlier appeared to be 10–30% greater than that of males without such previous experience, but a statistically significant increase was found only among males that had mated three to twenty-four hours earlier.

As can be seen from tables 22.3, 22.4 and 22.5, *Drosophila* is affected by the experience of being surrounded by older individuals, as well as by its own mating experience. We tried to determine whether all phases of the mating process occurred in flies that shortened their premating time

TABLE 22.3.

The Average Number of Matings ($\overline{X} \pm SE$) of Three-Day-old D. melanogaster
flies (Zero Hour), Grown Surrounded by Older Individuals and
24–168 Hours After Removal of These Individuals
(n = 10 Pairs in 15 Replicas, Compared with That Many Controls)

	Zero Hour	+24 hours	+72 hours	+120 hours	+168 hours
(A) Experimentals	8.7 ± .1	8.9 ± .2	7.5 ± .4	8.3 ± .2	8.7 ± .3
(B) Controls	7.9 ± .3	8.4 ± .1	6.6 ± .4	7.3 ± .4	8.4 ± .2
t test	2.5*	2.2*	1.6	2.2*	1.1
(df=28)					

(A) Groups with previous adult contact; (B) without such contact.
*$p < 0.05$.

TABLE 22.4.

Premating Time of D. melanogaster *Males, in Minutes/Seconds,*
(I) with and (K) Without Previous Mating Experience
(3–120 Hours Ago; 10 Pairs Observed in 6 × 20 Experiments)

Hours After Previous Mating Experience		Premating Time (min/sec)	I/K_{index} ($\overline{X} \pm SE$)	t Value (df=18)
3 Hours	I	11/12		
	K	(14/6)	0.79 ± .14	1.50*
24 Hours	I	10/18		
	K	(12/12)	0.84 ± .19	0.84
48 Hours	I	5/18		
	K	(10/30)	0.51 ± .16	3.06***
72 Hours	I	8/12		
	K	(10/9)	0.79 ± .09	2.33**
96 Hours	I	7/30		
	K	(11/6)	0.68 ± .10	3.20***
120 Hours	I	9/30		
	K	(14/12)	0.67 ± .26	1.27

*$p < 0.1$; **$p < 0.05$; ***$p < 0.01$.

TABLE 22.5.

Mating Success of D. melanogaster *Males, (I) with and (K) Without Previous Mating*
Experience (3–120 Hours Ago; 10-Pairs Observed in 5 × 20 Experiments)

Hours After Previous Mating Experience		No. of Matings ($\overline{X} \pm SE$)	I/K_{index} ($\overline{X} \pm SE$)	t Value (df=18)
3 Hours	I	6.5 ± .7		
	K	5.3 ± .6	1.3 ± .1	3.00**
24 Hours	I	7.5 ± .5		
	K	6.4 ± .6	1.2 ± .1	2.00*
48 Hours	I	7.0 ± .3		
	K	7.0 ± .4	1.1 ± .1	1.00
72 Hours	I	6.7 ± .5		
	K	6.5 ± .8	1.2 ± .2	1.00
120 Hours	I	6.7 ± .5		
	K	8.1 ± .3	0.8 ± .1	—

* $P < 0.1$; ** $P < 0.01$.

(tables 22.2 and 22.4), after having an earlier experience of being previously surrounded by active older flies. And indeed, we found that in some of these males the courtship ritual had shortened; i.e., they often just approached females directly, without having all the phases characteristic for the courtship of their species (Manning 1959; Welbergen, Spruijt, and van Dijken 1992).

In another set of 143 replicated experiments (table 22.6) we compared the matings among 10 pairs of young flies whose females and/or males were allowed, for the initial three days of their adult lives, to observe surrounding older individuals: (a) by all sensory means, (b) by auditory and olfactory means, (c) only visually, or (d) without such contacts and stimuli. In 20 synchronous experiments (table 22.6 A), young females allowed to use all stimuli in observing the surrounding flies were found to be reproductively more successful (6.2 matings on average, among 10 pairs), than those who used only visual, or only vibration and olfactory stimuli (4.8 matings). It appears that visual effects, as well as vibration and chemical stimuli, could be of importance in a female's biological surrounding, contributing to an increase of her mating success, in comparison with those individuals who did not experience the surrounding of older flies (3.9 matings among 10 pairs on average, in 20 replicas).

Somewhat different results have been obtained (table 22.6 B) when only males, in 23 additional experiments, were allowed to notice the surround-

TABLE 22.6.

Number of Matings Among 10 D. melanogaster *Pairs*

Combination of Postimaging Contacts	No. of Matings											n	$\overline{X}$ SE
	0	1	2	3	4	5	6	7	8	9	10		
(A) Females													
(a)		1	1	1	5	1	4	7				20	6.20 ± 0.22
(b)		1	4	4	4	3	3	1				20	4.85 ± 0.31
(c)			2	5	3	2	3	3	2			20	4.80 ± 0.34
(d)	1	3	3	2	3	3	1	3	1			20	3.85 ± 0.40
(B) Males													
(a)				1	0	2	6	7	0	3	4	23	7.17 ± 0.39
(b)					2	3	3	6	2	5	2	23	7.13 ± 0.38
(d)				3	1	2	6	5	2	4		23	6.34 ± 0.39
(C) Males and females													
(a)		1	0	3	2	8	15	21	21	19	10	100	7.34 ± 0.18
(b)		0	2	3	4	12	13	23	15	23	5	100	7.05 ± 0.19
(d)		1	0	4	8	10	19	22	16	14	6	100	6.77 ± 0.19

t Tests: A/ $t_{a/d, a/c} = 3.5 - 5.1(p<.001)$; $t_{b/d, c/d} = 1.9(p<.1)$
  B/ $t_{a/d, b/d} = 1.5 \ (p>.1)$
  C/ $t_{a/d} = 2.2 \ (p<.05)$; $t_{b/d} = 1.1 \ (p>0.1)$

NOTE: (A) females, (B) males, (C) both sexes allowed to observe the surrounding older flies during initial three days of their growth: (a) by all stimuli, (b) by vibration and olfactory means, (c) only visually, (d) without such contacts. In each of 143 experiments, 4 (in A), or 3 groups of 10 males × 10 females (in B and C) have been observed simultaneously.

ing flies from: (a) mildly closed (by cotton) glass test tubes, (b) test tubes covered by a dark paper, or (d) test tubes with no other flies outside. Among (a) and (b) males, an equal number of matings with virgin females was detected (7.1 on average), and among the males from (d) group only 6.3 mated with control females (the average from 23 × 10 pairs observed for one hour).

This experiment has been in addition repeated many times (n = 100; table 22.6 C), namely, when 3 groups of males and females, before being placed together, were allowed to observe the older flies: (a) by all stimuli, (b) without visual stimuli, or (d) without any surrounding of older individuals. The decrease in the (b) group was not significant, but the difference that was significant was between (a) and (d) groups, with 7.3 versus 6.8 matings per hour ($p < 0.05$).

The last 2 groups of experiments (B and C in table 22.6) demonstrate the importance of other stimuli, besides visual, for a verification of the importance of biological surroundings during the initial periods of young *Drosophila* adult life. However, the changes in response to all studied stimuli were significant when the females were exposed to these stimuli (table 22.6 A; $p < 0.05$). Here, one should consider that our experimental conditions (e.g., a glass tube mildly closed by cotton) could diminish in part the stimuli that may come from surrounding flies, implying that the above-determined influences of the biological environment could be even more important in nature.

Accepting Darwin's theory of natural selection as a basic theory of organic evolution, Dobzhansky (1937a, 1941, 1951a) emphasized the following: (1) *Populations* are the basic units of evolution. (2) *Fitness* of specific genotypes determines their chance to increase or decrease their frequency in future generations. (3) Differential *reproduction* rates are of crucial importance for determination of genetical constitution of a population. (4) *Balancing selection* is the basic force that maintains the genetic variability of a population. By taking into account these four postulates, one can explain the paradoxical effects of natural selection as the main evolutionary force, which results in an increase of biological diversity during the process of evolution, as well as in the maintenance of enormous genetic loads in natural populations.

Dobzhansky himself collected a huge amount of data, in the field as in the laboratory, demonstrating the basic mechanisms of evolutionary adaptations, using *Drosophila* as a model organism. His classical studies of ecology, behavior, chromosomal polymorphism, and fitness traits in *Drosophila* lasted almost fifty years and provided an incomparable amount of information about the synchronous contribution of genetical and environmental factors in numerous evolutionary adaptations of a species (see, e.g., Lewontin et al. 1981; Marinković in press). Dobzhansky's continuing influence reflects his ingenious perception of evolutionary principles. It is also reflected in his school of numerous students who were later the founders of new fields and laboratories in population and evolutionary genetics.

Reproductive behavior is certainly an important component of fitness of a genotype, population, or species (see Dobzhansky et al. 1977). That this behavior is primarily intrinsic and genetically controlled is well known, but how much it could be influenced by other biological factors, such as by the activities of older individuals of the parental generation, should be answered by studies in different organisms.

In natural conditions, individuals of a species may live almost solitarily or in more or less crowded groups. This happens also with *Drosophila*, which in the European climate start to appear in small numbers during the spring and become most abundant in the summer or in early autumn, reaching at some places (e.g., in *D. melanogaster*) many thousands of individuals in small areas of a territory. For a small and isolated group of such flies, the reproductive behavior must be primarily based on the intrinsic, i.e., instinctive events, which occur in a series of complex behavioral activities. These complex reproductive behaviors have been described by Sturtevant (1915) for 22 species and by Spieth (1968) for 200 species of *Drosophila*.

When surrounded by many flies, newly hatched adults may experience the reproductive behavior of older individuals, which may eventually contribute to a more pronounced ability of their own to copulate with the individuals of opposite sex. The question is how independent is reproductive behavior of an individual insect. Is it based primarily on inherited and instinctive events, or could this series of events be changed by environmental factors, giving rise to an increase, or decrease, of the reproductive success of a group of individuals?

The results presented in this paper show that older flies, as a biological surrounding environment, may influence the reproductive success of young individuals of *D. melanogaster*. The newly hatched males and females, if surrounded by older reproductively active flies, subsequently copulate more frequently and more quickly, than those that did not have such an experience. The number of their progenies was found to increase by a factor of two in our experimental conditions (table 22.2), which means that their fitness could be significantly increased. This may have great adaptive significance in the overcrowded conditions frequently found in nature, which lead to a shortening of generation time and contribute to an exponential growth of the population.

Table 22.6 shows that visual, vibratory, and chemical stimuli are all important aspects of the effects of surrounding older individuals. Females seem to be more sensitive than males in making the difference between studied stimuli.

Our experiments show that the observation of mating behaviors of surrounding flies can affect subsequent behavior for some time (table 22.3). During five days after the removal of such older reproductively active flies, young individuals had an increased reproductive activity, in comparison with a control group of the same age and origin. However, in the group tested after three days, this increase was not statistically significant. Recall that Folkers and Spatz (1981) showed that the memory of a visual learning behavior in *D. melanogaster* lasted only for fourteen hours.

We are not certain, though, whether the described contacts of young with older flies were imprinted as a new experience of some details of mating behavior of older individuals or as a result of an increased physiological level that may last for some time and be misinterpreted as the "memory" of a particular effect of the environment. Inherited (i.e., intrinsic) mating behavior certainly includes all the phases of the courtship. Modifications of this behavior, i.e., a shortening of prereproductive rituals, could be the result of an acquired influence of the 'parental generation' acting as a new experience of young *Drosophila* individuals.

A male's mating experience also contributes to a shortening of the time needed to find another female later and start mating with her (table 22.4). In males that forty-eight to ninety-six hours earlier had mated with a female, the premating time was significantly shorter than in those males that did not have such an experience before.

Such results could be further discussed from the point of view of evolutionary adaptations (Dobzhansky 1937b, 1970) and compared with relevant results of other authors (e.g., Kaul and Parsons 1966; Siegel and Hall 1979; Tompkins et al. 1983), who studied the relationships of intrinsic and acquired effects in reproductive behaviors of *Drosophila* species. Different rearing methods and developmental isolation do also affect subsequent adult behavior of *Drosophila* (Ehrman 1989, 1990; Kim, Ehrman, and Koepfer 1992, 1993).

Our results have shown that biological (social) environment definitely affects the future behavior of newly hatched adults in an insect species like *Drosophila melanogaster*. It leads to an increase of mating efficiency and, consequently, also to an increase in the fitness of such groups of young individuals. Such a "parental influence" could be present also in many other groups of invertebrates. It should be considered an important component of fitness, contributing significantly to their biological adaptations and to the enhancement of future generations.

REFERENCES

Dobzhansky, Th. 1937a, 1941, 1951a. *Genetics and the Origin of Species.* 1st, 2d, and 3d eds. New York: Columbia University Press.

———. 1937b. Genetic nature of species differences. *Am. Natur.* 71:404–20.

———. 1951b. Experiments on sexual isolation of *Drosophila*. X: Reproductive isolation between *D. pseudoobscura* and *D. persimilis* under natural and under laboratory conditions. *Proc. Natl. Acad. Sci.* 37:792–96.

———. 1970. *Genetics of the Evolutionary Process.* New York: Columbia University Press.

Dobzhansky, Th. and O. Pavlovsky. 1967. Experiments on the incipient species of the *Drosophila paulistorum* complex. *Genetics* 55:141–56.

Dobzhansky, Th., F. J. Ayala, G. L. Stebbins, and J. W. Valentine. 1977. *Evolution.* San Francisco: W. H. Freeman.

Dobzhansky, Th., L. Ehrman, and P. A. Kastritsis. 1968. Ethological isolation between sympatric and allopatric species of the *obscura* group of *Drosophila*. *Animal Behav.* 16:79–87.

Ehrman, L. 1989. Effects of lifelong experience on the rare-male mating advantage in *Drosophila pseudoobscura*. *Behav. Genet.* 19:755–56.

———. 1990. Developmental isolation and subsequent adult behavior of *Drosophila pseudoobscura*. *Behav. Genet.* 20:609–15.

Folkers, E. and H. Ch. Spatz. 1981. Visual learning behavior in *Drosophila melanogaster* wild type. *Insect Physiol.* 27/9:615–22.

Kaul, D. and P. A. Parsons. 1966. Competition between males in the determination of mating speed in *Drosophila pseudoobscura*. *Aust. J. Biol. Sci.* 19:945–47.

Kim, Y-K., L. Ehrman, and H. R. Koepfer. 1992. Developmental isolation and subsequent adult behavior of *Drosophila paulistorum*. I: Survey of the six semi-species. *Behav. Genet.* 22:545–56.

———. 1993. Different rearing methods affect subsequent adult behavior in *Drosophila paulistorum*. (Forthcoming.)

Kyriacou, C. P. and J. C. Hall. 1980. Circadian rhythm mutations in *Drosophila melanogaster* affect short-term fluctuations in the males courtship song. *Proc. Natl. Acad. Sci.* 77:6729–33.

Lewontin, R. C., J. A. Moore, W. B. Provine, and B. Wallace. 1981. *Dobzhansky's Genetics of Natural Populations.* I–XLIII. New York: Columbia University Press.

Manning, A. 1959. The sexual behaviour of two sibling species of *Drosophila*. *Behaviour* 15:123–45.

Marinković, D. (in press). Theodosius Dobzhansky and biological synthesis. In M. B. Adams and S. G. Ingevechtomov, eds., Proc. Int. Symp.: *Th. Dobzhansky and Evolutionary Synthesis*. Leningrad University Press.

Pruzan, A. and L. Ehrman. 1974. Age, experience, and rare male advantages in *Drosophila pseudoobscura*. *Behav. Genet.* 4:159–64.

Siegel, W. R. and J. C. Hall. 1979. Conditioned responses in courtship behavior of normal and mutant *Drosophila Proc. Natl. Acad. Sci.* 76:3430–34.

Spieth, H. T. 1968. Evolutionary implications of sexual behavior in *Drosophila*. In Th. Dobzhansky, M. K. Hecht, and W. C. Steere, eds., *Evolutionary Biology 2* New York: Appleton-Century-Croft.

Stanić, S. and D. Marinković. 1990. Intrinsic and acquired in reproductive behavior of *Drosophila*. *Arch. Biol. Sci.* 42:7–8.

———. 1992. Intrinsic and acquired in mating behavior of *Drosophila*. *Drosophila Info. Serv.* 71:246–47.

Sturtevant, A. H. 1915. Experiments on sex recognition and the problem of sexual selection in *Drosophila*. *Animal Behav.* 5:351–66.

Tompkins, L., R. W. Siegel, D. A. Gailey, and J. C. Hall. 1983. Conditioned courtship in *Drosophila* and its mediation by association of chemical cues. *Behav. Genet.* 13: 565–78.

Welbergen, P. H., B. M. Spruijt, and F. R. van Dijken. 1992. Mating speed and interplay between female and male courtship responses in *Drosophila melanogaster. J. Insect Behav.* 3:229–44.

PART SIX

Molecular Studies

23

Population Genetics in *Drosophila pseudoobscura:* A Synthesis Based on Nucleotide Sequence Data for the *Adh* Gene

Stephen W. Schaeffer

The studies of microevolutionary and macroevolutionary change in *Drosophila pseudoobscura* by Theodosius Dobzhansky established many of the fundamental questions addressed in population and evolutionary genetics. Dobzhansky in collaboration with Sturtevant used *D. pseudoobscura* as a model system because they could study the forces that alter gene frequencies within species and the mechanisms that lead to reproductive isolation between species (Provine 1980). Dobzhansky estimated genetic variation at the level of chromosomes, of lethal genes, and of proteins to determine the importance of natural selection in populations. His results should be viewed with caution because he used phenotypic measures of genetic diversity. If he saw two similar phenotypes in different populations, he did not know whether they were identical by descent or identical by kind. This made hypothesis testing difficult because alternative explanations could not be excluded. For instance, Dobzhansky observed extreme differences in inversion frequencies over space and time (Dobzhansky 1939, 1943; Dobzhansky and Queal 1938). Extreme differentiation among

The sequence data described in this paper were determined by my research technician Ellen L. Miller, without whose efforts these analyses would not have been possible. I would like to thank Wyatt W. Anderson and Richard C. Lewontin for thoughtful comments on this manuscript. This work was supported in part by a grant from the National Institutes of Health (GM42472).

different geographic populations may result from either neutral forces such as limited migration and random genetic drift in small isolated demes or selective forces where local environmental differences remove individuals from the population. Dobzhansky could not accurately determine whether the inversions in two populations were identical by descent, which would imply extensive migration between populations, or were identical by kind, which would be consistent with a model of limited migration.

Population genetics has entered a new phase of growth because of significant advances in experimental and theoretical approaches that use DNA to infer the evolutionary history of genes, of populations, and of species. The redundancy of the genetic code superimposes two histories on a nucleotide sequence, a selective history written in the amino acid sequences of proteins and a neutral history described by synonymous and noncoding sites (Lewontin 1985). The examinations of both types of sites along a chromosome allow alternative evolutionary histories to be discriminated because low recombination rates correlate the inheritance patterns of linked nucleotides (Chovnick, Gelbart, and McCarron 1977). Thus, selection acting on nonsynonymous sites can affect neutral variation in linked sites by increasing or decreasing levels of variation with either balancing selection or directional selection (Aquadro 1992, 1993; Kreitman 1987, 1991).

Population samples of nucleotide sequences are now being generated rapidly to estimate critical population parameters and test statistics that are capable of rejecting expectations of the neutral theory (Hudson, Kreitman, and Aguade 1987; Tajima 1989). The neutral theory of molecular evolution provides the null hypothesis that explains how nucleotide sequences will evolve when mutation, random genetic drift, migration, and recombination are the predominant forces acting in populations (Kimura 1983). Coalescent methods allow neutral gene genealogies to be simulated under various population genetic scenarios without extensive need for computer time (Hudson 1990; Hudson and Kaplan 1988; Kaplan, Darden, and Hudson 1988; Kingman 1982a,b; Slatkin 1991; Tajima 1983; Tavare 1984). Simulation results allow strong inferences to be made about the evolutionary history of populations.

A major strength of Dobzhansky's research was that he combined theoretical and empirical approaches to solve population genetics problems (Provine 1980; Smocovitis 1992). Provine (1980:59) suggests that "what Dobzhansky wanted to do was to take the theoretical structure of evolutionary theory, as represented by the work of Fisher, Haldane, and Wright, and flesh it out in accordance with what was actually known about populations of organisms." Dobzhansky used the theoretical framework of Wright even though he had difficulty understanding the mathematical theory; nevertheless, he helped to unify modern biology. Mayr (1993:32) disagrees and suggests that Dobzhansky's early training as a beetle taxonomist helped "to produce a synthesis between the views of naturalists (those who are really studying evolutionary phenomenon) and the geneticists." I think that Mayr underestimates the contribution of Wright's theoretical results to Dobzhansky's view of evolution relative to Dobzhansky's early training as a taxonomist and a naturalist. My graduate and postdoctoral training with Drs. W. W. Anderson and R. C. Lewontin emphasized the importance of comparing experimental data to the predictions of population genetics theory, which may in part be due to their training with Dobzhansky. I think that the current use of both experimental and theoretical approaches is a continuation of an approach begun by Dobzhansky and Wright.

My research program continues and extends the Dobzhansky tradition in three major ways: (1) I use *Drosophila pseudoobscura* as a model organism to study population genetics questions. (2) I determine levels of nucleotide diversity within populations and infer the major evolutionary forces that are responsible for the pattern and organization of nucleotide variation. (3) I combine both theoretical and experimental approaches to answer questions in molecular population genetics. This review highlights how nucleotide sequence data from the alcohol dehydrogenase (*Adh*) region in *Drosophila pseudoobscura* bears on the experimental work of Dobzhansky. I show that: (1) The two gene loci in the *Adh* region have evolved primarily by selectively neutral forces. (2) *D. pseudoobscura* has a large effective population size. (3) *D. pseudoobscura* is capable of extensive gene flow. (4) The Bogota population of *D. pseudoobscura* has diverged from all North American populations of *D. pseudoobscura*. I then discuss the possible mechanisms that explain the genetic differentiation of third chromosome inversion polymorphisms between populations.

The Alcohol Dehydrogenase Region of *Drosophila*

The alcohol dehydrogenase (*Adh*) region is composed of two genes, *Adh* and *Adh-Dup*, that were derived from an ancient duplication of an ancestral gene (figure 23.1) (Schaeffer and Aquadro 1987). The *Adh* gene is transcribed from two promoters during development, but the sequence that encodes the enzyme in the two transcripts is identical (Benyajati et al. 1983). The length of the enzyme is 254 or 256 amino acids, depending on

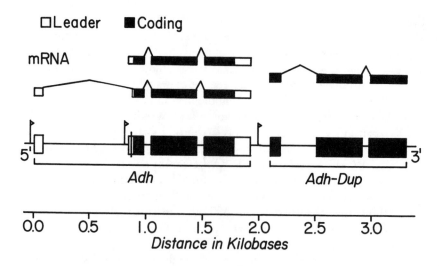

FIGURE 23.1.

Fine structure of the *Adh* region of *D. pseudoobscura*. The region is subdivided into 17 sequence domains from 5′ to 3′: 5′ Flanking, *Adh* Adult Leader, *Adh* Adult Intron, *Adh* Larval Leader, *Adh* Exon 1, *Adh* Intron 1, *Adh* Exon 2, *Adh* Intron 2, *Adh* Exon 3, *Adh* 3′ Leader, Intergenic, *Adh-Dup* Exon 1, *Adh-Dup* Intron 1, *Adh-Dup* Exon 2, *Adh-Dup* Intron 2, *Adh-Dup* Exon 3, and 3′ Flanking.

the *Drosophila* species that is studied (Sullivan, Atkinson, and Starmer 1990). The 5′ or distal promoter is used primarily in adults, whereas the 3′ or proximal promoter is used primarily in larvae (Savakis and Ashburner 1985; Savakis, Ashburner, and Willis 1986). In addition, two tissues, fat body and Malpighian tubules, show the same developmental utilization pattern of the two promoters (Savakis, Ashburner, and Willis 1986). Transcription factors responsible for the regulation of the distal and proximal promoters are currently being isolated (Benyajati et al. 1992; Heberlein, England, and Tjian 1985; Moses, Heberlein, and Ashburner 1990).

The *Adh* locus of *D. melanogaster* encodes two forms of the enzyme in natural populations (Grell, Jacobson, and Murphy 1964; Johnson and Denniston 1964; Ursprung and Leone 1965), and there is a latitudinal cline in gene frequency of the two allozymes formed in the northern and southern hemispheres (Oakeshott et al. 1982; Vigue and Johnson 1973). Nucleotide sequence data suggest that selection plays a prominent role in the maintenance of the two allozymes in natural populations (Aquadro et al. 1986; Hudson, Kreitman, and Aguade 1987; Kreitman 1983; Kreitman and Hudson 1991; Simmons et al. 1989). The *Adh* gene of *D. melanogaster* corresponds (Chr. 2L) to the *Adh-1* gene of *D. pseudoobscura* (Chr. 4) (Chambers et al. 1978), which will hereafter be referred to as *Adh*. The *Adh* gene in *D. pseudoobscura* encodes a single electrophoretic allele (Prakash 1977). The *Adh-Dup* gene was discovered when sequence comparisons of the region 3′ to *Adh* between *D. pseudoobscura* and *D. mauritiana* were found to be highly conserved (Schaeffer and Aquadro 1987). The sequence has been reported in numerous members of the subgenus *Sophophora* (Sullivan, Atkinson, and Starmer 1990) and has recently been detected in the *Scaptodrosophila* and *Drosophila* subgenera (Albalat and Gonzalez-Duarte 1993). All evidence gathered to date, e.g., synonymous to nonsynonymous ratio, codon bias, and cDNA clones, suggests that *Adh-Dup* produces a functional gene product, but what the protein does is unknown.

Nucleotide Sequence Data

The complete nucleotide sequence (3.5 kilobases) of the *Adh* region for 109 strains of *D. pseudoobscura* and its close relatives has been determined with either molecular cloning or PCR-mediated sequencing techniques (Higuchi and Ochman 1989; Saiki et al. 1988; Sambrook, Fritsch, and Maniatis 1989; Sanger, Nicken, and Coulson 1977; Schaeffer and Aquadro 1987; Schaeffer and Miller 1991, 1992a,b). The geographic range of *D. pseudoobscura* and the location of populations with sample sizes greater than three strains are shown in figure 23.2. The locations and numbers of the strains sequenced in each of 12 *D. pseudoobscura* populations were: Stemwinder Provincial Park, B. C., Canada (13); Rainbow Orchard, Apple Hill, Calif. (10); Gundlach-Bundschu Winery, Sonoma, Calif. (26); Bryce Canyon National Park, Utah (5); Mesa Verde National Park, Colo. (5); Kaibab National Forest, Ariz. (26); San Bernardino Mountains, Calif. (4); Mexico (5); Brookings, Oreg. (2); MacDonald Ranch, Calif. (1); Baja California, Mexico (1); and Gregory Canyon, Colo. (1). In addition, we have determined the sequences of 8 strains of the subspecies *D. pseudoobscura bogotana*, and 1 strain each of the sibling species *D. persimilis* and *D. miranda*. The GenBank/EMBL Data Library accession numbers for all sequences discussed in this paper have been previously published (Schaeffer and Miller 1993).

The nucleotide sequences have been aligned by minimizing the numbers of gaps and mismatches assumed in the sequences. All analyses considered in this paper used polymorphic nucleotide sites or segregating

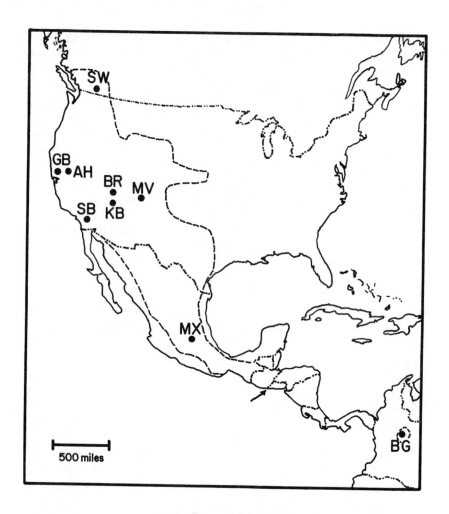

FIGURE 23.2.

Geographic range and collection localities of *Drosophila pseudoobscura*. The geographic range is designated by (– – – –). The arrow shows the southern boundary of the range, which is coincident with the southern border of Guatemala. The names of populations are: SW, Stemwinder Provincial Park, British Columbia, Canada; AH, Apple Hill, California; GB, Gundlach-Bundschu Winery, Sonoma Valley, California; BR, Bryce Canyon National Park, Utah; MV, Mesa Verde National Park, Colorado; KB, Kaibab National Forest, Arizona; SB, San Bernardino Mountains, California; MX, Mexico; and BG, Bogota, Colombia.

sites. A segregating site was defined as any aligned nucleotide site that had two or three nucleotides present. Any sites found within insertions or deletions inferred from the locations of gaps in the alignments were excluded from all analyses. Nine populations had sample sizes greater than 3: Stemwinder, Apple Hill, Gundlach-Bundschu, Bryce Canyon, Mesa Verde, Kaibab, San Bernardino, Mexico, and Bogota (figure 23.2).

Adh and the Neutral Theory

The selectionist view held by Dobzhansky was in direct contrast to the neutral theory of Kimura (1983). Dobzhansky assumed that all genetic variation in populations would affect an organism's fitness. He felt that genetic variation could protect a species from the consequences of an environmental change (Dobzhansky 1970). As a result, he predicted that most protein encoding genes within an organism's genome would be heterozygous (Dobzhansky 1955). Kimura, on the other hand, assumed that the vast majority of nucleotide changes were selectively neutral, i.e., functionally equivalent with respect to fitness. Thus, the neutral theory predicts that the vast majority of genetic variation within populations reflects a transient phase in molecular evolution. One of the newest goals in population genetics is to determine what proportion of genes within the genome is acted upon by positive Darwinian selection (Kreitman 1987, 1991).

A distinction should be made between positive and negative selection, for each force affects a newly arising mutation differently. New mutations are always rare and will have a frequency of 1/2N in a population of N individuals. In most cases, new mutations will be lost from the population but they will occasionally go to fixation. The probability that a new selectively neutral mutation will be fixed by random genetic drift is equal to its initial frequency, 1/2N (Kimura 1983). New deleterious mutations will be fixed with a probability less than the neutral expectation because of negative selection. Beneficial mutations will be fixed with a probability greater than a neutral mutation because of positive selection. The difference in the fixation rate for new selected mutations compared to a neutral gene depends on the strength of selection and the degree of dominance. Completely recessive alleles will behave as neutral genes until they have reached an appreciable frequency to be influenced by selection.

The controversy between selectionists and neutralists was about the role that positive Darwinian selection plays in the evolutionary process. Both groups acknowledged that individuals with deleterious traits will be removed from populations because of negative selection, but they disagreed on the importance of positive selection for the long-term evolution of populations. Kimura agreed that positive Darwinian selection occurs but suggested that fixations of beneficial alleles happen so quickly that they are unimportant in molecular evolution (Kimura 1983). Dobzhansky (1970:262) felt that positive selection actively maintains genetic diversity but acknowledged that "the strongest theoretical argument for the neutrality of some mutations is based on the degeneracy of the genetic code." The nucleotide sequence data from the *Adh* region of *D. pseudoobscura* will be discussed in considering Dobzhansky's selectionist viewpoint.

The sequences of 99 *Adh* genes reveal two forms of the inferred ADH enzyme. The major form is found in 98 strains of *D. pseudoobscura* and the minor form is found once. Two percent of the individuals in *D. pseudoobscura* populations would be expected to be heterozygous for the ADH enzyme. Two hypotheses may explain why the *Adh* gene in *D. pseudoobscura* encodes predominantly one enzyme form in natural populations, a neutral or selective hypothesis. The ADH enzyme may be nearly monomorphic because *Adh* has had strong selective constraints against amino acid replacements that result in a low neutral mutation rate for amino acid changes. In other words, most new amino acid changes are deleterious. Alternatively, a new protein form of ADH may have rapidly increased to near fixation by directional selection. Examination of synonymous variation in the *Adh* region can discriminate between these alternative hypotheses.

If the strong selective constraint hypothesis is true, then the substitution rates at linked neutral sites will be independent of the rates in amino acid replacement positions (Birky and Walsh 1988). Neutral synonymous variation would be expected to reflect the transient fixation process of an equilibrium neutral model. On the other hand, if the directional selection hypothesis is true, then the rapid fixation of a new enzyme allele would also reduce variation in linked neutral sites by the hitchhiking effect (Aguade, Miyashita, and Langley 1989; Begun and Aquadro 1991, 1992; Berry, Ajioka, and Kreitman 1991; Kaplan, Hudson, and Langley 1989; Maynard Smith and Haigh 1974). Under this scenario, the number of segregating linked sites would be expected to be small and the frequency spectrum of variable sites would have an excess of rare variants.

Two methods were used to test variable synonymous sites in *Adh* for departures from an equilibrium neutral model (Hudson, Kreitman, and Aguade 1987; Tajima 1989). Tajima's (1989) method computes a difference statistic, D, that is the difference of two estimates of the neutral mutation parameter, $M(S)$ and $M(k)$, that are determined from the number of segregating sites, S, and average number of pairwise differences, k, respectively. D, is distributed as Beta random variable and should not differ significantly from zero under an equilibrium neutral model. The Hudson, Kreitman, and Aguade (HKA) (1987) method is a modified test of statistical independence that determines whether regions that are highly polymorphic within species also diverge rapidly between species among several genetic loci. The HKA test statistic, X^2, is distributed as a chi-square random variable that assumes no recombination within locus and free recombination between loci. The ratio of within-species polymorphism and between-species divergence will be equivalent among loci under a neutral model. Any departures from statistical independence among the tested loci may indicate the action of positive Darwinian selection.

Estimates of $M(S)$ and $M(k)$ for 17 sequence domains in the *Adh* region in the 99 strains are shown in table 23.1. The Tajima test fails to reject an equilibrium neutral model for the 10 sequence domains of Adh ($M(S)=42.97$, $M(k)=23.97$, $D=-1.487$, $p>0.05$), for the 7 sequence domains of *Adh-Dup* ($M(S)=24.38$, $M(k)=12.26$, $D=-1.650$, $p>0.05$), and for the 17 sequence domains of the entire region ($M(S)=67.35$, $M(k)= 36.23$, $D=-1.563$, $p>0.05$).

The HKA test requires between-species divergence data and subdivision of the region into two or more loci, e.g., *Adh* versus *Adh-Dup*. the *D. miranda* rather than *D. persimilis* sequence was used for the between-species comparison because the *D. persimilis* sequence does not differ significantly from any of the *D. pseudoobscura* sequences (Schaeffer and Miller 1991). Excessive subdivision of the region for the HKA test could lead to false rejection of the neutral hypothesis because expected values of some cells may be less than 5. Alternatively, lumping too many domains together in the HKA test could lead to the loss of power to detect domains within the *Adh* region that may be influenced by selection. Thus, the *Adh* region was subdivided in two ways to avoid the problems listed above. The first HKA test subdivided the region into 17 subregions based on the fine structure of the gene (figure 23.1), and the second test partitioned the region into the genetic loci, *Adh* and *Adh-Dup*. The observed and expected number of polymorphic and divergent sites determined for the two HKA tests are shown in figure 23.2. Both HKA tests fail to reject a neutral model (17 domain test: $X^2=9.15$, df=16, $p>0.90$; 2 domain test: $X^2=0.27$, df=1, $p>0.50$) (table 23.2). The Tajima and HKA tests also fail to reject a neutral model in the 9 populations with sample sizes greater than 3 (Schaeffer and Miller 1992a,b). These data suggest that ADH is nearly monomorphic

TABLE 23.1.

*Estimates of the Neutral Mutation Parameter, 4Nμ, Based on
Synonymous Sites on a Per Locus and Per Nucleotide Basis
for the Adh Region in 99 Strains of* Drosophila pseudoobscura

Region	m	Per Locus Estimate		Per Nucleotide Estimate	
		$M(S)$	$M(k)$	L	π
Adh					
5' Flanking	39	1.16	0.43	0.030	0.011
Adult Leader	80	1.94	0.28	0.024	0.003
Adult Intron	700	23.42	12.35	0.033	0.018
Larval Leader	46	0.39	0.10	0.008	0.002
Exon 1	23	0.58	0.10	0.025	0.004
Intron 1	55	3.10	2.25	0.056	0.041
Exon 2	101	4.06	1.25	0.040	0.012
Intron 2	59	4.06	4.92	0.069	0.083
Exon 3	66	2.71	1.84	0.041	0.028
3' Leader	151	1.55	0.44	0.010	0.003
Intergenic	153	1.74	0.71	0.011	0.005
Adh-Dup					
Exon 1	24	0.58	0.06	0.024	0.003
Intron 1	205	4.84	1.66	0.024	0.008
Exon 2	101	6.39	3.39	0.063	0.034
Intron 2	53	2.90	2.16	0.055	0.041
Exon 3	83	6.77	3.93	0.081	0.047
3' Flanking	97	1.16	0.35	0.012	0.004
Summary					
Adh	1320	42.97	23.97	0.033	0.018
Adh-Dup	716	24.38	12.26	0.034	0.017
Total region	2036	67.35	36.23	0.033	0.018

m, effective number of synonymous sites; $M(S)$ and L, estimates of the neutral mutation parameter based on the number of synonymous sites (Tajima 1989, equation 5; Nei 1987, equation 10.3, respectively); and $M(k)$ and π, estimates of the neutral mutation parameter based on the average number of pairwise differences (Tajima 1989, equation 6; Nei 1987, equation 10.5).

because of strong selective constraints against amino acid substitutions rather than positive Darwinian selection.

The amino acid variation of ADH-DUP stands in sharp contrast to the polymorphism observed in the ADH protein. Nine of the 278 amino acid positions in *Adh-Dup* are polymorphic in *D. pseudoobscura* populations (S. W. Schaeffer and E. L. Miller, unpublished data). The variable amino acid sites generate 14 different protein alleles, and 79.2% of the individuals would be expected to be heterozygous within *D. pseudoobscura* populations, where heterozygosity is defined by $h = 1 - \Sigma p^2_i$, and p_i and p_i is the frequency of the ith protein allele. The observed and expected frequency distributions of ADH-DUP alleles were tested for departures from a neutral model with the Ewens-Watterson test (Ewens 1972; Watterson 1978). The Ewens-Watterson test fails to reject an infinite alleles model of molecular evolution (observed $\overline{F}$=0.208, expected $\overline{F}$=0.194, number of strains sampled=99, number of alleles=14, the 95% confidence interval =0.113 ≤ $\overline{F}$ ≤ 0.367 [see Hartl and Clark 1989; Whittam et al. 1986 for details on the test statistic and hypothesis testing]). Synonymous and nonsynonymous variation in ADH-DUP alleles is consistent with an equilibrium neutral model. The selective constraints on ADH-DUP appear to be relaxed compared to the extreme conservation of the ADH amino acid sequence.

Griffiths (1982) examined the joint distribution of the number of alleles and segregating sites in a random sample from the infinite alleles model

TABLE 23.2.

Results of the HKA Test for the Adh *Region Based on Sequence Comparisons of 99 Strains of* Drosophila pseudoobscura

Region	M(S)	Within-Species Polymorphism		Between-Species Divergence	
		Observed	Expected	Observed	Expected
Adh					
5' Flanking	1.2	6	6.0	2	2.0
Adult Leader	1.7	10	8.9	2	3.1
Adult Intron	24.5	121	126.7	49	43.3
Larval Leader	0.7	2	3.7	3	1.3
Exon 1	0.4	3	2.2	0	0.8
Intron 1	3.3	16	17.1	7	5.9
Exon 2	3.6	21	18.6	4	6.4
Intron 2	4.8	21	24.6	12	8.4
Exon 3	2.6	14	13.4	4	4.6
3' Leader	1.3	8	6.7	1	2.3
Intergenic	2.2	9	11.2	6	3.8
Adh-Dup					
Exon 1	0.7	3	3.7	2	1.3
Intron 1	4.9	25	25.3	9	8.7
Exon 2	5.8	33	29.8	7	10.2
Intron 2	2.6	15	13.4	3	4.6
Exon 3	5.6	35	29.1	4	9.9
3' Flanking	1.4	6	7.5	4	2.5
Two regions					
Adh	46.3	231	239.2	90	81.8
Adh-Dup	21.1	117	108.2	29	37.2

M(S) is the estimate of the neutral mutation parameter based on the number of synonymous sites (Watterson 1975, equation 1.4a). The observed values with species are the numbers of synonymous segregating sites within each region. The observed values between species are the number of nucleotide differences between a *D. pseudoobscura* and a *D. miranda* sequence. The estimate of the divergence time between *D. pseudoobscura* and *D. miranda* is 1.8.

and found a high correlation between the two variables. I examined this prediction of a neutral model by determining whether the number of unique polymorphisms or singletons in the *Adh* region was proportional to the number of alleles sampled from the 8 North American populations. A unique polymorphism (also referred to as an external site [Fu and Li 1993]) is defined as a segregating site where the low-frequency nucleotide is not shared with the outgroup species, *D. miranda*, and has a frequency of $1/n$ within *D. pseudoobscura*, where n is the sample size. I use unique polymorphisms rather than all segregating sites because I assume that singletons reflect the introduction of new mutations. The unique polymorphisms in the *Adh* region were determined from 94 sequences collected from North American populations. The singletons were assigned to a population based on which population had the sequence with the rare-frequency nucleotide. A χ^2 goodness of fit test was used to determine whether the observed numbers of unique polymorphisms in each population were proportional to that expected given each population's sample size. The number of singletons within a population is proportional to sample size (table 23.3) ($\chi^2 = 10.74$, $df = 7$, $p > 0.10$), and thus these data are consistent with an infinite alleles model where every new mutation generates a new allele (Griffiths 1982).

There is no evidence that positive Darwinian selection has acted to maintain protein diversity or fix beneficial enzyme forms in *Adh* or *Adh-Dup* in the recent history of these populations. The nucleotide data from *Adh* and *Adh-Dup* are not consistent with Dobzhansky's selectionist view of

TABLE 23.3.

Observed and Expected Number of Unique Polymorphisms in 8 Populations of Drosophila pseudoobscura Based on 94 Sequences of the Adh Region

Population	N	Observed	Expected
San Bernardino, Calif.	4	4	6.0
Mesa Verde, Colo.	5	4	7.6
Mexico	5	10	7.6
Bryce Canyon, Utah	5	13	7.6
Apple Hill, Calif.	10	15	15.2
Stemwinder, B.C.	13	15	19.8
Gundlach-Bundschu, Calif.	26	34	39.5
Kaibab, Ariz.	26	48	39.5
Total	94	143	143.0

N=sample size, Observed=observed number of segregating sites, Expected=expected number of segregating sites determined from the fraction of population sample size relative to the total number of strains.

the genome. The generality of these results should be viewed with caution because the studies of Schaeffer and Miller (1992a, b) consider nucleotide evolution in only two closely linked genes. Indeed, Begun and Aquadro (1992) suggest that positive selection has left its signature in the genome of *D. melanogaster*, especially when genes are located in regions with low recombination rates. Examination of more genes within the *Drosophila* genome is warranted to determine what proportion of the genome has been acted upon by positive Darwinian selection.

Effective Population Size, Gene Flow, and Migration

Dobzhansky and Wright used several approaches to estimate the effective population size (*N*) of *D. pseudoobscura* (Dobzhansky and Wright 1941; Wright, Dobzhansky, and Hovanitz 1942) because they wanted to explain the forces that generate the geographic differentiation of the third chromosomal inversion frequencies (Dobzhansky 1939; Dobzhansky and Queal 1938). Random genetic drift in small populations was thought to explain local differentiation of inversion frequencies if effective population sizes were small. Alternatively, if effective population sizes were large, then local adaptation was more likely to explain the geographic differentiation of inversion frequencies.

Estimates of the allelism of lethals within and between populations suggested that the effective population size of *D. pseudoobscura* was 1,000 to 80,000 individuals (Dobzhansky and Wright 1941; Wright, Dobzhansky, and Hovanitz 1942). Estimates of dispersion rates of flies from mark-recapture experiments indicate that the effective population size of *D. pseudoobscura* was approximately 80,000 (Crumpacker and Williams 1973; Dobzhansky and Wright 1943; Powell et al. 1976). These estimates of effective population size suggested that random genetic drift was the primary cause of differences in gene frequency between populations.

The nucleotide data from the *Adh* region of *D. pseudoobscura* may be used to estimate effective population size independently from the neutral mutation ($4N\mu$[Nei 1987, see equation 10.5]) and from the recombination ($4Nc$ [Hudson 1987, see equations 1, 2, and 4]) parameters, where μ is the neutral mutation rate per locus per generation and c is the neutral recombination rate per locus per generation. The *Adh* region is an ideal segment to use to estimate effective population size because the region has not been subject to natural selection that may reduce estimates of the neutral mutation and of the recombination parameters (Schaeffer and Miller 1992a, b). The $4N\mu$ and $4Nc$ were estimated to be 29.9 and 487.3, respectively, based on selectively neutral sites in the *Adh* region of *D. pseudoobscura* (Schaeffer and Miller 1993).

Li, Luo, and Wu (1985) estimated the average number of neutral mutations per year per base in pseudogenes to be 4.9×10^{-9}. The number of neutral mutation events expected in the 3.2 kilobase *Adh* region in *D. pseudoobscura:* $\mu \approx ([4.9 \times 10^{-9}$ events per base pair per year] $[3.2 \times 10^3$ base pairs])/(4 generations per year) $= 3.9 \times 10^{-6}$ events per locus, which assumes 4 generations of *D. pseudoobscura* per year. Thus, the estimate of the effective population size based on the neutral mutation rate is: $4N\mu /4\mu$ $=1.9 \times 10^6$.

Chovnick, Gelbart, and McCarron (1977) estimated the average number of recombination events per base pair in *D. melanogaster* females to be 1.7×10^{-8}. I may estimate the number of recombination events expected in the 3.2 kilobase *Adh* region in *D. pseudoobscura:* $c \approx (1.7 \times 10^{-8}$ events per base pair)(0.5)(3.2 $\times 10^3$ base pairs) $= 2.7 \times 10^{-5}$ events, which assumes that recombination occurs only in females. Thus, the estimate of the effective population size based on the neutral mutation rate is: $4Nc/4c = 4.5 \times 10^6$.

The estimates of effective population size of *D. pseudoobscura* based on the *Adh* sequence data are 24- to 56-fold greater than those of Dobzhansky and his collaborators based on their estimate of $N = 80,000$ (Dobzhansky and Wright 1941, 1943; Powell et al. 1976; Wright, Dobzhansky, and Horanitz 1942). These data suggest that the role of random genetic drift is not as strong in *D. pseudoobscura* as originally supposed by Dobzhansky and Wright. The large estimate of effective population size is consistent with the lack of amino acid polymorphism in *Adh* and higher levels of variation in synonymous sites. A population with a large effective size would harbor more variation in selective neutral sites, e.g., synonymous sites, but purifying selection would remove amino acid changes even though the phenotype was slightly deleterious (Aquadro, Lado, and Noon 1988; Ohta 1976).

The sequences from the *Adh* region have been used to estimate the migration parameter, *Nm*, among all pairs of the 9 populations collected from the geographic range of *D. pseudoobscura* (Schaeffer and Miller 1992a), where *m* is the migration rate per generation. The estimates of *Nm* show that 1 to 12 migrants are exchanged among North American populations per generation. The average value for *Nm* for all *D. pseudoobscura* populations is 2.38 migrants per generation with a lower limit of 2.26 and an upper limit of 3.36 for a 95% confidence interval. These data suggest that extensive gene flow occurs among *D. pseudoobscura* populations in North America. These values of *Nm* are consistent with direct estimates of dispersal (Coyne et al. 1982; Coyne, Bryant, and Turelli 1987; Jones et al. 1981) and other indirect estimates of genetic relatedness of populations based on allozyme and nucleotide data (Keith 1983; Keith et al. 1985; Riley, Hallas, and Lewontin 1989)

The movement of 2 individuals per generation is sufficient to homogenize gene frequencies among all geographic populations of *D. pseudoobscura* in the North America (Wright 1931). The range of *Nm* estimates shows that some populations may be exchanging more than 2 migrants per generation. The relationship between the number of migrants exchanged between populations and the interpopulation distance is consistent with an isolation-by-distance model (Wright 1943). Populations of *D. pseudoobscura* must, however, be separated by greater than 2000 kilometers before the number of migrants exchanged between populations falls below 1.

The Bogota Population of *D. pseudoobscura*

The time and mode of origin of the Bogota population of *D. pseudoobscura* have been the subject of moderate debate since its discovery by Dobzhansky and his colleagues (1963). The population was initially thought to be a recent introduction because extensive fly collections by researchers at the University of Texas failed to find any *D. pseudoobscura* in the lowlands and mountains of Central America. The isolated population of flies became more interesting when Prakash (1972) showed that the Bogota and United States flies were partially reproductively isolated, despite little genetic differentiation among these populations at the allozyme level (Prakash, Lewontin, and Hubby 1969). Prakash concluded that new species could form quickly because little genetic change was observed between populations. This is consistent with the suggestion of Hoenigsberg (1986) who stated that the Bogota population was founded recently by a few individuals followed by a rapid expansion of population size. Subsequent allozyme studies did find unique genetic differences in Bogota flies suggesting that the population was old enough to become reproductively isolated from United States populations (Ayala and Dobzhansky 1974; Coyne and Felton 1977; Singh, Lewontin, and Felton 1976).

Schaeffer and Miller (1991, 1992a) have used the nucleotide sequence data from the *Adh* region to estimate the time of origin of the Bogota population of *D. pseudoobscura*. They have shown that Bogota flies are genetically distinct from flies in North American populations and suggest that the Bogota population was isolated 155,000 years ago. Two hypotheses could account for the geographic isolation of Bogota flies, distance or environment. The homogeneity of North American populations suggests that flies can disperse distances that are greater than the distance that separates Guatemala and Bogota (Schaeffer and Miller 1992a); however, flies do not seem to be able to move between Central America and Bogota. Thus, an inhospitable environment in Central America rather than distance between Guatemala and Bogota is a better explanation of the geographic separation of North and South American populations of *D. pseudoobscura* (Schaeffer and Miller 1991, 1992a).

Schaeffer and Miller (1991) speculate on the biogeographical and environmental factors that led to the isolation of the Bogota population. They suggest that the Isthmus of Panama formed a land bridge in Central America 3–4 million years ago (Jones and Hasson 1985), which may have allowed *D. pseudoobscura* to expand its range from North to South America, or vice versa. The climate in present-day tropical regions could have been cooler and drier during periods of glacial maxima during the Pleistocene, allowing the flies to extend their distribution into Central and South America (Epling 1944). The limited knowledge of *D. pseudoobscura* ecology in Guatemala and Mexico supports this idea because flies are found only at higher elevations that are cooler and drier, even though flies in North America survive at lower elevations (Dobzhansky and Epling 1944). The tropical climate of the Isthmus of Panama now prohibits the movement of *D. pseudoobscura* between the two continents. The estimated date of the isolation of the Bogota population, 155,000 years ago, coincides with the most recent last glacial maximum (Van Campo et al. 1990).

Adh and the Inversion Polymorphisms

The geographic differentiation of inversion frequencies may be explained by two hypotheses, random genetic drift in small populations or a complex selection mechanism. The random genetic drift hypothesis assumes that *D. pseudoobscura* has either a small effective population size, a low migration rate, or both. Molecular studies of the *Adh* gene have shown that *Drosophila pseudoobscura* has a large effective population size that would allow even subtle fitness differences to be detected by selection. In addition, the average number of migrants is sufficient to genetically homogenize geographic populations (Schaeffer and Miller 1992a). Thus, the effects of random genetic drift in *D. pseudoobscura* are minimal, which disagrees with the initial conclusions of Dobzhansky (Dobzhansky 1939; Dobzhansky and Queal 1938; Koller 1939). Extensive gene flow should homogenize gene frequencies of all loci in the genome provided mutational and selective forces are equivalent (Cavalli-Sforza 1966; Slatkin 1985). Indeed, gene frequencies are spatially homogeneous on all chromosomes except the third, which is polymorphic for inversions (Aquadro et al. 1991; Keith 1983; Keith et al. 1985; Lewontin 1974; Prakash, Lewontin and Hubby 1969; Riley, Hallas, and Lewontin 1989).

The geographic differentiation of the inversions stands in sharp contrast to the expected genetic homogeneity predicted on the basis of the extensive gene flow (Riley, Hallas, and Lewontin 1989; Schaeffer and Miller 1992a; see Anderson 1989; Powell 1992 for reviews). Restriction map variation in the third chromosome gene amylase shows extreme differentiation between inversion types (Aquadro et al. 1991); however, amylase haplotypes within inversions tend to be genetically similar when chromosomes between geographic populations are compared. Thus, natural selection acting in varying environments would appear to counteract the homogenizing effects of migration (Endler 1973; Lewontin 1974). This explanation seems most parsimonious in considering inversion frequencies in the western United States. California populations are highly polymorphic for inversions, while populations in the southwestern desert are nearly monomorphic for the Arrowhead inversion (Anderson et al. 1991). The desert populations in Arizona and New Mexico should be highly polymorphic for inversion types, given the high migration rate and the diversity of California populations. If California flies are transported to Arizona and New Mexico, then only flies that carry Arrowhead survive after arriving from other parts of the geographic range. Similar reasoning has been used to support a selective explanation for the cline in *Adh* allozyme frequencies in *Drosophila melanogaster* in the eastern United States (Berry and Kreitman 1993).

Selection in varying environments does not explain why many inversions in California populations have intermediate frequencies. Does the polymorphism of inversions in California represent a stable balanced polymorphism? Studies of population cages have suggested that the inversions are maintained by balancing selection (see Anderson 1989 for a recent review of the factors that can change the stable equilibria). The homogeneity of inversion frequencies among California populations is consistent with a model of extensive migration, but it is not clear why so many inversions would be maintained across so many environments. Does this suggest that populations in California are fairly uniform? Seasonal fluctuations in inversion frequencies would suggest that the environments are not the same and do change over time (Dobzhansky 1943). Further studies of nucleotide diversity of genes on the third chromosome may provide valuable insight

into the forces acting on the inversion polymorphisms in California populations. For instance, are the inversions in different populations identical
by descent and is there any evidence for recombination between inversions
or is each inversion genetically isolated from all the others?

Theodosius Dobzhansky thought that natural selection was a pervasive
force in the evolution of organisms. Dobzhansky provided the current generation of population geneticists with fundamental evolutionary questions
to answer and a tradition of combining experimental and theoretical
approaches. I am not sure how Dobzhansky would react to the work that I
have done, because I have demonstrated that some genes, *Adh* and *Adh-
Dup*, are consistent with neutral evolution. On the other hand, the *Adh*
data provide strong support for the role selection plays in modulating the
frequencies of inversions between populations, which was one of
Dobzhansky's lifelong research goals.

REFERENCES
Aguade, M., N. Miyashita, and C. H. Langley. 1989. Reduced variation in the
yellow-achaete-scute region in natural populations of *Drosophila melanogaster.*
Genetics 122:607–15.
Albalat, R. and R. Gonzalez-Duarte. 1993. *Adh* and *Adh-dup* sequences of
Drosophila lebanonensis and *D. immigrants:* Interspecies comparisons. *Gene*
126:171–78.
Anderson, W. W. 1989. Selection in natural and experimental populations of
Drosophila pseudoobscura. Genome 31:239–45.
Anderson, W. W., J. Arnold, D. G. Baldwin, A. T. Beckenbach, C. J. Brown, S.
H. Bryant, J. A. Coyne, L. G. Harshman, W. B. Heed, D. E. Jeffrey, L. B.
Klaczko, B. C. Moore, J. M. Porter, J. R. Powell, T. Prout, S. W. Schaeffer, J.
C. Stephens, C. E. Taylor, M. E. Turner, G. O. Williams, and J. A. Moore.
1991. Four decades of inversion polymorphism in *Drosophila pseudoobscura.*
Proc. Natl. Acad. Sci. 88:10367–71.
Aquadro, C. F. 1992. Why is the genome variable? Insights from *Drosophila.*
Trends in Genet. 8:355–62.
———. 1993. Molecular population genetics of *Drosophila.* In J. Oakeshott and
M. J. Whitten, eds., *Molecular Approaches to Fundamental and Applied
Entomology,* pp. 222–66. New York: Springer-Verlag.
Aquadro, C. F., S. F. Deese, M. M. Bland, C. H. Langley, and C. C. Laurie-
Ahlberg. 1986. Molecular population genetics of the alcohol dehydrogenase
gene region of *Drosophila melanogaster. Genetics* 114:1165–90.
Aquadro, C. F., K. M. Lado, and W. A. Noon. 1988. The *rosy* region of
Drosophila melanogaster and *Drosophila simulans.* I: Contrasting levels of natu-

rally occurring DNA restriction map variation and divergence. *Genetics* 119:875–88.

Aquadro, C. F., A. L. Weaver, S. W. Schaeffer, and W. W. Anderson. 1991. Molecular evolution of inversions in *Drosophila pseudoobscura:* The amylase gene region. *Proc. Natl. Acad. Sci.* 88:305–9.

Ayala, F. J. and Th. Dobzhansky. 1974. A new subspecies of *Drosophila pseudoobscura. Pan-Pacific Entomol.* 50:211–19.

Begun, D. J. and C. F. Aquadro. 1991. Molecular population genetics of the distal portion of the *X* chromosome in *Drosophila:* Evidence for genetic hitchhiking of the *yellow-achaete* region. *Genetics* 129:1147–58.

———. 1992. Levels of naturally occurring DNA polymorphism correlate with recombination rates in *D. melanogaster. Nature* 356:519–20.

Benyajati, C., A. Ewel, J. McKeon, M. Chovav, and E. Juan. 1992. Characterization and purification of *Adh* distal promoter factor 2, Adf-2, a cell-specific and promoter-specific repressor in *Drosophila. Nucleic Acids Res.* 20:4481–89.

Benyajati, C., N. Spoerel, H. Haymerle, and M. Ashburner. 1983. The messenger RNA for alcohol dehydrogenase in *Drosophila melanogaster* differs in its 5'end in different developmental stages. *Cell* 33:125–33.

Berry, A. and M. Kreitman. 1993. Molecular analysis of an allozyme cline: Alcohol dehydrogenase in *Drosophila melanogaster* on the east coast of North America. *Genetics* 134:869–93.

Berry, A. J., J. W. Ajioka, and M. Kreitman. 1991. Lack of polymorphism on the *Drosophila* fourth chromosome resulting from selection. *Genetics* 129:1111–17.

Birky, C. W. and J. B. Walsh. 1988. Effects of linkage on rates of molecular evolution. *Proc. Natl. Acad. Sci.* 85:6414–18.

Cavalli-Sforza, L. L. 1966. Population structure and human evolution. *Proc. R. Soc. Lond. B.* 164:362–79.

Chambers, G. K., J. F. McDonald, M. McElfresh, and F. J. Ayala. 1978. Alcohol-oxidizing enzymes in 13 *Drosophila* species. *Biochem. Genet.* 16:757–67.

Chovnick, A., W. Gelbart, and M. McCarron. 1977. Organization of the Rosy locus in *Drosophila melanogaster. Cell* 11:1–10.

Coyne, J. A., I. A. Boussy, T. Prout, S. H. Bryant, J. S. Jones, and J. A. Moore. 1982. Long-distance migration of *Drosophila. Am. Natur.* 119:589–95.

Coyne, J. A., S. H. Bryant, and M. Turelli. 1987. Long-distance migration of *Drosophila.* 2: Presence in desolate sites and dispersal near a desert oasis. *Am. Natur.* 129:847–61.

Coyne, J. A. and A. A. Felton. 1977. Genic heterogeneity at two alcohol dehydrogenase loci in *Drosophila pseudoobscura* and *Drosophila persimilis. Genetics* 87:285–304.

Crumpacker, D. W. and J. S. Williams. 1973. Density, dispersion, and population structure in *Drosophila pseudoobscura. Ecol. Monogr.* 43:499–538.

Dobzhansky, Th. 1939. Genetics of natural populations. IV: Mexican and Guatemalan populations of *Drosophila pseudoobscura. Genetics* 24:391–412.

———. 1943. Genetics of natural populations. IX: Temporal changes in the composition of populations of *Drosophila pseudoobscura*. *Genetics* 28:162–86.

———. 1955. A review of some fundamental concepts and problems of population genetics. *Cold Spring Harbor Symp. Quant. Biol.* 20:1–15.

———. 1970. *Genetics of the Evolutionary Process.* New York: Columbia University Press.

Dobzhansky, Th. and C. Epling. 1944. Taxonomy, geographic distribution, and ecology of *Drosophila pseudoobscura* and its relatives. In T. Dobzhansky and C. Epling, eds., *Contributions to the Genetics, Taxonomy, and Ecology of Drosophila pseudoobscura and Its Relatives*, pp. 1–46. Baltimore: The Lord Baltimore Press.

Dobzhansky, Th., A. S. Hunter, O. Pavlovsky, B. Spassky, and B. Wallace. 1963. Genetics of natural populations. XXXI: Genetics of an isolated marginal population of *Drosophila pseudoobscura*. *Genetics* 48:91–103.

Dobzhansky, Th. and M. L. Queal. 1938. Genetics of natural populations. I: Chromosome variation in populations of *Drosophila pseudoobscura* inhabiting isolated mountain ranges. *Genetics* 23:239–51.

Dobzhansky, Th. and S. Wright. 1941. Genetics of natural populations. V: Relations between mutation rate and accumulation of lethals in populations of *Drosophila pseudoobscura*. *Genetics* 26:23–51.

———. 1943. Genetics of natural populations. X: Dispersion rates in *Drosophila pseudoobscura*. *Genetics* 28:304–40.

Endler, J. A. 1973. Gene flow and population differentiation. *Science* 179:243–50.

Epling, C. 1944. The historical background. In T. Dobzhansky and C. Epling, eds. *Contributions to the Genetics, Taxonomy, and Ecology of Drosophila pseudoobscura and Its Relatives*, pp. 145–83. Baltimore: The Lord Baltimore Press.

Ewens, W. J. 1972. The sampling theory of selectively neutral alleles. *Theor. Popul. Biol.* 3:87–112.

Fu, Y.-X. and W.-H. Li. 1993. Statistical tests of neutrality of mutations. *Genetics* 133:693–709.

Grell, E. H., K. B. Jacobson, and J. B. Murphy. 1964. Alcohol dehydrogenase in *Drosophila melanogaster:* Isozymes and genetic variants. *Science* 149:80–82.

Griffiths, R. C. 1982. The number of alleles and segregating sites in a sample from the infinite-alleles model. *Adv. Appl. Prob.* 14:225–39.

Hartl, D. L. and A. G. Clark. 1989. *Principles of Population Genetics.* Sunderland, Mass.: Sinauer.

Heberlein, U., B. England, and R. Tjian. 1985. Characterization of *Drosophila* transcription factors that activate the tandem promoters of the alcohol dehydrogenase gene. *Cell* 41:965–77.

Higuchi, R. G. and H. Ochman. 1989. Production of single-stranded DNA templates by exonuclease digestion following the polymerase chain reaction. *Nucleic Acids Res.* 17:5865.

Hoenigsberg, H. F. 1986. Population genetics in the American Tropics. XXII: The penetration of *Drosophila pseudoobscura* into Colombia. *Biol. Zentralbl.* 105:249–55.

Hudson, R. R. 1987. Estimating the recombination parameter of a finite population model without selection. *Genet. Res.* 50:245–50.

———. 1990. Gene genealogies and the coalescent process. *Oxf. Surv. Evol. Biol.* 7:1–44.

Hudson, R. R. and N. L. Kaplan. 1988. The coalescent process in models with selection and recombination. *Genetics* 120:831–40.

Hudson, R. R., M. Kreitman, and M. Aguade. 1987. A test of neutral molecular evolution based on nucleotide data. *Genetics* 116:153–59.

Johnson, F. M. and C. Denniston. 1964. Genetic variation of alcohol dehydrogenase in *Drosophila melanogaster*. *Nature* 204:906–7.

Jones, D. S. and P. F. Hasson. 1985. History and development of the marine invertebrate faunas separated by the Central American Isthmus. In F. G. Stehli and S. D. Webb, eds., *The Great American Biotic Interchange*, pp. 325–55. New York: Plenum.

Jones, J. S., S. H. Bryant, R. C. Lewontin, J. A. Moore, and T. Prout. 1981. Gene flow and the geographic distribution of a molecular polymorphism in *Drosophila pseudoobscura*. *Genetics* 98:157–78.

Kaplan, N. L., T. Darden, and R. R. Hudson. 1988. The coalescent process in models with selection. *Genetics* 120:819–29.

Kaplan, N. L., R. R. Hudson, and C. H. Langley. 1989. The "hitchhiking effect" revisited. *Genetics* 123:887–99.

Keith, T. P. 1983. Frequency distribution of esterase-5 alleles in two populations of *Drosophila pseudoobscura*. *Genetics* 105:135–55.

Keith, T. P., L. D. Brooks, R. C. Lewontin, J. C. Martinez-Cruzado, and D. L. Rigby. 1985. Nearly identical distributions of xanthine dehydrogenase in two populations of *Drosophila pseudoobscura*. *Mol. Biol. Evol.* 2:206–16.

Kimura, M. 1983. *The Neutral Theory of Molecular Evolution*. New York: Cambridge University Press.

Kingman, J. F. C. 1982a. The coalescent. *Stochast. Proc. Appl.* 13:235–48.

———. 1982b. On the genealogy of large populations. *J. Appl. Prob.* 19A:27–43.

Koller, P. C. 1939. Genetics of natural populations. III: Gene arrangements in populations of *Drosophila pseudoobscura* from contiguous localities. *Genetics* 24:22–33.

Kreitman, M. 1983. Nucleotide polymorphism at the alcohol dehydrogenase locus of *Drosophila melanogaster*. *Nature* 304:412–17.

———. 1987. Molecular population genetics. *Oxf. Surv. Evol. Biol.* 4:38–60.

———. 1991. Detecting selection at the level of DNA. In R. K. Selander, A. G. Clark, and T. S. Whittam, eds. *Evolution at the Molecular Level*, pp. 204–21. Sunderland, Mass.: Sinauer.

Kreitman, M. and R. R. Hudson. 1991. Inferring the evolutionary histories of the *Adh* and *Adh-Dup* loci in *Drosophila melanogaster* from patterns of polymorphism and divergence. *Genetics* 127:565–82.

Lewontin, R. C. 1974. *The Genetic Basis of Evolutionary Change*. New York: Columbia University Press.

————. 1985. Population genetics. *Ann. Rev. Genet.* 19:81–102.

Li, W.-H., C.-C. Luo, and C.-I. Wu. 1985. Evolution of DNA sequences. In R. J. MacIntyre, ed., *Molecular Evolutionary Genetics,* pp. 1–94. Plenum.

Maynard Smith, J. and J. Haigh. 1974. The hitch-hiking effect of a favorable gene. *Genet. Res.* 23:23–35.

Mayr, E. 1993. What was the evolutionary synthesis? *Trends in Ecol. Evol.* 8:31–35.

Moses, K., U. Heberlein, and M. Ashburner. 1990. The *Adh* gene promoters of *Drosophila melanogaster* and *Drosophila orena* are functionally conserved and share features of sequence structure and nuclease-protected sites. *Mol. Cell. Biol.* 10:539–48.

Nei, M. 1987. *Molecular Evolutionary Genetics.* New York: Columbia University Press.

Oakeshott, J. G., J. B. Gibson, P. R. Anderson, W. R. Knibb, D. G. Anderson, and G. K. Chambers. 1982. Alcohol dehydrogenase and glycerol-3-phosphate dehydrogenase clines in *Drosophila melanogaster* on different continents. *Evolution* 36:86–96.

Ohta, T. 1976. Role of very slightly deleterious mutations in molecular evolution and polymorphism. *Theor. Popul. Biol.* 10:254–75.

Powell, J. R. 1992. Inversion polymorphisms in *Drosophila pseudoobscura* and *Drosophila persimilis.* In C. B. Krimbas and J. R. Powell, eds., *Drosophila Inversion Polymorphism,* pp. 73–126. Ann Arbor, Mich.: CRC Press.

Powell, J. R., Th. Dobzhansky, J. E. Hook, and H. E. Wistrand. 1976. Genetics of natural populations. XLIII: Further studies on rates of dispersal of *Drosophila pseudoobscura* and its relatives. *Genetics* 82:493–506.

Prakash, S. 1972. Origin of reproductive isolation in the absence of apparent genic differentiation in a geographic isolate of *Drosophila pseudoobscura.* *Genetics* 72:143–55.

————. 1977. Genetic divergence in closely related sibling species *Drosophila pseudoobscura, Drosophila persimilis,* and *Drosophila miranda. Evolution* 31:14–23.

Prakash, S., R. C. Lewontin, and J. L. Hubby. 1969. A molecular approach to the study of genic heterozygosity in natural populations IV: Patterns of genic variation in central, marginal and isolated populations of *Drosophila pseudoobscura. Genetics* 61:841–58.

Provine, W. B. 1980. Origins of *The Genetics of Natural Populations* series. In R. C. Lewontin, J. A. Moore, W. B. Provine, and B. Wallace, eds., *Dobzhansky's Genetics of Natural Populations* I–XLIII, pp. 1–76. New York: Columbia University Press.

Riley, M. A., M. E. Hallas, and R. C. Lewontin. 1989. Distinguishing the forces controlling genetic variation at the *Xdh* locus in *Drosophila pseudoobscura. Genetics* 123:359–69.

Saiki, R. K., D. H. Gelfand, S. Stoffel, S. J. Scharf, R. Higuchi, G. T. Horn, K. B. Mullis, and H. A. Ehrlich. 1988. Primer-directed enzymatic amplification of DNA with a thermostable DNA polymerase. *Science* 239:487–91.

Sambrook, J., E. F. Fritsch, and T. Maniatis. 1989. *Molecular Cloning: A Laboratory Manual.* Cold Spring Harbor, N.Y.: Cold Spring Harbor Laboratory Press.

Sanger, F., S. Nicken, and A. R. Coulson. 1977. DNA sequencing with chain-terminating inhibitors. *Proc. Natl. Acad. Sci.* 74:5463–67.

Savakis, C. and M. Ashburner. 1985. A simple gene with a complex pattern of transcription: The alcohol dehydrogenase gene of *Drosophila melanogaster. Cold Spring Harbor Symp. Quant. Biol.* 50:505–14.

Savakis, C., M. Ashburner, and J. H. Willis. 1986. The expression of the gene coding alcohol dehydrogenase during the development of *Drosophila melanogaster. Dev. Biol.* 114:194–207.

Schaeffer, S. W. and C. F. Aquadro. 1987. Nucleotide sequence of the *Adh* gene region of *Drosophila pseudoobscura:* Evolutionary change and evidence for an ancient gene duplication. *Genetics* 117:61–73.

Schaeffer, S. W. and E. L. Miller. 1991. Nucleotide sequence analysis of *Adh* genes estimates the time of geographic isolation of the Bogota population of *Drosophila pseudoobscura. Proc. Natl. Acad. Sci.* 88:6097–6101.

———. 1992a. Estimates of gene flow in *Drosophila pseudoobscura* determined from nucleotide sequence analysis of the alcohol dehydrogenase region. *Genetics* 132:471–80.

———. 1992b. Molecular population genetics of an electrophoretically monomorphic protein in the alcohol dehydrogenase region of *Drosophila pseudoobscura. Genetics* 132:163–78.

———. 1993. Estimates of linkage disequilibrium and the recombination parameter determined from segregating nucleotide sites in the alcohol dehydrogenase region of *Drosophila pseudoobscura. Genetics* 135:541–52.

Simmons, G. M., M. E. Kreitman, W. F. Quattlebaum, and N. Miyashita. 1989. Molecular analysis of the alleles of alcohol dehydrogenase along a cline in *Drosophila melanogaster.* I: Maine, North Carolina, and Florida. *Evolution* 43:393–409.

Singh, R. S., R. C. Lewontin, and A. A. Felton. 1976. Genetic heterogeneity within electrophoretic "alleles" of xanthine dehydrogenase in *Drosophila pseudoobscura. Genetics* 84:609–29.

Slatkin, M. 1985. Gene flow in natural populations. *Ann. Rev. Ecol. Syst.* 16:393–430.

———. 1991. Inbreeding coefficients and coalescence times. *Genet. Res.* 58:167–75.

Smocovitis, V. B. 1992. Unifying biology: The evolutionary synthesis and evolutionary biology. *J. Hist. Biol.* 25:1–65.

Sullivan, D. T., P. W. Atkinson, and W. T. Starmer. 1990. Molecular evolution of the alcohol dehydrogenase genes in the genus *Drosophila. Evol. Biol.* 24:107–47.

Tajima, F. 1983. Evolutionary relationship of DNA sequences in finite populations. *Genetics* 105:437–60.

————. 1989. Statistical method for testing the neutral mutation hypothesis by DNA polymorphism. *Genetics* 123:585–95.

Tavare, S. 1984. Line-of-descent and genealogical processes, and their applications in population genetic models. *Theor. Popul. Biol.* 26:119–64.

Ursprung, H. and J. Leone. 1965. Alcohol dehydrogenases: A polymorphism in *Drosophila melanogaster. J. Exp. Zool.* 160:147–54.

Van Campo, E., J. C. Duplessy, W. L. Prell, N. Barratt, and R. Sabatier. 1990. Comparison of terrestrial and marine temperature estimates for the past 135 kyr off southeast Africa: A test for GCM simulations of paleoclimate. *Nature* 348:209–12.

Vigue, C. L. and F. M. Johnson. 1973. Isozyme variability in species of the genus *Drosophila*. VI: Frequency-property-environment relationships of allelic alcohol dehydrogenases in *D. melanogaster. Biochem. Genet.* 9:213–27.

Watterson, G. A. 1975. On the number of segregating sites in genetical models without recombination. *Theor. Pop. Biol.* 7:256–76.

————. 1978. The homozygosity test of neutrality. *Genetics* 88:405–17.

Whittam, T. S., A. G. Clark, M. Stoneking, and R. L. Cann. 1986. Allelic variation in human mitochondrial genes based on patterns of restriction site polymorphism. *Proc. Natl. Acad. Sci.* 83:9611–15.

Wright, S. 1931. Evolution in mendelian populations. *Genetics* 16:97–159.

————. 1943. Isolation by distance. *Genetics* 28:114–38.

Wright, S., Th. Dobzhansky, and W. Hovanitz. 1942. Genetics of natural populations. VII: The allelism of lethals in the third chromosome of *Drosophila pseudoobscura. Genetics* 27:363–94.

24

A Transposable Element in Natural Populations of *Drosophila melanogaster*

Ian A. Boussy

Theodosius Dobzhansky's overwhelming passion throughout his life was demonstrating the fundamental and dominant forces influencing the genetics and evolution of natural populations. For most of his career he used *Drosophila pseudoobscura* as his research organism. This allowed him to exploit known features of *Drosophila* genetics, a luxury unavailable to workers on the ecological genetics of most other organisms (see editorial discussions in Lewontin et al., 1981). The populations of *D. pseudoobscura* that Dobzhansky studied most intensely may not have been as free of human influence as he pretended (Coyne, Boussy, and Bryant 1984), but his life's work was to establish the importance of the major factors influencing evo-

This paper is an acknowledgment of my debt to my academic "grandfather," Professor Theodosius Dobzhansky. I also acknowledge my academic "father," Tim Prout, whose interest in all biological phenomena kept mine going, and who taught me a measure of personal and intellectual discipline for which I will always be grateful. I was also influenced by various academic "aunts" and "uncles," including Francisco Ayala, Jerry Coyne, John Gibson, Les Gottlieb, Mel Green, Subodh Jain, Margaret Kidwell, John McDonald, John Oakeshott, and Marty Tracey, some of whom are also academic "offspring" of Dobzhansky. I am especially grateful to Ron Woodruff for recollecting *D. melanogaster* along the eastern coast of Australia and for sharing the flies and data with me, and to Anita S.-F. Chong for continued ideas, support, and tolerance. I also thank Anthony E. Romanelli and Jennifer Soriano for their help in testing the flies.

lutionary genetics, especially natural selection, by evaluating them in natural populations.

Dobzhansky seemed to have little patience for odd, unexplained phenomena in genetics, including transposable elements and mutator systems. Louis Levine (personal communication) tells of sitting next to him during the historical presentation by Barbara McClintock at the Cold Spring Harbor Symposium of 1951 (McClintock 1951), in which she presented her pioneering work on the A*c* and D*s* transposable elements in maize. In the 1950s, the idea of "jumping genes" was thought by many to be either heretical or an oddity of the genetics of maize; it was not fully accepted by geneticists for almost thirty more years. Dobzhansky's own view at the time was skeptical. At the end of the talk, he turned to Levine and said: "There must be some other explanation!"

Dobzhansky's lifelong focus was thus on the major features of evolutionary genetics, especially on natural selection's effects on variability in chromosomes and genes. The major methodologies that Dobzhansky used included field collecting to determine patterns in nature, field experiments when possible, and analysis and experimentation in the laboratory with field-collected materials. Throughout his work, evolutionary and population genetic theory determined the design of collection and analysis of field and experimental data. His approach has shaped the study of modern evolutionary genetics and has proven to apply well to problems involving many nontraditional phenomena.

Dobzhansky also eschewed *D. melanogaster,* along with the other cosmopolitan species of *Drosophila,* characterizing them in print as "domestic, colonizing, species, or if you prefer the name—animal weeds" (Dobzhansky 1965:533). In conversation he was less circumspect, often referring to them as "garbage species." The genetics of such human commensals were not of as much interest to him as those of more "natural" species. I might not be studying natural populations of *Drosophila* if it were not for Dobzhansky; nevertheless, there are two reasons—his disinterest in oddities of genetics and his disdain of *D. melanogaster*—that Dobzhansky might have had little interest in this discussion, which concerns my studies of a transposable element in wild populations of *D. melanogaster.*

P-M Hybrid Dysgenesis

In the 1970s, *Drosophila melanogaster* workers began seriously examining a set of strange phenomena that came to be called "P-M hybrid dysgenesis" (Hiraizumi 1970; Sved 1976; Kidwell, Kidwell, and Sved 1977; Green

1977, 1979; Bregliano and Kidwell 1983). Occurring primarily in the off-spring of one-way crosses between different strains, these phenomena included: recombination in males (normally not present in *D. melanogaster*); chromosome transmission ratio distortion; high rates of generation of new mutants, which often proved unstable; generation of chromosomal breaks and rearrangements; and the occurrence of tempera-ture-dependent sterilities involving the nondevelopment of gonads (for reviews, see Bregliano and Kidwell 1983; Engels 1983, 1989). The phe-nomena occurred not at all or only at lower rates in the reciprocal crosses or within the strains.

The P-M hybrid dysgenesis system seemed to defy standard genetics. It yielded initially to a nearly quantitative genetic approach, in which strains were shown to have certain characteristics, and the determinants of these characteristics were shown to be distributed among the chromosomes. It was found that when certain lines were used as the male parent, in crosses to certain other lines, hybrid dysgenesis ensued; in the reciprocal cross, no dysgenic traits, or low levels of them, were manifested (Colgan and Angus 1978; Kidwell 1979; Schaefer, Kidwell, and Fausto-Sterling 1979; Engels and Preston 1979). The hypothesized determinants were called "P factors" (for "paternally contributing"), and strains of flies with active P factors were called P strains; their opposites were called M strains (for "maternally contributing").

Strongly interacting P and M lines were taken as standards. They were then used to characterize other lines by performing reference crosses and scoring the offspring for dysgenic traits. As diagrammed in figure 24.1, cross A (unknown males are crossed to females of a strong reference M strain) evaluates P activity potential (Kidwell, Kidwell, and Sved 1977), and

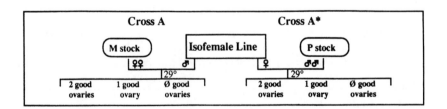

FIGURE 24.1.
Crosses A and A* are used to determine, respectively, the P activity potential and the suscep-tibility to P activity of a line. F_1 females from the crosses are dissected to determine the state of development of their ovaries.

cross A* (unknown females are crossed to males of a strong reference P strain) evaluates susceptibility to P activity (Engels and Preston 1980). Nondevelopment of gonads ("gonadal dysgenesis," or GD) among female of offspring reared at elevated temperature (29°C) is the trait most often tested. Populations with a high proportion of dysgenic offspring from cross A (GD ≥10%, up to 100%) and a low proportion from cross A* are called "P"; populations with low GD scores (≤10%) in both crosses are called neutral, or "Q"; and populations with low A GD and high A* GD (≥10%, up to 100%) are called "M" (Kidwell, Frydryk, and Novy 1983; Yamamoto, Hihara, and Watanabe 1984; Kidwell 1986; Boussy and Kidwell 1987; see table 24.1). Note that high GD in both crosses virtually does not occur; presumably, if it did, the population would have a high intrinsic degree of dysgenesis, an unstable condition that should evolve to loss of P activity or loss of susceptibility or both. Note also that these qualitative categories hide a continuum of phenotypes, from strong P to weak P to Q to weak M to strong M.

Various surveys in the 1980s demonstrated that wild populations of *D. melanogaster* around the world differed in their P-M properties from strong P to extreme M, with large variability within populations. North American, South American, and most of sub-Saharan African lines were primarily P (Engels and Preston 1980; Kidwell, Frydryk, and Novy 1983; Anxolabéhère et al., 1984; Kidwell and Novy 1985). French populations were predominantly Q, whereas populations in the rest of Europe, northern Africa, and across central Asia were M (Anxolabéhère, Nouaud, and Périquet 1982; Anxolabéhère et al., 1984; Anxolabéhère et al., 1985; Anxolabéhère et al., 1988). The overall pattern, then, was of continent-wide

TABLE 24.1.

Phenotypic Categorizing of Lines or Populations as P, Q, or M in the P-M Hybrid Dysgenesis System, Using as the Criterion of Dysgenesis the Degree of Gonadal Nondevelopment (GD) of Offspring of Reference Crosses

		Cross A GD	Cross A* GD
Phenotypic Category:	P	≥10%	≈0%
	Q	≤10%	≤10%
	M	≈0%	≥10%

The proportion of affected F_1 females (one or both ovaries undeveloped) is scored (or, by some authors, the proportion of affected ovaries). Som older laboratory lines completely lack *P* elements; they invariably test as strong M lines and are sometimes called "true M" lines. All other M lines carry at least some *P* elements, and are sometimes called "pseudo M" lines or "M" lines.

qualitative uniformity (with internal variability) or smooth transitions between P, Q, or M states over long distances (Anxolabéhère, Kidwell, and Périquet 1988).

The *P* Element

In the early 1980s, the hypothesized P factor that caused P-M hybrid dysgenesis was cloned and shown to be a 3 kilobase (kb) transposable element, thereafter called the *P* element (Bingham, Kidwell, and Rubin 1982; Rubin, Kidwell, and Bingham 1982; O'Hare and Rubin 1983; for a more recent review, see Engels 1989). With *P* element probes, *P* elements in the genomes of laboratory and wild lines can be visualized on autoradiograms of Southern blots. It was shown that *P* elements occur both as full-size (3 kb) elements and as internally deleted elements of various sizes and that there is often a preponderance of the deleted elements in a genome (O'Hare and Rubin 1983; Sakoyama et al., 1983; Todo et al., 1984; Sakoyama et al., 1985; Anxolabéhère et al., 1985).

A particular internally deleted *P* element of 1.2 kb is present at very high copy numbers per genome in European and Asian lines but is nearly absent in lines from North America; it was named the *KP* element after the line, Krasnodar, from which it was first cloned (Black et al. 1987). The high copy number of *KP* elements generated the hypothesis that they might have a repressor function that serves to control *P* element transposition, thus giving a selective advantage to individuals with more such elements per genome (Black et al. 1987; Jackson, Black, and Dover 1988). This interpretation is controversial, however, since lines with many *KP* elements may show little or no repressor ability (Boussy et al., 1988; Heath and Simmons 1991; Raymond et al., 1991). Other internally deleted *P* elements have been implicated as repressors as well (Nitasaka, Mukai, and Yamazaki 1987; Robertson and Engels 1989).

Studies have shown that the genomes of all *D. melanogaster* from all parts of the world collected since the mid-1970s contain *P* elements, regardless of their P, Q, or M status in reference crosses (Anxolabéhère, Kidwell, and Périquet 1988). North American flies were found to carry many copies of full-size *P* elements and many copies of various size internally deleted elements, but virtually no *KP* elements. Flies from Europe and Asia carry a majority of *KP* elements, as well as other deleted elements of various sizes, with many to few or no copies of full-size elements.

Various laboratory stocks, however, completely lack *P* elements. These were designated "true M" (Kidwell 1985), to differentiate them from "pseudo M" or "M'," lines of M phenotype that bear *P* elements (Engels 1984). Testing of lines collected in different years and kept in the laboratory since revealed a remarkable fact: lines lacking *P* elements were more common among the older lines, and all lines collected since 1966 in North America, since 1969 in western Europe, and since 1974 in central Asia carried *P* elements (Anxolabéhère, Kidwell, and Périquet 1988). This suggested that *P* elements had invaded (or become common) in *D. melanogaster* only since about 1950 and had spread around the world since (the "recent introduction" hypothesis; Kidwell 1983, 1986; Anxolabéhère, Kidwell, and Périquet 1988).

Australian *D. melanogaster*

My initial studies were undertaken to see whether there might be variability with latitude in the P-M system among *D. melanogaster* populations in Australia. Latitudinally clinal patterns had previously been discovered in chromosomal inversion frequencies (Knibb, Oakeshott, and Gibson 1981; Knibb 1982) and in frequencies of various isozymes (Oakeshott et al. 1981, 1982, 1983). It was thought that the inversion and isozyme clinal patterns were determined by natural selection (especially since similar latitudinal clines had been found in North America and Europe). Much of Australia is uninhabitably dry for *D. melanogaster*, but along the eastern coast there is a relatively humid region that extends from the tropics in the north to cool temperate climes in the south. Tasmania is completely cool temperate. *D. melanogaster* occurs at least seasonally throughout this range. The eastern coast of Australia then provides a nearly linear habitat across about 30° of latitude, in which temperatures at different times of the year vary virtually linearly with latitude. It seemed likely that if P-M hybrid dysgenesis (or any other character) were to vary with ecological parameters such as temperature, it would be more accessible to study in such a linear habitat gradient than in more diverse or two-dimensional realms.

I tested individuals from isofemale lines (strains each started with one gravid female collected in the wild, a means of sampling a population used by Dobzhansky as early as 1937 [Dobzhansky and Queal 1938]). The lines were collected from widely separated localities along the eastern coast of Australia. The results demonstrated that, indeed, there were large differences between populations from different parts of the coast of Australia (Boussy 1987). In the north, from Cairns (16.9°S lat) to Coff's Harbour (30.3°S lat), the populations tested were phenotypically P. A single population tested from Bateman's Bay in the southeast (35.7°S lat) was Q. A population from the south (Cann River, 37.6°S lat) and one from Tasmania (Cygnet, 43.2°S lat) were both M. All the populations showed a great deal of variability between individual lines tested within the population.

Collections from more localities were made, by myself and A. S.-F. Chong in 1985 along the eastern coast north of Sydney (34°S lat) to Coff's Harbour, and in 1986 by Dr. Phil Anderson south of Sydney to Cann River. Testing of these lines showed that in fact the overall pattern was discontinuous, with P populations north of Sydney, Q populations south of Sydney to Eden (37.1°S lat), and M populations at Genoa (37.5°S lat) and Cann River (Boussy and Kidwell 1987). Indeed, the discontinuities were dramatic.

The P activity of the P region populations declined from north to south, down to 20% cross A GD at Ourimbah (33.4°S lat), just north of Sydney; it then dropped to 6% at Wollongong (34.4°S lat), only 150 km south of Ourimbah. Cross A scores from Wollongong south to Eden were similar, ranging from 1 to 8% GD; so were the cross A* scores, which ranged from 1 to 10%. These results define that region as Q. The transition from Q populations to M populations between Eden and Genoa was even more dramatic than that from P to Q. The mean cross A* GD score for Eden was only 1.2%; that for Genoa was 48.2%. These collection localities are only about 50 km apart! Thus the P-Q-M latitudinally clinal pattern was partially a step-cline, with a P-to-Q transition over 150 km or less and a Q-to-M transition over only 50 km or less.

Molecular analysis of lines from the cline revealed that all populations carried many *P* elements per genome (Boussy et al., 1988). The number of full-size *P* elements per genome declined from north to south, and the number of *KP* elements increased from north to south. Crude quantification of the numbers of full-size *P* elements and *KP* elements per genome by densitometry of autoradiographs allowed comparison to cross A and cross A* GD sterilities. There appeared to be a threshold of full-size *P* element numbers above which lines tended to show P activity in cross A. Likewise, there was another threshold, below which a line tended to show susceptibility to P activity in cross A*. The thresholds were not strongly defined, but the Spearman's rank correlation coefficients were significant ($r_s = 0.55$, $p = 0.02$, and $r_s = -0.65$, $p = 0.01$, respectively). The numbers of *KP* elements per genome also correlated with P activity and susceptibility in a way involving a threshold. Cross A GD sterility was manifested only by lines with numbers of *KP* elements below a certain level, and cross A* GD sterility was manifested only by lines with numbers of *KP* elements above a certain level. Among Q and M lines, the relationship of *KP* numbers to cross A* GD sterility appeared to be random ($r_s = 0.14$, $p \approx 0.65$), indicating that *KP* elements were not a prime determinant of susceptibility to P activity in these lines.

Resampling the Cline

In 1991, Ron Woodruff (Bowling Green State University, Ohio) collected flies from 27 sites along the eastern coast of Australia, including many of the localities I had sampled in 1983–86. The goal was to determine whether the clinal pattern were still present and whether it had qualitatively or quantitatively changed since 1986. Investigation of the lines he collected is still under way, but preliminary results are available from GD testing of 37 lines from 18 localities.

The formerly P and Q regions are still qualitatively P and Q, respectively, based on 8 lines tested from the P region and 7 from the Q region. A chi-square test of the proportions of individual lines from the 1983 to 1986 collections and from the 1991 collections, grouped by whether their GD scores were >10% or ≤10%, shows no statistically significant differences between these time periods for the 2 regions. There appears to be a quasi-stable clinal pattern over most of the eastern coast.

In contrast, 2 populations in the previously "M" region, Genoa and Cann River, are now Q or nearly Q (cross A GD ranged from 0 to 3%; see table 24.2 for A* GD means). A chi-square test of the proportions of individual lines from the 1983–1986 collections from Cann River and Genoa, respectively, and from the 91 collections, grouped by whether their A* GD scores were ≥10% or <10%, shows that the collections are statistically different (table 24.3). It seems likely that this change is the result of a migration of flies (or of chromosomes) from the Q region into Genoa and Cann River, although it is possible that there is an innate tendency for M populations to evolve to a Q phenotype.

TABLE 24.2.

Mean A* Gonadal Dysgenesis Scores Among Isofemale Lines Collected in 1983 and 1986 in Genoa and Cann River and of Lines Collected in 1991 from Both Localities; Each Mean Is Followed by the Standard Error of the Mean (s.e.) and the Number of Lines Tested (n)

	1983–86 A* GD ± s.e. (n)	1991 A* GD ± s.e. (n)
Genoa (1986)	48.2 ± 11.7 (5)	12.8 ± 6.7 (6)
Cann River (1983 and 1986)	25.6 ± 4.1 (40)	2.0 ± 1.2 (6)

TABLE 24.3.

Numbers of Lines with Cross A* GD Sterilities Above and Below 10%
in the 1983 and 1986 Collections from Cann River and Genoa,
and in the 1991 Collections from the Same Localities

	1986 Genoa, 1983 and 1986 Cann River	1991 Genoa and Cann River
Cross A* GD sterility: ≥10%	28	2
<10%	17	10

A contingency test yields chi-square = 7.89, with $p < .005$.

Southern blots of the tested 1991 lines show that all carry many *P* elements of different sizes. All have at least a few putative full-size *P* elements, and most carry many copies of *KP* elements. The largest number of *KP* elements per genome is seen in the southernmost lines tested, but lines from as far north as Coff's Harbour have many *KP* elements per genome. Only the northernmost lines differ from this in having very few copies of *KP* elements per genome. This is consistent with results from the 1983–1986 collections (Boussy et al., 1988 and other data not shown).

Inheritance of Repressor Function

The ability of a P line to repress P activity was originally called "P cytotype" because it was at least partly maternally inherited (Engels and Preston 1979), but some lines have been shown to have repressor ability that is strictly chromosomally inherited (Kidwell 1985). Lines from North America seem to predominantly show at least some maternal inheritance of repressor function, whereas those from Europe show only the chromosomal type. The correlation with the high numbers of *KP* elements in European lines suggested that *KP* elements had an active role in repressing P activity and that this activity was only chromosomally inherited (Black et al., 1987; Jackson, Black, and Dover 1988), although, as noted above, this is controversial. The maternally and chromosomally inherited repressor functions can be differentiated by crosses as shown in figure 24.2. The F_1 females from crosses 1 and 2 are chromosomally identical, so if their female offspring differ in the GD sterility they manifest, the difference must be due to maternal inheritance of repressor function.

Some preliminary testing of 1991 lines from Australia has shown that 3 lines from P and Q region localities all had up to 100% maternal inheritance of repressor function; in contrast, testing of 1 phenotypically Q line from Cann River showed that it had strictly chromosomal inheritance of its repressor function (table 24.4). Since the molecular bases for repressor function are only partially understood, it is currently difficult to ascribe biological significance to this observation. If it holds for more lines, such a discrepancy between the M region lines and those from the P and Q regions would provide another biological distinction between them and may relate to the origins of the differences (see below).

The overall patterns described above are summarized in figure 24.3. The figure shows the eastern coast of Australia, with the sites referred to in

TABLE 24.4.

GD Sterilities in F_2 Females from Crosses 1 and 2 with Four 1991 Lines, Along with a Designation of the Type of Region from Which the Line Came (P, Q, or M, as Designated from 1983–1986 Results), and the Cross A and A* GD Scores for the Line. GD Scores Are as Percent Affected Females

Line:	Region:	Cross A GD	Cross A* GD	Cross 1 GD	Cross 2 GD
BB 22	P	88	0	39	99
Lau 6	P	14	5	1	59
Ull 6	Q	2	6	0	93
CR 9	M	1	0	81	63

Note that the cross 1 and 2 GD results are very different for the first three lines tested, indicating maternal inheritance of repressor function, but are similar for the fourth line, from Cann River, indicating only chromosomal inheritance of repressor function in that line.

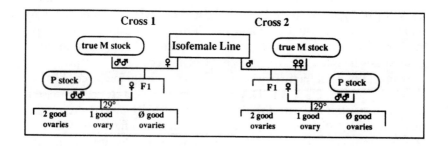

FIGURE 24.2.

Crosses 1 and 2 test whether repressor function has a maternally inherited component (Black et al., 1987). If the GD sterility observed in the offspring females from cross 2 is greater than that from cross 1, there is a maternally inherited component of repressor function. If not, the inheritance of repressor function is strictly chromosomally inherited.

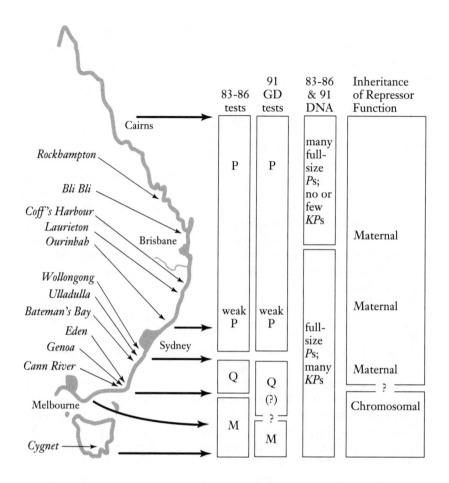

FIGURE 24.3.

Summary of the experimental results, arrayed geographically. The east coast of Australia is shown, with sites referred to in the text indicated. The first two columns classify regions as P, weak P, Q, or M, based on gonadal dysgenesis (GD) tests in crosses A and A* of lines collected in 1983–1986 and in 1991. The next column describes the qualitative results of Southern blot analyses of genomic DNAs. The last column refers to preliminary tests of inheritance pattern of repressor function (maternal or chromosomal). Question marks indicate phenotype and boundary determinations that are uncertain as yet.

the text indicated. Next to that are columns representing the results of gonadal dysgenesis testing for P, Q, or M phenotype (results for 1983–1986 lines and for 1991 lines are presented separately), for Southern blot analysis for presence of full-size *P* elements and *KP* elements in the genomes, and a tentative geographical interpretation of inheritance of repressor function.

Interpretation of the Clinal Pattern

As noted above, the occurrence of *P* elements in populations of *D. melanogaster* differs worldwide (Anxolabéhère, Kidwell, and Périquet 1988; Ronsseray, Lehmann, and Anxolabéhère 1989). In central Asia, genomes have few *P* elements, and those are typically internally deleted forms, such as the *KP* element. In western Europe, genomes are full of *P* elements, with a predominance in frequency of *KP* elements. In North America, genomes are replete with *P* elements, full-sized and deleted, but *KP* elements are virtually not found among them.

Not all regions outside Australia are so uniform over large areas. Japanese populations have been found to be mostly weak P or Q, but some M populations have been described (Ohishi, Takanashi, and Ishiwa Chigusa 1982; Takada et al., 1983; Yamamoto, Hihara, and Watanabe 1984), and variability ranging from P to Q to M has also been described in South Africa (Getz and van Schaik 1988). The variability does not form a clinal pattern in either region. If the recent introduction hypothesis is true for *P* elements, then the variabilities from continent to continent, within areas such as Japan and South Africa, and in the clinal pattern in Australia probably represent a transient state that will tend to uniformity. Variability over short geographic distance as in Australia may be the result of the dynamics of *P* element invasion of populations, or it may represent the results of migration and admixture of flies with different initial characteristics.

I suggest as a hypothesis that the pattern found in eastern Australia could be the result of two (or more) introductions from parts of the world such as North America and Europe that differ in several aspects of the P-M system. Flies from the northernmost "P" region populations are most like North American flies: they have few *KP* elements, are P in phenotype, and their repressor function is largely maternally inherited (P cytotype). In the southern part of the P region, flies are like North American flies in having P cytotype and a (weak) P phenotype, but they carry many *KP* elements per genome as European flies do. Flies from the Q region are like North American flies in having P cytotype but like European flies in their Q phenotype and in the numbers of *KP* elements they carry. Flies from the southern "M" region are most like European flies: they have many *KP* elements, are Q or M in phenotype, and their repressor function is chromosomally inherited.

The changes from M to Q phenotypes I report here for the Genoa and Cann River populations between 1983 and 1986 and 1991 may represent the results of further migration and admixture between M and Q populations. Perhaps the biggest mystery, in Australia and worldwide, is not that changes have occurred, but that more change has not occurred and that populations have not gone to uniformity. One would have expected that the fly migration and rates of *P* element spread that drove *P* elements around the world in less than forty years should have been sufficient to lead to homogenization within geographic domains in fairly short order. Perhaps after the initial introduction phase, repressor function in a population prevents rapid movement of the *P* elements, leading to much slowed evolution.

Would Dobzhansky Have Approved?

Dobzhansky's lifelong study of the population genetics of *D. pseudoobscura* began with careful field studies of the occurrence of chromosomal polymorphisms in populations in western North America. The large variability over time and space that he documented provided the material for many of his studies and those of his students, and his early work especially underlines the importance of careful surveys in describing natural populations.

Description of the dramatic latitudinal cline in *P*-element-related properties in eastern Australian *D. melanogaster* emerged from such a survey. The close geographical juxtaposition of populations differing greatly in their P-M characteristics is across a virtually linear habitat with a correlated range of temperatures. Study of the properties of the cline and its change over time will provide unique data for analysis of the population dynamics of a transposable element. A gigantic natural experiment has been set in motion, and we are currently observing it.

The ubiquity of transposable elements was not appreciated by the vast majority of geneticists until about 1980. Whether Dobzhansky would have considered them of sufficient evolutionary interest to warrant investigation is a moot point. It has been hypothesized that hybrid dysgenesis caused by transposable elements may act to generate postzygotic isolation between populations differing in their transposable element complements (Bingham, Kidwell, and Rubin 1982; Kidwell 1983; Ginzburg, Bingham, and

Yoo 1984). Experimental studies have not borne out this hypothesis as important for three pairs of closely related *Drosophila* species (Coyne 1985, 1986) or for four interfertile species and semispecies of the *D. affinis* subgroup (Hey 1989). Speciation by hybrid dysgenesis remains hypothetical.

The primary evolutionary significance of transposable elements may be as mutators. Transposable elements certainly form part of the genetic makeup of even such species as *D. pseudoobscura*. It has been estimated that as many as 80% of "spontaneous" mutations in *D. melanogaster* are due to transposable element insertions (Green 1988). The evolutionary impact of transposable elements in causing episodic elevations of mutation rates could be enormous (e.g., Berg 1982), especially if their transpositions could be triggered by environmental or "genomic" shocks (McClintock 1984) and occur in clusters (Woodruff and Thompson 1992), and may thus be involved when species invade new habitats or niches.

Finally, transposable elements may for the most part be simply "selfish" or "parasitic DNA" that exists as a genomic parasite (Dawkins 1989). Certainly the known dynamics of transposable elements are consistent with their having no immediate positive role for their hosts (Charlesworth and Langley 1989). Nonetheless, given their potential for strong effects, a complete understanding of the mode and tempo of evolutionary genetics would seem to require understanding of the population genetics and evolution of transposable elements.

I feel that Dobzhansky would have approved of the observations discussed here. His own observations, starting in the 1930s, on the frequencies and distributions of inversions in *D. pseudoobscura* set the stage for several decades of work that has been continued by his intellectual offspring (Anderson et al., 1991). Louis Levine has commented to me that these inversions might well have been considered oddities themselves when Dobzhansky was beginning his studies of them. Dobzhansky himself initially viewed the inversions as neutral markers that varied between localities owing to genetic drift, only several years later changing his views to believe that they were acted upon by natural selection (see discussions in Lewontin et al., 1981). My hope is that the observations of the studies summarized here can eventually be analyzed in the light of models of transposable element population and genomic dynamics.

On the Nature of Nature

Most people who worked with Dobzhansky have at least one personal Dobzhansky story; I am no exception. This story serves to keep things in perspective for me when the secrets of nature do not seem to unfold as simply as I would like. I once went to Dobzhansky's office to inquire about the relationships and differing degrees of sterilities and partial sterilities in crosses within the *paulistorum* complex. Dobzhansky explained to me the populations, races, subspecies, semispecies, species *in statu nascenti*, sibling species, and full species in the group at great length, with details of distributions, sexual isolation, and reciprocal crosses. After half an hour of explanation, he finally finished. I said "Whew! Professor Dobzhansky! It is all extremely *complex!*"

He replied: "As it should be!"

REFERENCES

Anderson, W. W., J. Arnold, D. G. Baldwin, A. T. Beckenbach, C. J. Brown, S. H. Bryant, J. A. Coyne, L. G. Harshman, W. B. Heed, D. E. Jeffery, L. B. Klaczko, B. C. Moore, J. M. Porter, J. R. Powell, T. Prout, S. W. Schaeffer, J. C. Stephens, C. E. Taylor, M. E. Turner, G. O. Williams, and J. A. Moore. 1991. Four decades of inversion polymorphism in *Drosophila pseudoobscura. Proc. Natl. Acad. Sci.* 88:10367–71.

Anxolabéhère, D., L. Charles-Palabost, A. Fleuriet, and G. Périquet. 1988. Temporal surveys of French populations of *Drosophila melanogaster:* P-M system, enzymatic polymorphism and infection by the sigma virus. *Heredity* 61:121–31.

Anxolabéhère, D., K. Hu, D. Nouaud, G. Périquet, and S. Ronsseray. 1984. The geographical distribution of *P-M* hybrid dysgenesis in *Drosophila melanogaster. Génétique, Sélection et Évolution* 16:15–26.

Anxolabéhère, D., M. G. Kidwell, and G. Périquet. 1988. Molecular characteristics of diverse populations are consistent with the hypothesis of a recent invasion of *Drosophila melanogaster* by mobile *P* elements. *Mol. Biol. Evol.* 5:252–69.

Anxolabéhère, D., D. Nouaud, and G. Périquet. 1982. Cytotype polymorphism of the *P-M* system in two wild populations of *Drosophila melanogaster. Proc. Natl. Acad. Sci.* 79:7801–03.

Anxolabéhère, D., D. Nouaud, G. Périquet, and P. Tchen. 1985. P-element distribution in Eurasian populations of *Drosophila melanogaster:* A genetic and molecular analysis. *Proc. Natl. Acad. Sci.* 82:5418–22.

Berg, R. L. 1982. Mutability changes in *Drosophila melanogaster* populations of Europe, Asia, and North America and probable mutability changes in human populations of the U.S.S.R. *Jpn. J. Genet.* 57:171–83.

Bingham, P. M., M. G. Kidwell, and G. M. Rubin. 1982. The molecular basis of P-M hybrid dysgenesis: The role of the P element, a P-strain-specific transposon family. *Cell* 29:995–1004.

Black, D. M., M. S. Jackson, M. G. Kidwell, and G. A. Dover. 1987. KP elements repress P-induced hybrid dysgenesis in *Drosophila melanogaster. EMBO J.* 6:4125–35.

Boussy, I. A. 1987. A latitudinal cline in P-M gonadal dysgenesis potential in Australian *Drosophila melanogaster* populations. *Genet. Res.* 49:11–18.

Boussy, I. A. and M. G. Kidwell. 1987. The P-M hybrid dysgenesis cline in eastern Australian *Drosophila melanogaster:* Discrete P, Q and M regions are nearly contiguous. *Genetics* 115:737–45.

Boussy, I. A., M. J. Healy, J. G. Oakeshott, and M. G. Kidwell. 1988. Molecular analysis of the P-M gonadal dysgenesis cline in eastern Australian *Drosophila melanogaster. Genetics* 119:608–15.

Bregliano, J. C. and M. G. Kidwell. 1983. Hybrid dysgenesis determinants. In J. A. Shapiro, ed., *Mobile Genetic Elements*, pp. 363–410. New York: Academic Press.

Charlesworth, B. and C. H. Langley. 1989. The population genetics of *Drosophila* transposable elements. *Ann. Rev. Genet.* 23:251–87.

Colgan, D. J. and D. S. Angus. 1978. Bisexual hybrid sterility in *Drosophila melanogaster. Genetics* 89:5–14.

Coyne, J. A. 1985. Genetic studies of three sibling species of *Drosophila* with relationship to theories of speciation. *Genet. Res.* 46:169–92.

———. 1986. Meiotic segregation and male recombination in interspecific hybrids of *Drosophila. Genetics* 114:485–94.

Coyne, J. A., I. A. Boussy, and S. H. Bryant. 1984. Is *Drosophila pseudoobscura* a garbage species? *Pan Pacific Entomol.* 60:16–19.

Dawkins, R. 1989. *The Selfish Gene.* 2d ed. Oxford: Oxford University Press.

Dobzhansky, Th. 1965. "Wild" and "domestic" species of *Drosophila.* In H. G. Baker and G. L. Stebbins, eds., *The Genetics of Colonizing Species*, pp. 533–46. New York: Academic Press.

Dobzhansky, Th. and M. L. Queal. 1938. Genetics of natural populations. I: Chromosomal variation in populations of *Drosophila pseudoobscura* inhabiting isolated mountain ranges. *Genetics* 23:239–51.

Engels, W. R. 1983. The P family of transposable elements in *Drosophila. Ann. Rev. Genet.* 17:315–44.

———. 1984. A *trans*-acting product needed for P factor transposition in *Drosophila. Science* 226:1194–96.

———. 1989. P elements in *Drosophila melanogaster.* In D. E. Berg and M. M. Howe, eds., *Mobile DNA*, pp. 437–84. Washington, D.C: American Society for Microbiology.

Engels, W. R. and C. R. Preston. 1979. Hybrid dysgenesis in *Drosophila melanogaster:* The biology of female and male sterility. *Genetics* 92:161–74.

———. 1980. Components of hybrid dysgenesis in a wild population of *Drosophila melanogaster. Genetics* 95:111–28.

Getz, C. and N. van Schaik. 1988. Mobile genetic P elements in South African populations of *Drosophila melanogaster. S. Afr. J. Sci.* 84:908–12.

Ginzburg, L. R., P. M. Bingham, and S. Yoo. 1984. On the theory of speciation induced by transposable elements. *Genetics* 107:331–41.

Green, M. M. 1977. Genetic instability in *Drosophila melanogaster:* De novo induction of putative insertion mutations. *Proc. Natl. Acad. Sci.* 74:3490–93.

———. 1979. Genetic instability in *Drosophila melanogaster.* Dosage and mutator activity of an *MR* chromosome. *Mutation Res.* 62:529–31.

———. 1988. Mobile DNA elements and spontaneous gene mutation. In M. E. Lambert, J. F. McDonald, and I. B. Weinstein, eds., *Eukaryotic Transposable Elements as Mutagenic Agents*, pp. 41–50. Cold Spring Harbor, N.Y.: Cold Spring Harbor Laboratory Press.

Heath, E. M. and M. J. Simmons. 1991. Genetic and molecular analysis of repression in the P-M system of hybrid dysgenesis in *Drosophila melanogaster. Genet. Res.* 57:213–26.

Hey, J. 1989. Speciation via hybrid dysgenesis: Negative evidence from the *Drosophila affinis* subgroup. *Genetica* 78:97–104.

Hiraizumi, Y. 1970. Spontaneous recombination in *Drosophila melanogaster* males. *Proc. Natl. Acad. Sci.* 68:268–70.

Jackson, M. S., D. M. Black, and G. A. Dover. 1988. Amplification of *KP* elements associated with the repression of hybrid dysgenesis in *Drosophila melanogaster. Genetics* 120:1003–13.

Kidwell, M. G. 1979. Hybrid dysgenesis in *Drosophila melanogaster:* The relationship between the P-M and I-R interaction systems. *Genet. Res.* 33:205–17.

———. 1983. Evolution of hybrid dysgenesis determinants in *Drosophila melanogaster. Proc. Natl. Acad. Sci.* 80:1655–59.

———. 1985. Hybrid dysgenesis in *Drosophila melanogaster:* Nature and inheritance of *P* element regulation. *Genetics* 111:337–50.

———. 1986. Molecular and phenotypic aspects of the evolution of hybrid dysgenesis systems. In S. Karlin and E. Nevo, eds., *Evolutionary Processes and Theory*, pp. 169–98. New York: Academic Press.

Kidwell, M. G. and J. B. Novy. 1985. The distribution of hybrid dysgenesis determinants in North American populations on *D. melanogaster. Drosophila Info. Serv.* 61:97–100.

Kidwell, M. G., T. Frydryk, and J. B. Novy. 1983. The hybrid dysgenesis potential of *Drosophila melanogaster* strains of diverse temporal and geographical natural origins. *Drosophila Info. Serv.* 59:64–69.

Kidwell, M. G., J. F. Kidwell, and J. A. Sved. 1977. Hybrid dysgenesis in *Drosophila melanogaster:* A syndrome of aberrant traits including mutation, sterility, and male recombination. *Genetics* 86:813–33.

Knibb, W. R. 1982. Chromosome inversion polymorphisms in *Drosophila melanogaster* II: Geographic clines and climatic associations in Australasia, North America and Asia. *Genetica* 58:213–21.

Knibb, W. R., J. G. Oakeshott, and J. B. Gibson. 1981. Chromosome inversion polymorphisms in *Drosophila melanogaster.* I: Latitudinal clines and associations between inversions in Australasian populations. *Genetics* 98:833–47.

Lewontin, R. C., J. A. Moore, W. B. Provine, and B. Wallace, eds. 1981. *Dobzhansky's Genetics of Natural Populations I–XLIII.* New York: Columbia University Press.

McClintock, B. 1951. Chromosome organization and genic expression. *Cold Spring Harbor Symp. Quant. Biol.* 16:13–47.

———. 1984. The significance of responses of the genome to challenge. *Science* 226:792–801.

Nitasaka, E., T. Mukai, and T. Yamazaki. 1987. Repressor of P elements in *Drosophila melanogaster:* Cytotype determination by a defective P element carrying only open reading frames 0 through 2. *Proc. Nat. Acad. Sci.* 84:7605–8.

Oakeshott, J. G., G. K. Chambers, J. B. Gibson, W. F. Eanes, and D. A. Willcocks. 1983. Geographic variation in *G6pd* and *Pgd* allele frequencies in *Drosophila melanogaster.* *Heredity* 50:67–72.

Oakeshott, J. G., G. K. Chambers, J. B. Gibson, and D. A. Willcocks. 1981. Latitudinal relationships of esterase-6 and phosphoglucomutase gene frequencies in *Drosophila melanogaster.* *Heredity* 47:385–96.

Oakeshott, J. G., J. B. Gibson, P. R. Anderson, W. R. Knibb, D. G. Anderson, and G. K. Chambers. 1982. Alcohol dehydrogenase and glycerol-3-phosphate dehydrogenase clines in *Drosophila melanogaster* on different continents. *Evolution* 36:86–96.

O'Hare, K. and G. M. Rubin, 1983. Structures of P transposable elements and their sites of insertion and excision in the *Drosophila melanogaster* genome. *Cell* 34:25–35.

Ohishi, K., E. Takanashi, and S. Ishiwa Chigusa. 1982. Hybrid dysgenesis in natural populations of *Drosophila melanogaster* in Japan. I: Complete absence of the P factor in an island population. *Jpn. J. Genet.* 57:423–28.

Raymond, J. D., T. A. Ojala, J. White, and M. J. Simmons. 1991. Inheritance of P-element regulation in *Drosophila melanogaster. Genet. Res.* 57:227–34.

Robertson, H. M. and W. R. Engels. 1989. Modified P elements that mimic the P cytotype in *Drosophila melanogaster. Genetics* 123:815–24.

Ronsseray, S., M. Lehmann, and D. Anxolabéhère. 1989. Copy number and distribution of P and I mobile elements in *Drosophila melanogaster* populations. *Chromosoma* 98:207–14.

Rubin, G. M., M. G. Kidwell, and P. M. Bingham. 1982. The molecular basis of P-M hybrid dysgenesis: The nature of induced mutations. *Cell* 29:987–94.

Sakoyama, Y., T. Todo, S. Ishiwa, H. Ryo, S. Kondo, and T. Honjo. 1983. Movable P elements in Q type strain: Cloning and gene structure of Q type P elements. *Jpn. J. Genet.* 58:678.

Sakoyama, Y., T. Todo, S. Ishiwa-Chigusa, T. Honjo, and S. Kondo. 1985. Structures of defective P transposable elements prevalent in natural Q and Q-derived M strains of *Drosophila melanogaster. Proc. Natl. Acad. Sci.* 82:6236–39.

Schaefer, R. E., M. G. Kidwell, and A. Fausto-Sterling. 1979. Hybrid dysgenesis in *Drosophila melanogaster:* Morphological and cytological studies of ovarian dysgenesis. *Genetics* 92:1141–52.

Sved, J. A., 1976. Hybrid dysgenesis in *Drosophila melanogaster:* A possible explanation in terms of spatial organization of chromosomes. *Aust. J. Biol. Sci.* 29:375–88.

Takada, S., M. Murai, E. Takanashi, K. Ohishi, K. Fukami, N. Hagiwara, Y. Satta, and S. Ishiwa. 1983. On the P-M system in natural populations of *Drosophila melanogaster* in and around Japan. *Jpn. J. Genet.* 58:686.

Todo, T., Y. Sakoyama, S. I. Chigusa, A. Fukunaga, T. Honjo, and S. Kondo. 1984. Polymorphism in distribution and structures of P elements in natural populations of *Drosophila melanogaster* in and around Japan. *Jpn. J. Genet.* 59:441–51.

Woodruff, R. C. and J. N. Thompson, Jr. 1992. Have premeiotic clusters of mutation been overlooked in evolutionary theory? *J. Evol. Biol.* 5:457–64.

Yamamoto, A., F. Hihara, and T. K. Watanabe. 1984. Hybrid dysgenesis in *Drosophila melanogaster:* Predominance of Q factor in Japanese populations and its change in the laboratory. *Genetica* 63:71–77.

Contributors

Chapter 1
Chia-Chen Tan
Institute of Genetics
Fudan University
Shanghai, 200433 China

Chapter 2
G. Ledyard Stebbins
Department of Genetics
University of California, Davis
Davis, CA 95616

Chapter 3
Howard Levene
Department of Statistics
Columbia University
New York, NY 10027

Chapter 4
Costas B. Krimbas
Department of Genetics
Agricultural University of Athens
Iera Odos 75
Athens 11855 Greece

Chapter 5
Bruce Wallace
Department of Biology
Virginia Polytechnic Institute and State University
Blacksburg, VA 24601

Chapter 6

Timothy Prout
Center for Population Biology
University of California, Davis
Davis, CA 95616

Chapter 7

Hampton L. Carson
Department of Genetics and Molecular Biology
John A. Burns School of Medicine
1960 East-West Road
University of Hawaii
Honolulu, HI 96822

Chapter 8

Jeffrey R. Powell
Department of Biology
Yale University
P.O. Box 6666
New Haven, CT 06511

Chapter 9

Richard C. Lewontin
Museum of Comparative Zoology
Harvard University
Cambridge, MA 02138

Chapter 10

Max Levitan
Department of Cell Biology and Anatomy
Mt. Sinai School of Medicine
New York, NY 10029
William J. Etges
Department of Biological Sciences
University of Arkansas
Fayetteville, AK 72701

Chapter 11

Louis Levine and Robert F. Rockwell
Department of Biology
City College of New York
New York, NY 10031

Olga Olvera, María Esther de la Rosa, and Judith Guzmán
Departmento de Radiobiología y Genética
Instituto Nacional de Investigaciones Nucleares
Sierra Mojada #447-2º piso
México, D. F., 11010 Mexico

Víctor M. Salceda
Facultad de Ciencias
Universidad Autónoma de Baja California
Apartado Postal 1880
Ensenada, 22800, B. C. Mexico

Jeffrey R. Powell
Department of Biology
Yale University
P. O. Box 6666
New Haven CT 06511

Wyatt W. Anderson
Department of Genetics
University of Georgia, Athens
Athens, GA 30602

Chapter 12

Louis Bernard Klaczko
Departamento de Genética e Evolução
Instituto de Biologia
Universidade Estadual de Campinas
Cx. Postal 6109
Campinas, 13081 SP Brazil

Chapter 13

Danko Brncic
Department of Cell Biology and Genetics
Faculty of Medicine
University of Chile

Independencia 1027
Santiago (7) Chile

Chapter 14
Anssi Saura
Department of Genetics
University of Umeå
S-901 87 Umeå Sweden

Chapter 15
Michael Golubovsky
Division of Evolutionary Biology
Institute of the History of Science and Technology
St. Petersburg 199034 Russia
Leonid Kaidanov
Deparment of Genetics and Selection
St. Petersburg State University
St. Petersburg 199034 Russia

Chapter 16
Antonio Fontdevila
Departament de Genèticai de Microbiologia
Universitat Autònoma de Barcelona
Bellaterra 08193 Spain

Chapter 17
Laurence D. Mueller
Department of Ecology and Evolutionary Biology
University of California, Irvine
Irvine, CA 92717

Chapter 18
Lee Ehrman and Jan R. Factor
Division of Natural Sciences
SUNY
Purchase, NY 10577
Ira Perelle
Mercy College
Dobbs Ferry, NY 10522

Chapter 19

Antonio R. Cordeiro
Departamento de Genética
Universidade Federal do Rio de Janeiro
Cx. Postal 68011
Rio de Janeiro 21944 RJ Brazil

Helga Winge
Departamento de Genética
Universidade Federal do Rio Grande do Sul
Cx. Postal 15053
Porto Alegre 91-501-970 RS Brazil

Chapter 20

Agustí Galiana and Andrés Moya
Departament de Genètica
Facultat de Biología
Universitat de Valencia Spain
Burjassot 46100 Spain

Francisco J. Ayala
Department of Ecology and Evolutionary Biology
University of California, Irvine
Irvine, CA 92717

Chapter 21

Seppo Lakovaara and Jaana O. Liimatainen
Department of Genetics
University of Oulu
SF 90570 Oulu
Finland

Chapter 22

Dragoslav Marinković
Faculty of Biology
University of Belgrade
Belgrade 11000 Yugoslavia

Snežana Stanić
Faculty of Biology
University of Kragujevac
Kragujevac, Yugoslavia

Chapter 23
Stephen W. Schaeffer
Department of Biology
The Pennsylvannia State University
University Park, PA 16802

Chapter 24
Ian A. Boussy
Department of Biology
Loyola University Chicago
6525 N. Sheridan Road
Chicago, IL 60626

Species Index

Cereus validus, 201

Drosophila, 4, 5, 7, 9, 10, 11, 17, 18, 31, 32, 33, 53, 57, 59–60, 140, 149, 162, 167, 200, 201, 212, 214; alcohol dehydrogenase (Adh) region of, 331–33, 344; body size, 227; crossover frequencies, 7; colonization, 282; distal promoter, 332; DNA-DNA hybridization, 80; fecundity, 227; genetic variability, long-term studies, 189; Hawaiian, 53; genomes, 193, 194; inbreeding, 191–92, 294; inversions, 62, 114, 143; karyotypes, 80, 81; lethal mutations, 193; low activity (LA) strains, 191–94; mobile elements (ME), 189, 191–94, 195; niche separation, 214; phylogenies, 79–81; population stability, theoretical models of, 230–33; proximal promoter, 332; recombination analysis, 192; seasonal variation, 16; sex ratio, 145; viability, 227, 228; viruses, 191, 195.

Drosophila alpina, courtship behavior in: female, role of, 302–4; field observations, equipment used, 301–2; territorial behavior, 302, 310

Drosophila buzzatii: body size, 206, 212; chromosome rearrangement, 205, 210; cladodes, 209, 210, 211, 213; collection sites, 201–2; colonization of, 201, 203; comparative approach, 203; fecundity, 202, 203, 211; fitness components, 202, 205, 212; genetic markers, 209; genotype, 206, 208; heterokaryotypes, 202; heterozygote deficiency, 210, 213; host plants, 201; inbreeding, 210; karyotype, 203, 206, 208; larval viability, 202, 205; longevity, 202, 203, 206, 212; mating, 209, 213; origination of, 201; parasitism, 205 215; phenotypic plasticity, 205; pupal viability, 205; sexual selection, 206, 209; yeast heterogeneity, 210–11

Drosophila mediopunctata: adaptation, 151; allele frequencies, 143, 144, 151; *altitudinal differences*, 145; chromosome inversions, 142–45, 151; coadaptation, 146, 151; collection sites, 142, 145; genetic markers, 142, 144; haplotypes, 144; heterokaryotes, 143; heterozygotes, 143, 151; homokaryotes, 143; karyotype, 143; linkage disequilibrium, 144, 145; morphological variation, 146–51; polytene chromosomes, analysis of, 144; sex ratio, 145–46, 151; wings, ellipses analysis of, 148–50

Drosophila melanogaster, 4, 57, 59, 66, 67, 142, 160, 161, 179, 201, 211, 212, 341; alcohol dehydrogenase (Adh) region, 332–33, 345; Aus-

Name Index

Subject Index

Adaptation, 58, 73; *see also* Coadaptation
Adh-Dup gene, 331, 333, 338, 340–41, 346
Alcohol dehydrogenase (Adh), 331–33, 336, 337, 340, 341, 342, 343, 344, 345, 346; tests for, 336–37
Allegheny plateau site, 115
Alleles, 50, 52, 94, 95, 336; electrophoretic, 333; frequencies, 52, 143, 144, 151, 157, 177; "wild type," 13
Allozygotes, 216
Allozymes, 54, 209, 343; frequency changes in, 36; heterozygosity, 61
Amazonian sites, 243, 249, 263
Amecameca site, 129, 131, 134, 135, 136
Amylase, 51
Andean-Brazilian sites, 243
Andreas Canyon (California) site, 9
Arkansas sites, 108, 110, 112, 114, 117; *see also* Ozark sites
Armenia site, 190
Arroyo Escobar site, 201, 205, 206
Australian sites, 358–60, 364, 365
Autozygotes, 216

Bacteria, 244, 246, 259–60
Balanced lethal stock (VA/BA) technique, 178, 181

Balanced selection, *see* Selection
"Balance" theory, 13, 98, 175, 176
Bergsonism, 26
"Biological species concept" (BSC), 71, 83
Blowflies, 230, 233
Blue Ridge Mountain site, 115
Body size, 206, 212, 227
Bogata sites, 343–45
Brazil sites, 141, 266
Bryce Canyon National Park site, 284
BSC ("biological species concept"), 71, 83

California Institute of Technology (Cal. Tech.), 3, 4, 5, 7, 9, 14, 19, 83
Canary Island sites, 177
Carboneras site, 201, 205, 206, 209, 211, 212, 213, 214
Caucasus region, 190, 191, 192
Central American sites, 243, 249, 263, 269
Chernobyl event, 184
Chile sites, 155, 159–61
Chromosomal inversions, *see* Gene arrangements; Inversions
Chromosomes, 5, 17, 57, 60, 61, 96, 194, 205; acrocentric, 174, 176; B, 194; C, 67; "coupling," 115; euchromatic regions of, 62; heterozygosity, 63, 184; interpopulation crosses, 97;